Crafting Great APIs with Domain-Driven Design

Annegret Junker • Fabrizio Lazzaretti

Crafting Great APIs with Domain-Driven Design

Collaborative Craftsmanship of Asynchronous and Synchronous APIs

Apress®

Annegret Junker
Munich, Germany

Fabrizio Lazzaretti
Zurich, Switzerland

ISBN-13 (pbk): 979-8-8688-1456-3   ISBN-13 (electronic): 979-8-8688-1457-0
https://doi.org/10.1007/979-8-8688-1457-0

Copyright © 2025 by Annegret Junker and Fabrizio Lazzaretti

This work is subject to copyright. All rights are reserved by the Publisher, whether the whole or part of the material is concerned, specifically the rights of translation, reprinting, reuse of illustrations, recitation, broadcasting, reproduction on microfilms or in any other physical way, and transmission or information storage and retrieval, electronic adaptation, computer software, or by similar or dissimilar methodology now known or hereafter developed.

Trademarked names, logos, and images may appear in this book. Rather than use a trademark symbol with every occurrence of a trademarked name, logo, or image we use the names, logos, and images only in an editorial fashion and to the benefit of the trademark owner, with no intention of infringement of the trademark.

The use in this publication of trade names, trademarks, service marks, and similar terms, even if they are not identified as such, is not to be taken as an expression of opinion as to whether or not they are subject to proprietary rights.

While the advice and information in this book are believed to be true and accurate at the date of publication, neither the authors nor the editors nor the publisher can accept any legal responsibility for any errors or omissions that may be made. The publisher makes no warranty, express or implied, with respect to the material contained herein.

Managing Director, Apress Media LLC: Welmoed Spahr
Acquisitions Editor: Smriti Srivastava
Development Editor: Laura Berendsom
Coordinating Editor: Jessica Vakili

Distributed to the book trade worldwide by Springer Science+Business Media New York, 1 New York Plaza, New York, NY 10004. Phone 1-800-SPRINGER, fax (201) 348-4505, e-mail orders-ny@springer-sbm.com, or visit www.springeronline.com. Apress Media, LLC is a Delaware LLC and the sole member (owner) is Springer Science + Business Media Finance Inc (SSBM Finance Inc). SSBM Finance Inc is a **Delaware** corporation.

For information on translations, please e-mail booktranslations@springernature.com; for reprint, paperback, or audio rights, please e-mail www.bookpermissions@springernature.com.

Apress titles may be purchased in bulk for academic, corporate, or promotional use. eBook versions and licenses are also available for most titles. For more information, reference our Print and eBook Bulk Sales web page at http://www.apress.com/bulk-sales.

Any source code or other supplementary material referenced by the author in this book is available to readers on GitHub (https://github.com/Apress/Crafting-Great-APIs-with-Domain-Driven-Design). For more detailed information, please visit https://www.apress.com/gp/services/source-code.

If disposing of this product, please recycle the paper

Dedicated to Stefan and my beloved brother Bernd.

For Hannah Li – my wife, my best friend, my everything.

Contents

About the Authors xv

About the Technical Reviewer xvii

Acknowledgements xix

Preface xxi

Acronyms xxiii

Introduction xxix

Part I The Importance of API Design

1 Transforming Problematic Application Programming Interfaces ... 3

    Complexity in a Volatility, Uncertainty, Complexity, and
        Ambiguity (VUCA) World 3
        Complexity 3
        VUCA World 5
    The Term API 6
    APIs Nobody Wants to Use 7
        An API Taken Directly from a Database 8
        An API Formulated Solely from the Backend Developer's
            Point of View 10
        A Purely Technical API Without Business Relevance 11
        An API from the Ivory Tower 11
        Overloaded World-Domination API 12
        APIs Without Documentation and Unexpected Behavior 13
    Introduction to Modern Development Processes 14
    Collaborative Approaches Improve Processes 17
    Why a Modern Development Process Is Necessary and Valuable 19
    API-First vs. Code-First 20
    Points to Remember 22
    Review Questions 23
    References 23

2	**Communication Categories**	25
	The Term API	25
	Communication in General	29
	Messages as the Central Building Block	30
	Communication Styles and Mechanisms	30
	Communication Types: Synchronous and Asynchronous Communication	32
	Communication and APIs	35
	Communication Mechanisms, Architectural Patterns, and Suitable Protocols	37
	Request- and Response-Based Mechanisms	38
	Messaging vs. Eventing	40
	Messaging	40
	Eventing	41
	Points to Remember	42
	Review Questions	43
	References	43
3	**Quality Requirements for APIs**	45
	Functional Suitability	46
	Functional Completeness	47
	Functional Correctness	47
	Functional Appropriateness	47
	Performance Efficiency	47
	Time Behavior	48
	Resource Utilization	48
	Capacity	48
	Compatibility	48
	Coexistence	49
	Interoperability	49
	Interaction Capability	49
	Appropriateness Recognizability	50
	Learnability	50
	Operability	50
	User Error Protection	50
	User Engagement	50
	Inclusivity	51
	User Assistance	51
	Self-Descriptiveness	51
	Reliability	52
	Faultlessness	52
	Availability	52
	Fault Tolerance	53
	Recoverability	53
	Security	53

	Maintainability	54
	Modularity	54
	Reusability	55
	Analyzability	55
	Modifiability	55
	Testability	55
	Flexibility	56
	Adaptability	56
	Scalability	56
	Installability	56
	Replaceability	57
	Safety	57
	Summary	57
	Review Questions	58
	References	59

Part II Domain-Driven API Design

4 Online Library 63
   First Requirements Gathering 63
   First Implementation Ideas 65
   Start of Prioritization 68
   Points to Remember 69
   Review Questions 70
   References 70

5 API Design Supported by Domain-Driven Design 71
   Qualifying Business Ideas 71
      Business Model Canvas 71
      Capability Map 74
      Wardley Map 74
      Prioritization of Capabilities 79
   Gathering Business Requirements 82
      Domain Storytelling 82
      Visual Glossary 90
      Event Storming 96
      Context Map 104
      Discussion of Data Exchange Using API-First Approach 109
      Discussion of Synchronous APIs 111
      Discussion of Events 113
   Generation of API Definitions with AI 114
   Conclusion 115
   Points to Remember 117
   Review Questions 117
   References 118

6   Interface Definitions ... 121
    Components of an Interface Definition ... 121
        Short History of Interface Definitions ... 122
        WSDL, RAML, Swagger, and OpenAPI ... 123
        AsyncAPI ... 126
        Necessary Information for Interfaces ... 126
        Components of an Interface Definition for Synchronous and
            Asynchronous Interfaces ... 127
    Data Formats and Their Schemas ... 130
        JSON ... 133
        YAML ... 136
        XML ... 137
        Protobuf ... 141
        Avro ... 143
        Comparison of Data Formats ... 146
        Conclusion ... 146
    Idempotency and Guarantees ... 147
        Problems in Transmission ... 148
        Mitigating Problems ... 149
    Definition of Synchronous Interfaces ... 153
        Definition as gRPC ... 154
        Definition as GraphQL ... 156
        Definition as REST with OpenAPI ... 158
        Comparison of gRPC, GraphQL, and HTTP REST ... 169
    Definition of Asynchronous Interfaces ... 170
        AsyncAPI ... 171
        What AsyncAPI Misses ... 181
        CloudEvents ... 182
    Advantages and Disadvantages of Different Schemas and Protocols ... 183
        Schemaless Versus Schemafull with Respect to the Single
            Data Formats ... 185
        Read and Write Functions and Their Advantages and
            Disadvantages ... 185
        Tooling for Linting ... 186
        Compression During Transit ... 187
        North–South and East–West Communication ... 187
        Decision Tree Use of Schemas and Protocols ... 188
    Typical Antipatterns ... 190
        Synchronous Events ... 190
        Asynchronous Commands ... 191
        Messages as Events ... 191
        Huge Events ... 192
    AI Generation of API Definitions: Chances and Limitations ... 192
    Conclusion ... 195
    Points to Remember ... 195

	Review Questions	196
	References	197
7	**Defining the Online Library Interfaces**	**205**
	Introduction	205
	Bounded Context Canvas	207
	Architecture Communication Canvas	208
	API Product Canvas	210
	Inventory Management	212
	Definition of AsyncAPI	212
	Definition of OpenAPI	216
	Catalog Management	216
	Management Application	219
	Search Application	221
	Lending	224
	Reading	232
	Appointment Management	236
	Teaching and Reading to Kids	239
	Member Management	240
	Conclusion	245
	Points to Remember	247
	Review Questions	248
	References	249

Part III Enabling Transformation

8	**Developer Experience and API Implementation**	**253**
	Versioning of APIs	253
	What Is a Breaking Change?	254
	Backward, Forward, and Full Compatibility	257
	Consider If a New Version Number Is the Solution	257
	Synchronous APIs	258
	Asynchronous APIs	259
	Schema Registry	260
	Testing	264
	Types of Testing	264
	Concepts	270
	Conclude Testing	271
	Continuous Integration	272
	Git Flow	272
	Continuous Deployment	273
	Benefits and Use of CI/CD with APIs	274
	Conclusion	275
	Publishing APIs	276
	API Management Platform	276
	Backends for Frontends (BFF)	277

	Service Mesh	278
	API Discovery	279
	ApiOps as a Process	280
	Developer Experience When Integrating an API	280
	Sandbox	281
	Observability	281
	Ability to Get Help	282
	Software Development Kits (SDKs)	282
	Common Standards	283
	Data Mesh Concept	283
	Analytics and Monitoring	284
	Error Discovery	284
	Usage Insights into APIs	285
	Billing with Metrics	285
	Analyzing Performance	285
	Conclusion	285
	Points to Remember	285
	Review Questions	286
	References	287
9	**Collaborative Design and Agility**	291
	Collaborative Design Process	291
	Business Model	292
	Strategic Design	292
	Tactical Design	295
	Overview	295
	Necessary Mindset for API Design	297
	Necessary Mindset	297
	Additional Methodologies in the Process	297
	Conway's Law	301
	Team Topologies	303
	Collaborative Processes in Classical Project Environments	303
	Short Feedback Loops and Stable APIs	306
	Conclusion	306
	Points to Remember	307
	Review Questions	307
	References	308
10	**Iterative Enhancements**	311
	Adding New Features	311
	Adding a Small Feature Extension	311
	Enhancement of Online Library by Completely New Features	313
	Integration of External Interfaces	314
	Standardization	316
	Collaboration Between Partners	318
	Publishing Events to External Partners	318

	Publishing External Events	319
	Security Concerns	322
	Conclusion	322
	Points to Remember	322
	Review Questions	322
	References	323
11	**Brownfield Project**	325
	Hypothetical Insurance Legacy System	325
	Legacy System	326
	Business Analysis in a Brownfield	326
	Domain Storytelling of Current Process	327
	Intended Process	329
	Architectural Approach to Modernization for the Insurance Application	331
	Business Analysis Using Wardley Map Method	332
	Architectural Transition	335
	Technical Analysis	335
	Decoupling Layer	335
	Modernization Layer	337
	Developed Architecture	338
	Conclusion	339
	Points to Remember	340
	Review Questions	340
	References	341
12	**Shortcuts in the Process**	343
	North Star	343
	Definition of REST APIs Directly After Domain Storytelling	345
	Definition of Events Directly from Event Storming	345
	Tactical Design	345
	Conclusion	345
	Points to Remember	346
	Review Questions	346
	References	346
13	**APIs and Events in a Serverless World**	347
	New Requirements and Challenges in the Cloud	347
	The CAP Theorem and Eventual Consistency	348
	Using the Cloud's Benefits	350
	What Is Serverless?	353
	How to Implement APIs Without a Server	353
	Points to Remember	354
	Review Questions	354
	References	354

Part IV Summarizing

14 Avoiding Mistakes in the Definition of Events and APIs 359
    Revisiting the Examples from the Beginning 359
        An API Directly from the Database 359
        An API Solely Formulated from the Backend Developer's
            Point of View 360
        A Purely Technical API Without Business Relevance 360
        An API from the Ivory Tower 360
        The Overloaded World-Domination API 360
        APIs Without Documentation and Unexpected Behavior 361
    Other Problems 361
    Conclusion 362
    Points to Remember 362
    Review Questions 362

15 A Couple of Beautiful APIs 363
    What Is a Beautiful API in the Eyes of the Authors? 363
    Some Beautiful Real-World APIs 363
        Synchronous APIs 363
        Asynchronous APIs 365
    Conclusion 365
    Points to Remember 365
    Review Questions 366
    References 366

16 Summary 367
    References 370

Glossary 373

Solutions 381

Index 391

About the Authors

Annegret Junker is Chief Software Architect at codecentric AG in Germany. She has worked in software development for over 30 years. She worked in quite different roles and domains, e.g., insurance, car manufacturer, logistics, financial services, etc. She is interested in DDD, microservices, and everything along with it. Currently, she is helping a customer on a transition project in health insurance.

Fabrizio Lazzaretti is a Managing Consultant specializing in software architecture at Wavestone. He has over a decade of expertise in software architecture, development, and DevOps. He is a maintainer of the Cloud Native Computing Foundation (CNCF) project CloudEvents. Recognized for his proficiency as both a team player and leader in software development, architecture, and Scrum, he excels in dynamic work settings. His primary areas of expertise lie in event-driven architecture and microservice design within cloud environments. Currently, he is actively engaged in enterprise architecture within the insurance sector.

About the Technical Reviewer

Linus Basig is a seasoned software engineer and technology leader with over 10 years of experience in the industry. He currently serves as the Head of Engineering at CARU AG in Zurich, overseeing the development of the cutting-edge home emergency solution CARU care. Prior to his role at CARU, Linus held positions as a software engineer at both CARU AG and Siroop AG. He loves to use domain- and event-driven architecture patterns to tame complex systems. In addition to his passion for code and architecture, he's engaged in the medical unit of his local militia fire service.

Acknowledgements

We wish to express our deepest gratitude to our families and friends. We especially wish to thank Hannah Li for her patience and support and for letting us discuss the book in our free time. We wish to thank Bernd, Stefan, and Stefan for their unwavering support and encouragement.

We are grateful to Apress for their guidance and expertise in bringing this book to fruition. Revising the book was a long journey that involved the invaluable participation and collaboration of reviewers, technical experts, and editors. We wish to thank Linus Basig and Dominik Keller for their constant support and expert suggestions, which greatly improved the quality of the book.

We would also like to acknowledge the valuable contributions of our colleagues and coworkers during our many years in the tech industry. They have taught us so much and provided helpful feedback on our work. Moreover, we would like to thank our colleagues at codecentric and Wavestone for their support and the freedom they gave us to write this book. Finally, we would like to thank all those readers who expressed an interest in the book and supported its production. Your encouragement has been invaluable.

Preface

Around 2010, Annegret started using Domain-Driven Design in her day-to-day work. At the time she was working at a large financial technology (fintech) company and saw flaws in the software caused by a lack of domain knowledge. She created, together with colleagues, an extensive glossary of the company's work. It grew, and, as she was forced to acknowledge, no one understood it because it was simply too large to comprehend; there were just too many terms. Applying smaller portions to it and calling them subdomains helped in explaining the company's business. Later, while working in the automotive industry, she realized that only strategic design helped in coming up with well-tailored services. And designing services requires well-designed Application Programming Interface (API).

From 2012, Fabrizio worked in numerous business application contexts in various industries. A major problem he detected was wrong API designs and a lack of collaboration between teams, IT, and business. He tried to resolve this issue through collaboration and various design approaches, including API designs. Over time, Fabrizio's methodological approaches have become increasingly sophisticated through mutual collaboration with Annegret and others, with all parties refining their processes together. His decade-long fascination with asynchronous communication has been an area of focus.

Annegret and Fabrizio met each other while working in a large insurance company a few years ago and found in each other someone who was similarly passionate about APIs and event-driven architecture. At the time, Annegret was already an acknowledged expert in Domain-Driven Design and collaborative modeling and consulted her company in that area. Fabrizio brought a profound knowledge of modern web technologies and API implementation to their discussions. Annegret told Fabrizio about her ideas for a book that would bring together Domain-Driven Design and API design. The very first sketches of the book piqued Fabrizio's interest, and they joined forces on this book, combining their passions for well-designed APIs and well-tailored services.

Munich, Germany Annegret Junker
Zurich, Switzerland Fabrizio Lazzaretti
February, 2025

Acronyms

AaaP	API as a Product, 280, 285, 360
ACID	atomicity, consistency, isolation, durability, 349, 350
ACK	acknowledgment, 33, 40
ACL	anti-corruption layer, 233, 238, 248, 316, 335, 338
ADR	Architecture Decision Record, 295, 345
AI	artificial intelligence, 64, 114, 121, 147, 192, 193, 254, 334, 338, 352
AMQP	Advanced Message Queuing Protocol, xxxi, 31, 38, 41, 171, 181, 187, 189, 195, 319, 378
ANSI	American National Standards Institute, 29
API	Application Programming Interface, xxi, xxx–xxxiii, 3, 6–14, 16–18, 20, 22, 23, 25–32, 35–39, 42, 45–58, 68, 71, 105, 108, 114–116, 121, 125–131, 134, 146, 147, 153, 156, 157, 159, 163, 166, 168, 170, 171, 176, 181, 186, 190, 192, 193, 195, 205, 207, 210–213, 215, 216, 219, 220, 229, 233, 236, 238, 239, 244–249, 253–255, 257–259, 262–266, 268–272, 274–287, 291, 295, 297, 298, 301, 302, 305–308, 311, 315–318, 320, 325, 326, 329, 336, 337, 339, 340, 343, 345–347, 350–354, 359–367, 369, 370, 373, 378, 381, 384–388
ASCII	American Standard Code for Information Interchange, 374
ATAM	architecture tradeoff analysis method, 335
BASE	basically available, soft state, and eventually consistent, 350
BBoM	Big Ball of Mud, 105, 108
BDD	behavior-driven development, 271, 286, 300, 301, 368, 369, 385
BFF	Backends for frontends, 277, 278, 286, 287
BPMN	Business Process Model Notation, 298, 369
CD	continuous deployment, 273
CDC	Change Data Capture, 336
CDN	content delivery network, 351
CI	continuous integration, 264–266, 272, 273, 275, 286

CI/CD	continuous integration and continuous delivery/deployment, 266, 273, 275, 338
CLI	command line interface, 146, 166, 187
CNCF	Cloud Native Computing Foundation, 182, 375
COM/DCOM	Component Object Model / Distributed Component Object Model, 122, 123
CORBA	Common Object Request Broker Architecture, 123, 158
CQRS	Command and Query Responsibility Segregation, 217–219, 221, 223, 295, 319
DaaP	Data as a Product, 375
DCE	Distributed Computing Environment, 122
DDD	Domain-Driven Design, 3, 6, 18, 20, 55, 102, 121, 193, 212, 221, 226, 232, 234, 235, 247, 271, 292, 303, 305, 322, 326, 335, 337, 339, 360, 362, 378, 388
DTO	data transfer object, 146
e2e	End-to-end, 269, 272
eBPF	Extended Berkeley Packet Filter, 279
ECMA	European Computer Manufacturers Association, 133
EDA	event-driven architecture, 31, 35, 39, 41, 126, 171, 175, 191, 212, 318, 376
ERP	Enterprise Resource Planning, 8
ESB	Enterprise Service Bus, 39
FaaS	Functions as a Service, 353
gRPC	gRPC Remote Procedure Calls, xxxi, 31, 39, 141, 153, 154, 156–158, 169, 170, 184, 185, 187, 189, 190, 195, 211, 369, 377
GUID	Global Unique Identifier, 11, 23, 381
gZIP	Gnu ZIP, 187
HATEOAS	hypermedia as the engine of application state, 159
HTML	Hypertext Markup Language, 125, 137
HTTP	Hypertext Transfer Protocol, 7, 8, 28, 30–32, 36, 38, 124, 125, 127, 129, 130, 150, 158, 163, 169–171, 175, 181, 182, 187, 195, 196, 210, 211, 220, 258, 268, 278, 315, 317, 319, 320, 322, 351, 377–380

IaaS	Infrastructure as a Service, 353
IaC	Infrastructure as Code, 136
IAM	Identity and Access Management, 240
IDE	integrated development environment, 57, 171
IDL	interface definition language, 22, 122, 123, 125, 142, 143, 156, 158, 159, 178, 195, 318, 369
IETF	Internet Engineering Task Force, 134
IP	internet protocol, 31, 282
ISO	International Organization for Standardization, 29, 137
JDK	Java Development Kit, 378
JSON	JavaScript Object Notation, xxxi, 125, 130–138, 143, 146, 147, 162, 165, 166, 169, 176, 178, 182, 187, 190, 195, 206, 254, 255, 259, 261, 262, 265, 370
JWT	JSON Web Token, 229, 230
KPI	key performance indicator, 269, 285
LAN	local area network, 25
LASR	Lightweight Approach for Software Reviews, 335
LLM	large language model, 193, 352
MQTT	message queuing telemetry transport, xxxi, 31, 41, 171, 182
MTBF	mean time between failures, 53
mTLS	mutual TLS, 167, 278
MTTR	mean time to recovery, 53, 274
OAS	OpenAPI specification, 50, 134
OASIS	Organization for the Advancement of Structured Information Standards, 39
OAuth	open authorization, 240
OCI	Open Container Initiative, 268
OData	Open Data Protocol, 31, 39
OIDC	OpenID Connect, 240, 242
OKR	objectives and key results, 285
ONC RPC	Open Network Computing Remote Procedure Call, 122
OODA	observe, orient, decide, act, 74
OS	operating system, 26–28
OSI	Open Systems Interconnection, 277, 278
OWASP	Open Worldwide Application Security Project, 54, 138, 269

PaaS	Platform as a Service, 353	
PDF	Portable Document Format, 192	
PDP	policy decision point, 276	
PoC	Proof of Concept, 20, 22, 189, 249	
PR	pull request, 273, 275	
QoS	Quality of Service, 182	
RAML	RESTful API Modeling Language, 121, 123–125	
RDBMS	relational database management system, 348–350, 354	
REST	Representational State Transfer, xxxi, xxxii, 8, 38, 153, 156–159, 169, 170, 184–186, 189, 190, 195, 196, 205, 221, 255, 258, 269, 317, 326, 336, 337, 353, 360, 363, 365, 378	
RPC	remote procedure call, 31, 38, 39, 143, 154	
SaaS	Software as a Service, 29	
SDK	software development kit, 28, 57, 182, 282, 283, 364, 365	
SGML	Standard Generalized Markup Language, 137	
SIEM	security information and event management, 379	
SOA	service-oriented architecture, 31, 39	
SOAP	Simple Object Access Protocol, 39, 123–125	
SOM	System Object Model, 122, 123	
SQL	Structured Query Language, 29, 49, 57, 217	
SSE	Server-Sent Events, 31	
SSRF	Server Side Request Forgery, 138	
TCP	Transmission Control Protocol, 28, 31–33, 149, 319, 380	
TDD	Test-Driven Development, 266, 270, 271, 275, 286, 370, 385	
TLS	Transport Layer Security, 167, 279, 320	
UDP	user datagram protocol, 31, 32	
UI	user interface, 26, 51, 146, 269, 281, 347	
UML	Unified Modeling Language, 90	
URL	Universal Resource Locator, 128, 138, 159, 192, 258, 320	
UUID	universal unique identifier, 153, 155, 360	
UX	User Experience, 293	
VCS	version control system, 272	
VPN	virtual private network, 53	

VUCA	Volatility, Uncertainty, Complexity, and Ambiguity, 3, 5, 22, 291
W3C	World Wide Web Consortium, 137
WADL	Web API Description Language, 125
WAF	web application firewall, 48
WAN	wide area network, 25
WSDL	Web Services Description Language, 121, 123–125, 265
XML	Extended Markup Language, xxxi, 123, 124, 130–132, 137, 138, 146, 169, 182, 195, 265
XPCOM	Mozilla's Component Object Model, 123
XSD	XML schema definition, 124, 138, 265
YAML	Yet Another Markup Language, 125, 131, 132, 136, 137, 146, 165, 166, 178, 195, 264

Introduction

This introduction explains how to read this book and gives an overview of its chapters.

Conventions Used

Every chapter ends with some "Review questions," which will serve to verify that the chapter's contents have been understood. The solutions to the questions appear at the end of the book in Solutions on page 381.

Important Information

Some important things to look at are highlighted in boxes with shaded borders, as shown below.

> **Important information**

This is a box with important notes.

Background Information

Sometimes we have exciting topics that are not the focus of our attention and not necessary to understand the key messages. These parts are also presented in information boxes with darker shading, like the following:

Background information

This box contains background information.

Code

The code presented in the book is published on GitHub at https://github.com/Apress/Crafting-Great-APIs-with-Domain-Driven-Design.

In the following section, we present an overview of upcoming chapters for the reader's orientation.

Chapter Overview

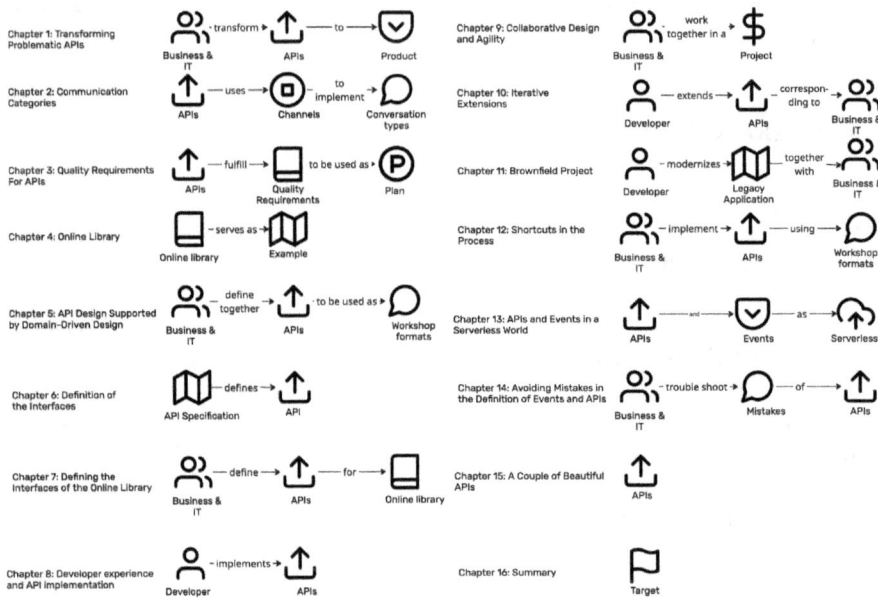

Figure 1 Chapter overview

The figure shows a book chapter in a visual format.

Chapter 1, "Transforming Problematic Application Programming Interfaces"

In this chapter, you will obtain a first impression of how bad and ugly APIs can affect development speed and time to market. You will also get a first glance at a modern and collaborative development process. Also, you will understand how elegant and attractive APIs will help in these areas.

Introduction   xxxi

Chapter 2, "Communication Categories"

This chapter discusses communication categories and how those communication categories are implemented by different API approaches and protocols. It defines the term API and how it is used in this book. From remote APIs to asynchronous communication via events, the terms are explained and categorized so that you can determine which form fits best in your application development model.

Chapter 3, "Quality Requirements for APIs"

All software projects must meet quality requirements, but APIs need to fulfill specific requirements when it comes to resilience, security, scalability, and even performance. In this chapter, we discuss these and additional aspects of maintainability and compatibility.

Chapter 4, "Online Library"

Throughout the book will use examples to illustrate the different approaches to defining and implementing APIs. This chapter contains an introduction to the online library that serves as an example.

Chapter 5, "API Design Supported by Domain-Driven Design"

This chapter discusses in great detail the collaborative development process already briefly introduced in the first chapter. Here, you will learn how to collaboratively model and define APIs.

Chapter 6, "Interface Definitions"

The definition of message payload in Extended Markup Language (XML), JavaScript Object Notation (JSON), Protobuf, and Apache Avro poses a challenge an API designer will need to face. You will learn how to define synchronous interfaces such as Representational State Transfer (REST) using OpenAPI, gRPC Remote Procedure Calls (gRPC), or GraphQL. Furthermore, you will learn how to define asynchronous APIs using AsyncAPI for different protocols like Apache Kafka, message queuing telemetry transport (MQTT), and Advanced Message Queuing Protocol (AMQP). You will see that asynchronous communication can be done via a broker, but it is possible to go brokerless. The advantages and disadvantages of the multiple schemas, protocols, and data formats are discussed. Certain anti-patterns in synchronous and asynchronous communication are explained, and a discussion of how to avoid them is presented.

Standards for communication patterns like REST and CloudEvents are introduced and discussed.

At the end of the chapter, a decision matrix that shows when to use which protocol and communication approach is given.

Chapter 7, "Defining the Online Library Interfaces"

In this chapter, we will apply the theory discussed in Chapter 6 to reality. The APIs of the online library will be designed, and a discussion about why a certain protocol or communication pattern is used will be presented.

Chapter 8, "Developer Experience and API Implementation"

Defining APIs represents a major step. But they need to be tested, versioned, and integrated as well. All three aspects are emphasized continuously. We will discuss the use of API Management applications, including API Marketplace and API Gateways. The concepts of service mesh and data mesh are discussed. Analytics and monitoring are important during API operation, and these will be discussed in the chapter as well.

Chapter 9, "Collaborative Design and Agility"

This chapter discusses how the presented collaborative design approach can be embedded in modern, agile software development processes.

Chapter 10, "Iterative Enhancements"

The original development of an API can be enhanced step by step. API development allows iterative enhancements to APIs as enhancements of the entire business process or as smaller improvements to single APIs.

Chapter 11, "Brownfield Project"

This chapter will discuss how to apply the methodologies and techniques discussed in preceding chapters to a brownfield project. The brownfield project will be introduced, and we will discuss how to improve a legacy application step by step to a maintainable and stable application.

Chapter 12, "Shortcuts in the Process"

In the preceding chapters, we discussed collaborative processes, which can usually be applied to large and medium-sized projects. But how can the approaches be applied to smaller projects, or even in a day-to-day work environment? We will discuss how to use the introduced methodologies and workshop formats and when certain steps can be skipped.

Chapter 13, "APIs and Events in a Serverless World"

In a serverless world, APIs play a crucial role. Without APIs, a serverless world is simply not possible. We will discuss the role of APIs in a serverless world and how this leads to APIs as a product. The chapter includes a discussion of the role of APIs and events in the future of software applications.

Chapter 14, "Avoiding Mistakes in the Definition of Events and APIs"

Avoiding mistakes is a major topic. In this chapter, the authors will explain certain mistakes they have encountered in their professional experience and how such mistakes can be avoided. Thus, you do not need to make mistakes yourself to learn from them.

Chapter 15, "A Couple of Beautiful APIs"

Toward the end of the book, the authors will present a selection of exquisite APIs and explain their appeal. We hope that by the time you finish the book, your views will align with the authors'. However, it is important to remember that the notion of beauty is subjective.

Chapter 16, "Summary"

APIs and events are THE drivers in a digital and cross-linked world. Innovation is driven by collaboration, not by competition. We will present our outlook on how collaboratively designed APIs create a real business advantage.

We hope that you will enjoy reading the book as much as we enjoyed working on it. Let us start with the first chapter and see how we can transform problematic APIs.

Part I
The Importance of API Design

Why is API design important?

Transforming Problematic Application Programming Interfaces 1

In this chapter, we consider problematic Application Programming Interfaces (APIs) and take a first look at possible solutions. Using modern and collaborative processes allows us to avoid problematic APIs and how we can design beautiful APIs.

Complexity in a Volatility, Uncertainty, Complexity, and Ambiguity (VUCA) World

Software development is intrinsically complex. It handles a wide variety of problems, from controlling washing machines to complex stock trading using artificial intelligence. *Eric Evans* introduced Domain-Driven Design (DDD) for the express purpose of managing this complexity. It explains the subtitle of his groundbreaking work: *Tackling Complexity in the Heart of Software* [1].

To understand why software development needs DDD, we need to understand, first, what complexity means in the context of software development.

Complexity

Complexity is something where cause and effect can only be perceived in hindsight [2].

This definition is taken from the *Cynefin* framework introduced by Snowden [2]. It aids decision-making using the clarity or even lack of clarity in the relationship between cause and effect. *Cynefin* is a Welsh word that means habitat. The different categories are shown in Figure 1-1.

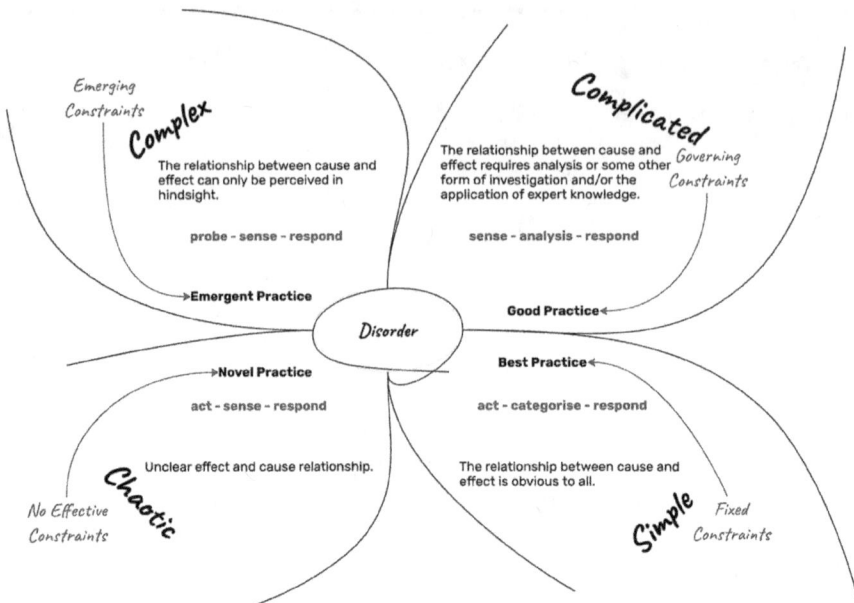

Figure 1-1 Cynefin framework, see [2]

To be able to make decisions, we can differentiate five different environments:

- **Simple**
  We can make simple decisions when best practices are available. This is comparable to a recipe. Your cake will burn if you bake it for too long. You act, and the result is obvious. In development, it is comparable to using a programming language. If you do not use the syntax correctly, your program cannot be compiled or interpreted. You will get an error.
- **Complicated**
  The result of a decision is not obvious. However, an expert can predict it. In software development, we use security libraries. Only bona fide experts can predict what will happen when implementing a cipher algorithm. Therefore, it is a governance constraint to use cipher libraries and not to implement them themselves when you are not a security expert.
- **Complex**
  As stated earlier, in a complex environment, you cannot predict the outcome of a process. The relationship between cause and effect can only be perceived in hindsight. Therefore, we must probe, perceive, and respond. The process of the Synergetic Blueprint is a way to proceed through development in small steps to find emerging constraints like business requirements and context boundaries.

- **Chaotic**
  The relationship between cause and effect is completely unclear. There are no practical constraints. One must act and subsequently observe the effect on a larger scale, even though the cause might still be unclear. But in critical situations, you cannot wait. We experienced such a chaotic environment during the Covid crisis. Governments need to act in the absence of knowledge. Micromanagement and other anti-patterns of good project management might bring about such an environment in software development. Applying good practices in project management avoids those situations.
- **Disorder**
  A disordered environment describes an area where no other properties can be applied. Applying random decisions brings experience so that other areas can be used.

Eric Evans launched his groundbreaking work tackling complexity in software development by stressing the domain [1]. Since then, the emphasis on functional tailoring software has become one of the cornerstones of software development. Additionally, we must adapt our approaches quickly in face of the VUCA world, which will be discussed in the following subsection.

VUCA World

A volatile, uncertain, complex, and ambiguous world leads to constantly changing software requirements in both business and technology [3]. A constantly changing world requires constant adaption in the development process and the use of technology. We must face such a world with good practices and constant adaptation of our development process. Let us look at what a volatile, uncertain world means for software development [4].

- **Volatility**
  Volatility means that contextual conditions are constantly changing. What is true today will not necessarily be true in the future. It is not possible to change the selected technology with each commit. Therefore, the software must be modularized to enable changing technologies for single modules without disturbing other parts of the application. Microservices represent a great approach to modularization. However, the costs of the increased complexity in operations must be balanced carefully against the advantages of a flexible architecture [5]. Even more, business models are changing as well, though probably not at the same speed as technology. Software needs to be flexible to be able to react to these changes.
- **Uncertainty**
  Uncertainty means that business requirements or even technology requirements cannot be completely known. We do not know everything. However, we

will learn. Completeness would mean that the complex system is completely described. We know software systems are intrinsically complex [1], and we know that cause and effect can only be detected in hindsight [2]. That means, among other things, that software systems cannot be described completely [4]. Therefore, we need a development process that includes a collaborative ideation process and a step-by-step method that incorporates new knowledge robustly.

- **Complexity**
  We have already discussed the complex nature of software systems. Many paths will take us to our goal, but we do not know them beforehand. We can only face it using emerging constraints [2]. The development process must be collaborative and adaptable. Later in this book we will show how the process can be adapted in brownfield processes (Chapter 11) or where shortcuts can be taken (Chapter 12).
- **Ambiguity**
  Ambiguity in the context of software development means that different attendees in a process have different perspectives. Those perspectives lead to different uses of terms and misunderstandings when using different languages. What's more, language is ambiguous. A customer in a sales context means something different than a customer in delivery. In a sales context, I want to know as much as possible about a prospective customer. In delivery, the only information I need is the delivery address. The use of term *customer* in both contexts is correct. However, it can lead to confusion. To avoid ambiguity, we need to define a ubiquitous language section "Gathering Business Requirements" and document it in a Visual Glossary [6]. The ubiquitous language is tied to a context to prevent ambiguity [1]. The development process must support the concept of bounded contexts.

One way to tackle complexity is by using APIs. They allow us to subdivide a large problem into smaller ones.

The Term API

APIs are used to connect different functions of software applications.

This book can be read without any prior knowledge about designing APIs or specifying and implementing them. This informative read will teach you how to craft elegant Application Programming Interfaces (APIs) that will align seamlessly with business and architectural requirements. It will also guide you on developing modern applications using Domain-Driven Design DDD, with a strong emphasis on APIs.

The main idea behind interfaces dates back to the earliest days of computing. In 1947, Goldstine and von Neumann published the idea of "libraries" [7].

The term APIs first appeared in 1968 [8]. It describes how an interface and its accompanying implementation can be separated from each other. Thus, it is possible to change the implementation without changing the interface and correspondingly affect the API user.

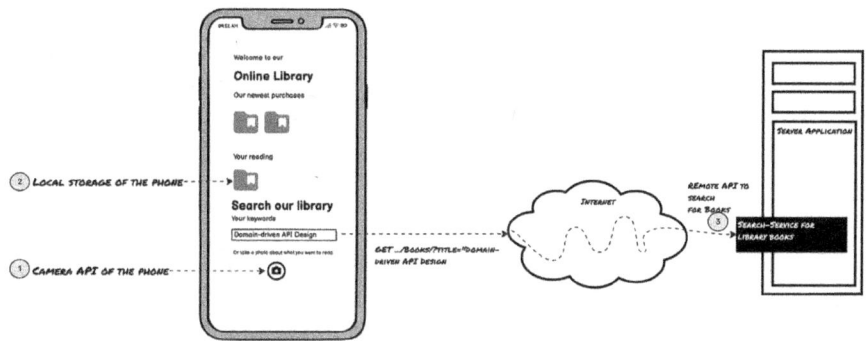

Figure 1-2 Different API types used in a mobile application adapting an idea of Lauret [9]

APIs themselves can be used for quite different purposes - granting access to the hardware of a smartphone like its camera, functions or libraries inside an application like access to a database, and so forth [9], as is shown in Figure 1-2. However, this book focuses on remote access to functions and data distributed over a network.

The different kinds of APIs, as shown in Figure 1-2, are discussed in detail in Chapter 2.

As already described in 1947, when an API is used, no knowledge about the implementation behind it is necessary. Such APIs are especially necessary if architectures evolve over several years. In such a way, the implementation can be changed without changing the implementation of clients using the interface. Obviously, that is only possible when the specification and behavior of the interface remain unchanged.

APIs are forever – code is for the moment.

APIs can be specified in very different ways. The most commonly known one is the OpenAPI specification standard [10] to specify Hypertext Transfer Protocol (HTTP) interfaces, which will be described in more detail in Chapter 6.

If you have no knowledge about the implementation of an API, you will rely on the clarity of the API's specification, which will need to describe the outcome. However, APIs in the real world often contradict this requirement. They are not understandable, and their behavior is unexpected. These are bad APIs. In the next sections, we will discuss some examples of bad APIs.

APIs Nobody Wants to Use

If you work as a programmer, you must use APIs. In the modern, complex world, it is no longer possible to create software without using APIs, especially APIs that are acquired from outside your own organization. Usually, it is not possible to simply go to the inventor of an API and ask its developer what they meant by something.

The following are examples of what the authors consider to be poor APIs. Some cite SAP-based systems. These are Enterprise Resource Planning (ERP) systems often used in Europe. This does not mean that all SAP systems are bad. On the contrary, most implemented SAP systems are well designed and offer APIs that can be understood well.

Let us now dig into some examples of what badly designed APIs mean to their users. The following examples are based on REST and HTTP because, for most readers, they should already be known at a high level. Moreover, the interface definitions are shown in a kind of pseudocode for simplicity.

In section "Why a Modern Development Process Is Necessary and Valuable", we will show how the process described in this book can avoid these bad APIs.

An API Taken Directly from a Database

Imagine a business partner's table of SAP [11] slated to be published one to one as an API.

Again, the sample is used because SAP systems are widely known. It might not be as bad if the database properties were named well. But database types only slightly correspond to well-designed API types.

The corresponding API would look like Listing 1-1.

Looking at it, you might not understand what is meant by CLIENT. CLIENT means a tenant of a system, which is clear if you know the system. However, it is unclear when you want to use the system from the outside. You probably won't know that you need to provide such tenant information.

The property PARTNER represents the business partner. That might be clear when it is well described in the API definition. But it seems rather overengineered when dealing with an API that only refers to customers and not to suppliers.

The reason you have to provide a flag BPKIND that indicates the type of business partner when only customers can be indicated is not clear. In such cases, the API implementation needs to hide the database schema. The API user should only see what they are interested in, for example, the customer resource.

The definition shown is barely understandable. However, the payload of the sample response doesn't improve the situation, as shown in Listing 1-2.

A user of the API needs to know what *business partner* means in the database: a customer, a delivery partner, a supplier, a consultant, or even all of the above. The terms client, partner, type, and so forth are not part of a well-defined ubiquitous language in this case. How to define a ubiquitous language will be described in Chapter 5.

APIs should be defined from an external viewpoint, correlating to the functionality they signify. Defining them from an internal perspective results in an interface that is challenging to comprehend.

```
 1  schemas:
 2    BusinessPartner:
 3      description: An existing business partner
 4      properties:
 5        CLIENT:
 6          type: CLNT
 7        PARTNER:
 8          type: string
 9          minLength: 10
10          maxLength: 10
11        TYPE:
12          type: string
13          minLength: 1
14          maxLength: 1
15        BPKIND:
16          type: string
17          minLength: 4
18          maxLength: 4
19        ...
20    CLNT:
21      description: Number of a tenant in a multitenant environment
22      type: string
23      length: 3
24      ...
```

Listing 1-1 Example of API definition taken directly from database definition

```
1  {
2    "BusinessPartner": {
3      "CLIENT": "ABC",
4      "PARTNER": "1234567890",
5      "TYPE": "A",
6      "BPKIND": "CUST",
7      ...
8    }
9  }
```

Listing 1-2 Example: Payload for API taken directly from database definition

We have observed that an API developed from a database definition is difficult to grasp. However, API providers typically cannot anticipate all of their customers. Thus, they need to formulate their API based on their limited understanding.

The shown API would be great inside an SAP system, where the ubiquitous language is defined by the use of the SAP system and understood by the developers of the system.

```
schemas:
  BusinessPartner:
    description: Company business partner
    properties:
      tenant:
        description: Tenant number in multitenant environment
        type: string
        minLength: 3
        maxLength: 3
      partnerNumber:
        description: Number of business partner
        type: string
        minLength: 10
        maxLength: 10

      ...

      name:
        description: Name of person
        type: string
        minLength: 2
        maxLength: 40
      firstName:
        description: First name of person
        type: string
        minLength: 2
        maxLength: 40
      ...
```

Listing 1-3 Example: API definition purely from provider's point of view

An API Formulated Solely from the Backend Developer's Point of View

Let us assume we need an API where end customers can change their own master data, in particular their name. Such an API could look like Listing 1-3. However, as one might expect, you will not think of a customer when you see the term `BusinessPartner` as defined in the SAP system.

The corresponding API might be better understood than the previous example. However, the internal controlling numbers (`tenant`, `partnerNumber`) must not be visible to the outside world. A partner number is not understandable to a developer who wants to change a customer's name. A customer identifier as a path variable would be the better choice because that is what an external developer expects in such an API, as shown in Listing 1-4. There are certain best practices for formulating APIs. We will discuss them in Chapter 6. Such a definition meets the expectations of developers because developers expect APIs to be defined using best practices. If an API is defined using those best practices, it can be used elegantly.

```
1  paths:
2    /customers/{customer-id}:
3      get:
4        ...
5
6  schemas:
7    Customer:
8      ...
9
```

Listing 1-4 Example: API definition purely from provider's point of view

Obviously, internally, those identifiers need to be mapped to the numbers used in the database.

A Purely Technical API Without Business Relevance

An API, which is defined completely based on technical factors, might be comparable to the previous examples. The designer of the earlier API has decided that a Global Unique Identifier (GUID) might be better in an API than a database number.

An incremental database number is usually easy to guess. It can give insights into how many customers a company has. It might be used to gather information about the number of customers in the enterprise or the increase in customers over time.

In contrast, in an elegant API like the one in Listing 1-4, a GUID is the key to accessing customer information. This is in contrast to legacy applications that typically provide only a database number, as discussed earlier.

A purely technical solution might be to define an API where a client can obtain customer identifiers using corresponding partner numbers. In such a case, the schema described earlier (Listing 1-3) does not need to be changed, but the GUID can be used as a parameter (Listing 1-5).

Such a technical workaround is not only annoying; it even implies that technical implementation details spoil the defined API.

However, APIs defined from a purely architectural point of view without any technical background lead in the wrong direction as well, as we can see in the next example.

An API from the Ivory Tower

Let us assume that a large organization wants to offer its products to the outside world. Let us assume it is a shoe retailer. An ivory tower architect working in his office who is not associated with an implementation teams defines the API. The API should provide products as a product catalog so that customers can order the products.

```
paths:
  /customer-identifier
    get:
      description: delivers a customer identifier by a business
      ↪ partner number
      parameters:
        - ref: '#/components/parameters/BusinessPartnerNumber'
      responses:
        200:
          description: Succesful operation
          content:
            application/json:
              schema:
                type: string
                format: uuid
...
components:
  parameters:
    BusinessPartnerNumber:
      name: business-partner-number
      type: string
      in: query

```

Listing 1-5 Example: Technically poor API

The API would look like Listing 1-6. The API returns all the organization's products. There might be organizations where this would work. But in most cases, it would not work. A shoe retailer would have thousands of shoes to sell. Without any filtering, the collection of shoes to be created on the client side will be too large for most browsers or mobile applications. Moreover, the database on the server side would be overloaded. Even if the database load is handled by some caches, the client object tree would be too large for most clients.

Even if an API is formulated from the consumer's and business's point of view, technical restrictions on the server and client sides need to be taken into consideration.

Our last example is one of the authors' favorites. Let us look at the world domination API.

Overloaded World-Domination API

An inexperienced API designer might be forced to put everything that the application can do in a single API. Especially in legacy systems, which are not elegantly tailored, all functionality might occur in the API. Imagine an accounting system where customers can manage their transactions. Such a world domination API would look like Listing 1-7.

```
paths:
  /products:
    description: Returns all products from catalog
    responses:
      200:
        description: Successful operation
        content:
          application/json:
            schema:
              type: array
              items:
                $ref: '#/components/schemas/Product'
              ...
```

Listing 1-6 Example: Ivory tower API

```
paths:
  /customers:
    ...
  /customers/{customer-id}:
    ...
  /customers/{customer-id}/addresses:
    ...
  /customers/{customer-id}/addresses/{address-id}:
    ...
  /customers/{customer-id}/accounts:
    ...
  /customers/{customer-id}/account/{account-id}:
    ...
  /customers/{customer-id}/account/{account-id}/transactions:
    ...
```

Listing 1-7 Example: World domination API

Even though the API should only handle accounts and accompanying transactions, it contains customer functionality as well. The API is not tailored for a bounded context (which will be explained in Chapter 4). This is how world domination APIs were created.

APIs are abstractions, as will explained in more detail in Chapter 2. They are usually used outside a department or even organization. Because the user of your API cannot just go to your office and ask you how to use your API, you need to provide documentation for it.

APIs Without Documentation and Unexpected Behavior

As a developer, you will eventually deal with APIs that are documented badly or not at all. Badly documented APIs can be found across the internet. Imagine an API where individual properties are not explained or where no error responses are defined.

APIs evolve over time, and certain properties may no longer be pertinent. Imagine an API that can be used to onboard customers. Consider a person's middle name as a mandatory attribute. Over time, the corresponding company may wish to expand into Spanish-speaking countries, where middle names are not customary, but a second family name is. If the API provider decides to retain the existing API, using the `middleName` property for the second family name, this could be perceived as unconventional, though not uncommon. However, it greatly diminishes the API's legibility. Users would be confused as to why `middleName` was being interpreted as a second family name.

Another example might be `gender`. A couple of years ago, the mandatory attribute *gender* was defined as an enumerator with two entries, `MALE` and `FEMALE`. With the recognition of diverse identities as an important aspect of our community, other entries need to be allowed. However, changing the enumerator of an attribute means publishing a new version of the API, which might cause troubles on the consumer side, even though it is a nonbreaking change. Therefore, an additional attribute, `enhancedGender`, was introduced as optional. Thus, one needs to select a mandatory gender attribute as male or female and, probably, `enhancedGender`. The reader can imagine that such an API would not be well received.

Other examples of bad APIs include violations of standards (see RFC 9110 [12]), for example, to returning a "200 OK" message for successful operation in HTTP, even though an error occurred. Other examples are specifications that define a content type like `application/json`, but the API returns a simple text.

To avoid such bad APIs, try to imagine how you would implement an API and what features you would wish for. Collaborative and iterative development processes where business and IT experts work closely together can help to avoid APIs no one wants to use. This book will look at methodologies and best practices to successfully apply those processes.

Introduction to Modern Development Processes

Collaborative workshop formats and iterative processes characterize modern collaborative development processes, as shown in Figure 1-3. The picture shows different steps of the development process, especially the API development process. This process constitutes the core of this book and is presented below. It will be described in more detail later (Chapter 5). The significant terms are introduced here, along with a brief overview.

Let us look into the individual steps.

1. A businessman creates a business plan.
    A businessman develops a business plan. The idea can be implemented, at least partly, with software. For example, say someone wants to create an online library where people can lend books out to read. Or they want to create a bicycle rental service where people can book and return bicycles using a mobile app. The

Introduction to Modern Development Processes

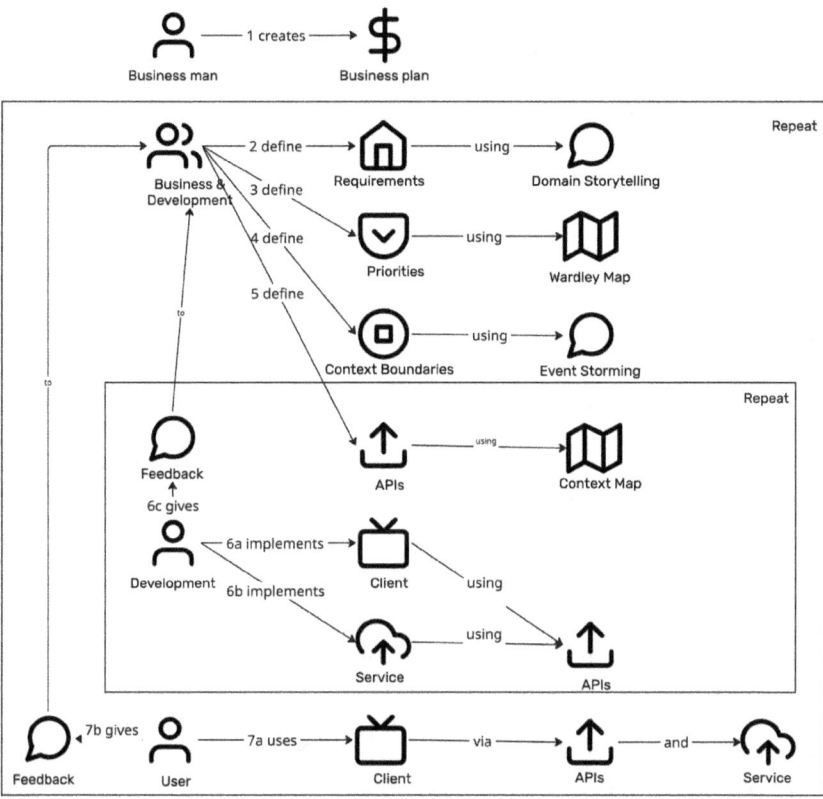

Figure 1-3 Modern, collaborative development process

business plan is documented using a Business Model Canvas [13]. The business plan defines the business capabilities in the first step.

2. Business and development define the business requirements using Domain Storytelling.

   Business experts and IT specialists together define the business requirements for the product to be implemented with respect to the business plan. For this they use the Domain Storytelling workshop format [14]. The story is told sentence by sentence, with a moderator drawing each sentence. The business expert can determine immediately whether or not they were correctly understood. Figure 1-3 is an example of such a domain story.

   The sentences of the story contain an actor represented by a corresponding icon, an action represented by a named arrow, and a working object on which the action is applied. Sentences can be enhanced by adverbial phrases describing certain conditions, for example, "using Wardley map."

The terms used by business experts can be explained in a Visual Glossary [6]. Using Domain Storytelling, business capabilities can be detailed and defined. The initial concepts of bounded contexts can be defined. During this state of development, a business capability can be mapped to a bounded context in a one-to-one relation.

3. Business and development define the priorities using a Wardley map.

   Business experts and IT specialists together define priorities for the development of the product. They use a Wardley map for prioritization during the evolution of necessary software components [15]. A Wardley map offers a way to sort business capabilities as the market matures. The degrees of maturity are genesis, custom, product, and commodity.

   Genesis means that a development is forthcoming.

   Custom means that the corresponding business capability needs to be developed by the enterprise itself.

   Product means the corresponding capability is available on the market but needs high customization to fit into the business.

   Commodity means that the corresponding capability can be purchased on the market and can be used without high customization.

4. Business and development define the bounded contexts using Event Storming.

   Business experts and IT specialists define the boundaries for individual software components together in a collaborative workshop format. The workshop format is called Event Storming [16].

   Business experts and IT specialists search for domain events. They mark business-relevant status changes in business objects. The domain events are enhanced by commands, which trigger them, by roles (human users) or technical processes (technical users), which state the command, and the data that are necessary or are changed when the event happens. The changed data are candidates for aggregates of an accompanying bounded context. The data that are necessary to fulfill the event but are not changed are read models or views.

   At the end of the workshop, the bounded contexts can be defined. When either a role or an aggregate changes, it can be a bounded context. Workshop attendees need to define whether or not the change indicates a bounded context.

5. Business and development define APIs.

   Business experts and IT specialists together define API specifications. This can be done using graphical tools or specification languages like OpenAPI and AsyncAPI. A context map can be used as graphical support.

   To create a context map for each bounded context found, a rectangle or other figure is drawn. Based on the Event Storming workshop, the necessary data exchanges from the read models or aggregates can be identified. These are shown as arrows in the direction of the data flow using the names read models and aggregates found in the Event Storming.

   Based on the data flow, it can be decided whether an asynchronous or synchronous flow is necessary, which will be discussed in more detail in Chapter 2.

   The necessary paths or messages can be defined based on the communication patterns defined in the context map. The schemas of the necessary data can be

specified using the created Visual Glossarys. In this way, it is guaranteed that business and IT will use the same language – the ubiquitous language – in a bounded context.
6. Developement teams implement client and server applications using the API definitions.
The team can implement the client and server applications. Because the APIs were defined earlier by business and development together, the server and client implementations can be done independently of each other. Client and service implementations use the same language – the ubiquitous language. The business is well defined and can be elegantly represented in the implementation.
During the implementation, the development team can give feedback to business management and the IT experts about the API implementation.
7. User use the business functionality supported by a client, API, and server implementation.
Users can use the application via a client, and they access the server functionality via an API. Users can also give feedback to the business and development teams so that the API and its implementation can be continuously improved.

The presented development process is collaborative and iterative. The feedback loops during the implementation of an application, as well as during the usage of the application, allow for constant improvements and iterative development. APIs can maintain backward compatibility while their implementation and behavior are enhanced.

Collaborative Approaches Improve Processes

As we saw earlier in section "Introduction to Modern Development Processes", modern development processes can be collaborative and iterative.
However, let us dive deeper into why they make it possible to avoid bad APIs, as we saw in section "APIs Nobody Wants to Use".
Recall that the API in the first example in section "An API Taken Directly from a Database" was defined based on the database. It could not be understood because the database was designed in a much more generic way than was necessary for the API. If you apply the process introduced earlier in section "Introduction to Modern Development Processes", you can avoid such an API. Business and IT experts define the necessary data flows together using their ubiquitous language so that the API can be defined before the database is defined. The terms in the API are understandable, and the internal implementation of the database is hidden behind a facade.
The second example in section "An API Formulated Solely from the Backend Developer's Point of View" was easier to understand than the first example. However, a consumer without more knowledge would not be able to understand the API. Therefore, it is recommended that the Domain Storytelling and Event Storming workshop formats be carried out together, not only with business experts from the producer's side but also with business and technical experts from the

consumer's side. The methodologies are explained in sections Domain Storytelling and Event Storming. Despite the apparent difficulty in engaging consumers, the Business Model Canvas proves instrumental by identifying key partners. These key partners typically have a vested interest in the application's success and are thus more prone to participate in workshops. Emphasizing the consumer's perspective during API design yields substantial value.

The third example (section "A Purely Technical API Without Business Relevance") is harder to avoid because it does not arise when software is designed on a green field. It reflects the difficulties that crop up when an application evolves over years. A DDD development process is not only a front-up design. It likewise requires a certain amount of governance. The same mechanisms and methodologies used to design a software application on a green field can be used to evaluate design decisions using a brownfield approach. We will discuss those in Chapter 11.

An API from the ivory tower is only a theory (section "An API from the Ivory Tower"). Usually, architects are heavily involved in all stages of the development process. However, it is not just the architects who need to be involved; the business experts and developers implementing the APIs need to be as well. A well-designed architecture cannot be achieved without involving business experts and the involved developers in the workshop formats Domain Storytelling and Event Storming.

A world domination APIs, as in section "Overloaded World-Domination API", can be avoided by well-tailored services. Bounded contexts can be detected using DDD, and services can be tailored to business requirements. A well-tailored system allows for evolution over a long time period and easy maintenance even after years of use.

APIs without documentation or unexpected behavior (section "APIs Without Documentation and Unexpected Behavior") should not appear at all, even if no collaborative or iterative development process is used. But they do indeed appear because of misunderstood business or development requirements under some sort of time constraints. You can be sure that any workshop you conduct with people outside business and IT will repay you several times over thanks to the use of an understandable language, well-tailored systems, and good and understandable APIs.

Applying an API-First approach, the specification is written first. The documentation is explicitly in the specification. Using API-First, it is easier not to "forget" the documentation (section "API-First vs. Code-First").

Only understandable, well-tailored, and well-documented APIs can be offered to access business capabilities. As far as marketing the business functionalities is concerned, it should be done not only via user interfaces but also via APIs. The entire process is explained in Chapter 5.

Keeping these improvements in mind, let us dig deeper into why those processes are valuable and necessary.

Why a Modern Development Process Is Necessary and Valuable

Unsuccessful projects are often unsuccessful not because they are badly planned or underbudgeted or have a lack of resources. They are unsuccessful because we are all humans.

Humans can be overconfident or inexperienced or try to avoid revealing their mistakes. It is our nature to be sometimes overconfident and to start tasks we are not actually capable of performing [17]. We think we are experienced enough to lead a large project when we are not. It is difficult for us to tell someone that we made a mistake.

People are not robots. And software development is a people business [18]. Therefore, it is necessary to build trust between all parties involved in a project.

Trust you can build to involve everyone affected in your decisions:

- Build a common understanding between business and IT using Domain Storytelling.
- Build a ubiquitous language using a Visual Glossary.
- Build a reliable and stable project setup using Event Storming and a context map.

Using models and methodologies leads a project to trust in the necessary collaboration. Trust is not only created within a project among all project members; it is established even between project members and customers because even customers and sponsors are involved in workshops. Trust in a project makes it possible to manage mistakes jointly and pair overconfident with inexperienced members. In this way, everyone benefits.

What's more, in this way, one can see the project from a product point of view. A product point of view is beneficial because it handles the unpredictable world instead of stating forecasts as project management requires. In product mode, teams are funded, not projects. In this way, teams can stay longer and are not only responsible for solutions; they are responsible for building, running, and iterating solutions. Most importantly, creativity, building trust, and operation are combined in one team instead of across several teams. Teams working in product mode are able to reorient themselves quickly. They can reduce the end-to-end cycle time. They are able to iterate, which is not possible in a forecast-requiring project mode [19].

Using product mode is beneficial because the end-to-end cycle time is reduced, and failures are early and cheaper in comparison to a classic project mode. The product is highly supported by the collaborative Domain Storytelling and Event Storming workshop formats.

Being faster and creating more value in product development or even in an agile project requires a collaborative and modern development process.

API-First vs. Code-First

APIs are facades to behind-the-scenes implementation. They are used to abstract implementation from usage. When you perform such an abstraction, you can do it in two ways:

- Define the abstraction by the implementation – "Code-First" or
- Define the abstraction by the usage – "API-First."

Both ways are valuable and can be used.

Different variations of the "Code-First" and "API-First" approaches are shown in Figure 1-4.

Backend first means that the backend service is implemented first. The resulting API is pushed to be used by other services and frontend applications.

Frontend first means that the frontend application is first implemented. The API definition is pushed to the backend to be implemented.

A third variation, **database first**, means the database model defines the APIs used by the backend and the frontend. In this scenario, the database schema is designed first and will then be incorporated into the backend logic. From the backend it will then heavily influence the API design to the frontend, where it will then shape the frontend logic.

API-First means that the contract is defined before any implementation.[1]

In a Code-First approach, as the name suggests, developers write the Code-First. They document the code after the fact and let the API specification generate. Developers with a clear understanding of API design might see certain advantages in this approach [20]:

- Quick time to market and
- Suitable for Proof of Concepts (PoCs) and rapid prototyping.

It might not be suitable to involve stakeholders, testers, and technical writers because they need to wait for the API specification until the code is ready. Usually, such specifications are difficult to read because they are generated out of quite technical code.

On the other hand, it allows an API-First approach to involve stakeholders and project members. An architect or a developer writes an API specification, for example, an OpenAPI specification first, and subsequently, implementation is performed. Even though at first glance it seems time-consuming, it has a couple of advantages:

- All project members involved are informed about the API definition [20].

---

[1] The contract is closed between backend service and other services. The API contract is not valid for communication to the database. The DDD pattern "Repository" needs to be used to secure the domain model from the database model [1].

API-First vs. Code-First

Backend first

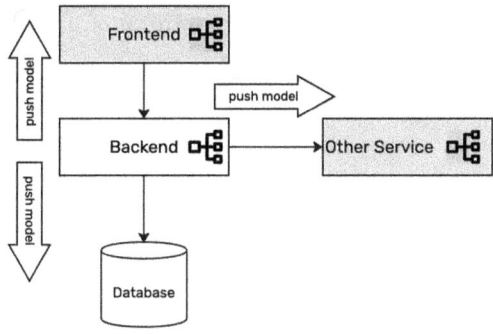

Frontend first

Database first

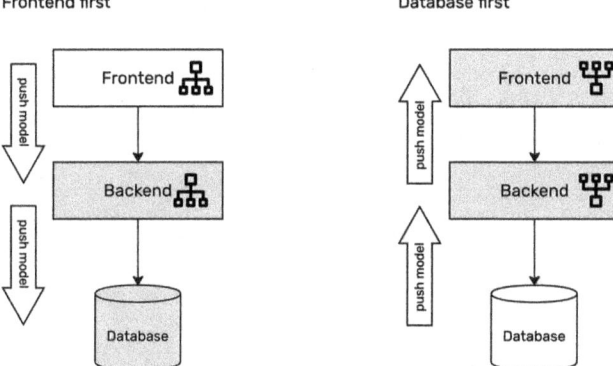

API-First

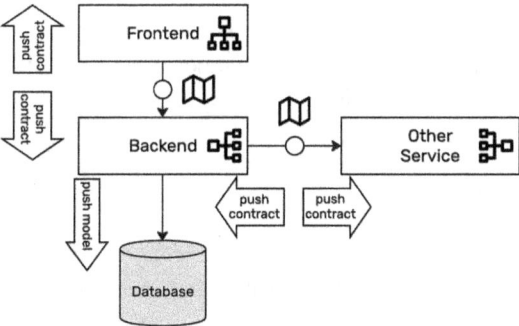

Figure 1-4 API-First vs. Code-First

- It allows parallel work of service and client implementation, testers, and technical writers [20].
- The ubiquitous language represented in the API specification can be reliably used in the implementation code.

Moreover, the implementation code can be generated. The generators are available for multiple programming languages and API specification languages (interface definition languages (IDLs)).

The advantages of the approaches are as follows:

- Code-First: fast development, fast time to market
- API-First: reliable service and client implementation, reliable application of ubiquitous language, stakeholder involvement

The disadvantages of the approaches are as follows:

- You must wait for API specification until the service implementation is done; the specification is not available to be discussed before implementation.
- Implementation development must wait until all stakeholders agree upon the specification.

Which approach can be used for which project nature?

- PoCs
- Product development and agile projects

Because API design is critical for the success of software products or even for software projects, a design-first approach or, as it is called here, the API-First approach, is preferred. It is fast and efficient. In this book, we will show how an API-First approach can be successfully applied in projects and how to design elegant APIs with it.

To understand the function of APIs better in modern software development, let us look at the communication categories first.

Points to Remember

- APIs are everywhere.
- Most projects run in a VUCA world and therefore need to be able to quickly adapt to changes.
- Code-First is great for quick time to market and PoCs.
- API-First is better for projects that require a well-defined API with a ubiquitous language, multiple informed stakeholders, and parallel implementation.

Review Questions

1.1 Ugly APIs Consider the poor and ineffective APIs you have encountered. Which of the following characteristics would you attribute to a subpar API? Multiple responses are applicable.

(a) There are well-described endpoints in a REST API.
(b) A property is named `employee` and used for external stakeholders.
(c) An integer database key is revealed in the API.
(d) The API makes it possible to obtain a GUID based on a database key.
(e) A customer resource is accessed via a customer identifier in the form of a GUID.
(f) The term "voltage adapter" is used as defined in a ubiquitous language.
(g) One API reveals all possible resources of an enterprise.

1.2 Feedback loops Reflect on a modern and collaborative development process. What are the minimum essential feedback loops required?

(a) A modern development process does not need any feedback loops. Through the collaboration of business experts and IT specialists, no mistakes are made.
(b) Feedback is necessary only between IT specialists. Business experts are only involved in definitions.
(c) Feedback is necessary between developers and business experts and IT specialists – but only during implementation.
(d) Customers should give feedback to business managers when using the application.
(e) Even though feedback should always be given, feedback from development to the API definition is necessary. Other feedback can be given, but it is not essential.

1.3 "API-First" or "Code-First" Reflect on the development of a new web application where a production-ready version needs to be implemented. Which development approach would you choose?

(a) I would choose API-First.
(b) I would choose Code-First.
(c) I would wait until my manager said what to do.

References

1. Evans E (2004) Domain-Driven Design: Tackling Complexity in the Heart of Software. Addison-Wesley, Boston
2. Snowden D (2010) The cynefin framework. [Online] Available: https://www.youtube.com/watch?v=N7oz366X0-8. Visited on 21 Jan 2025
3. Bennett N, Lemoin GJ (2014) What vuca really means for you. [Online] Available: https://hbr.org/2014/01/what-vuca-really-means-for-you. Visited on 26 Jan 2025

4. Junker A (2023) Integrationsarchitekturen in einer vuca world. In: OOP Munich 2023
5. Lilienthal C, Schwentner H (2023) Domain-Driven Transformation. dpunkt.verlag. ISBN: 978-38-64908-84-2
6. Zörner S (2015) Softwarearchitekturen dokumentieren und kommunizieren, Entwürfe, Entscheidungen und Lösungen nachvollziehbar und wirkungsvoll festhalten, 2., überarbeitete und erweiterte Auflage. Hanser, München, 277 pp. Literaturverz. S. [269]–272. ISBN: 978-34-46443-48-8
7. Goldstine HH, von Neumann J (1947) Planning and Coding of Problems for an Electronic Computing Instrument. Report on Mathematical and Logical Aspects of an Electronic Computing Instrument, Part II, vol 1–3. Institute for Advanced Study, Princeton
8. Cotton IW, Greatorex FS (1968) Data structures and techniques for remote computer graphics. In: Fall Joint Computer Conference
9. Lauret A (2019) The Design of Web APIs. Manning Publications Co. LLC, New York, 1324 pp. Description based on publisher supplied metadata and other sources. ISBN: 9781-63-8351-19-1
10. OpenAPI Initiative (2021) Openapi specification v.3.1.0. [Online] Available:https://spec.openapis.org/oas/latest.html. Visited on 14 July 2024
11. LeanX (2024) Sap table but000. [Online] Available: https://leanx.eu/en/sap/table/but000.html. Visited on 29 Jun 2024
12. Fielding RT, Nottingham M, Reschke J (2022) HTTP Semantics, RFC 9110. https://doi.org/10.17487/rfc9110. [Online] Available: https://www.rfc-editor.org/info/rfc9110. Visited on 29 July 2024
13. Strategyzer (2024) The business model canvas. [Online] Available: https://www.strategyzer.com/library/the-business-model-canvas. Visited on 21 July 2024
14. Hofer S, Schwentner H (2022) Domain Storytelling, a Collaborative, Visual, and Agile Way to Build Domain-Driven Software. Pearson International, London. ISBN: 978-01-37458-91-2
15. Wardley S (2022) Wardley Maps, 2. überarbeitete und erweiterte Auflage. Simon Wardley, Heidelberg
16. Brandolini A (2024) Event storming. [Online]. Available: https://www.eventstorming.com/. Visited on 30 Jun 2024
17. Kravcenko V (2024) Why software projects fail. [Online] Available: https://vadimkravcenko.com/shorts/why-software-projects-fail/. Visited on 10 Aug 2024
18. Welch M (2019) Software: A people business. [Online] Available: https://www.odyodel.com/software-people/. Visited on 10 Aug 2024
19. Narayan S (2018) Products over projects. [Online] Available: https://martinfowler.com/articles/products-over-projects.html. Visited on 10 Aug 2024
20. Tucci M (2023) Code-first vs. design-first: Eliminate friction with api exploration. [Online] Available: https://swagger.io/blog/code-first-vs-design-first-api/. Visited on 10 Aug 2024

Communication Categories 2

In this chapter, we look at communication channels – which communication channels exist, and how to distinguish them. This will help us later in deciding which communication strategy to use between components.

In this book, we focus on network-based APIs, and therefore, we look at network-based communication. These APIs are sometimes called remote APIs or web APIs [1, 2]. All APIs we define in this book are specifically designed to be called over a network including the internet, local area networks (LANs), and wide area networks (WANs). However, for completeness, we note here that more types of APIs and corresponding communications exist. In the next section, we will define the term API and show the broad and varied uses of it.

The Term API

In Chapter 1, we already looked at some examples of APIs. Now let us define the term API. An application programming interface is a programmable interface to some functionality of an application. This functionality is hidden behind an interface. The simpler way to use functionality would be directly without an interface. But without an interface, there is no abstraction, and you have tight coupling between components or just long, unstructured code in the most extreme. If there were no abstraction whatsoever with some kind of interface, there would simply be copy-pasted code everywhere, as no functionality would ever be abstracted in a reusable manner. For that reason – and many others which we will go over later – interfaces are used to abstract. In general, every API is an abstraction of the functionality behind it, which can be used programmatically. We see how an API works in Figure 2-1.

APIs are already pervasive throughout the software industry for different reasons. We will examine the details of the smartphone example given in Chapter 1. The smartphone camera application uses a wide variety of APIs (this is also visualized

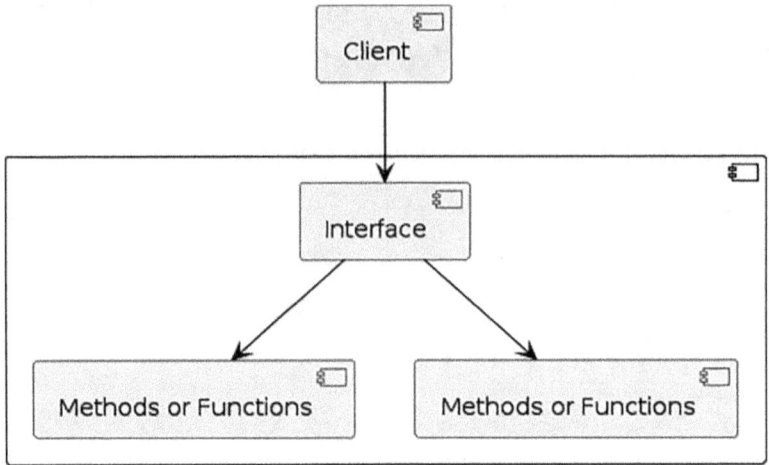

Figure 2-1 The interface is the glue between the caller and the backend code

in Figure 2-2). We go over a small user journey from taking a picture with the camera app and to sharing the photo with a chat:

This section is inspired by "The Design of Web APIs" by Lauret, where a more detailed example of all types of APIs can be found [3].

1. When the camera application starts, it gets invoked by the operating system (OS) with some **lifecycle hook**. These are functions that the application needs to implement and are called by the OS.[1] With these API endpoints, the OS can start the app and know when the app will be ready for use in a standardized way. The OS does not need any knowledge about the application.
2. The app calls **internal functions** of its code base. These are modules that are internally abstracted. Using modules makes the testing of components simpler. It allows different developers to implement different modules independently.
3. The app loads some user interface (UI) elements. These elements can be loaded from a **library** and be called using an API. In a mobile OS, those components are changing constantly with each version of the OS. However, the API to load those components is stable and does not change with each version update.
4. The app tries to **access the camera feed**. Fortunately, the app can use an OS API to do so. The camera can also be accessed abstractly. The app does not need to know details about the vendor and model of the camera.[2]

---

[1] Lifecycle hooks in Android: https://developer.android.com/guide/components/activities/activity-lifecycle.

[2] Access camera in Android: https://developer.android.com/media/camera/camerax/preview.

The Term API 27

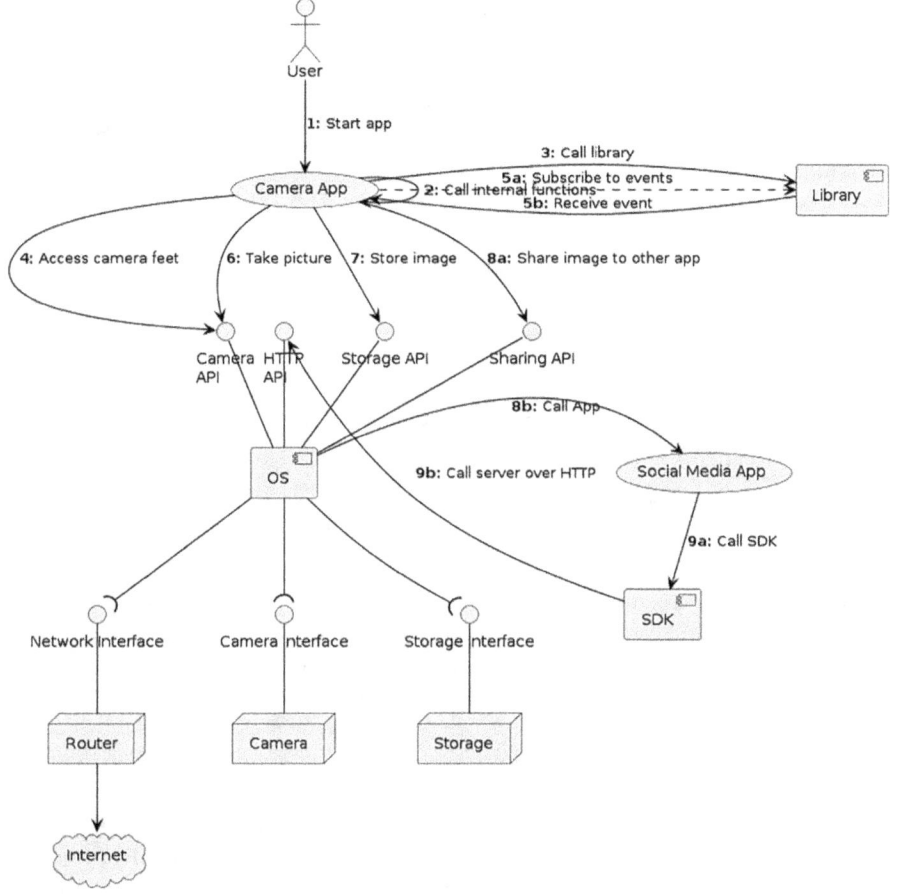

Figure 2-2 API usage in camera app example

5. The app **subscribes to events** to receive the events when the shutter button is pressed. The underlying OS itself receives the event by a system interrupt and then sends the event to the subscribers.[3]
6. The event is received. The app can now **take a picture** using the camera API.[4]

---

[3] In Android: https://developer.android.com/develop/ui/views/components/button#HandlingEvents.

[4] Take a picture in Android: https://developer.android.com/media/camera/camerax/take-photo.

7. The picture taken needs to be **stored** using an API that enables file streaming for storage. This is a great example of a wrapped API on Android: You have a media abstraction of the API over the normal storage API.[5]
8. Next, the user wants to **share the image** taken. This action involves APIs, too. The camera app needs to call the sharing API of the OS. The receiving app needs to implement the API to be called with the shared data.[6]
9. The receiving app publishes the photo in a chat. The app calls an API from the server over the **internet**.[7] Maybe the API will be called again over a **software development kit (SDK)**, which abstracts the internet call away and looks like a function call of a library.

   Naturally, the internal call via HTTP must be further directed through OS APIs in order to be transmitted over a Wi-Fi or phone router module.[8]

To summarize, here are the benefits of APIs (most of them were already shown in the camera app example [2, 4–6]):

Loose coupling and reusability   APIs are forever; code is for the moment. Code can change without affecting the associated components (in runtime and in development). This results in better reusability of code, which can be vastly improved using a good API design. Such improved code can be used for more cases like this.

Separate versioning   APIs allow separate releases and versioning. When an API is available in multiple versions, it allows smooth updates that do not have to be made instantly.

Separate coding languages   APIs make it possible to communicate between different architectures and programming languages.

Abstraction of underlying layers   APIs allow for abstracting underlying layers, for example, the use of SDKs that use an HTTP interface that uses Transmission Control Protocol (TCP). All these layers can, however, be changed since they are abstracted by APIs.

Abstraction for testing   An API can have multiple implementations. This feature is particularly useful in testing scenarios. It allows for simple testing with mock or alternative implementations, while the code under test does not need to know about that. The code under test is still referred to by the same API.

---

[5] Storage APIs in Android: Media API: https://developer.android.com/training/data-storage/shared/media#add-item; underlying APIs (kernel level): https://source.android.com/docs/core/architecture/android-kernel-file-system-support.

[6] Share API in Android: https://developer.android.com/training/sharing.

[7] Simple Hypertext Transfer Protocol (HTTP) API in Android: https://developer.android.com/reference/java/net/HttpURLConnection.

[8] Android Telephony API: https://developer.android.com/reference/android/telephony/TelephonyManager.

Clean architecture APIs allow for clean divisions between components and modules. This is used to separate bounded contexts but can also be used for libraries, for example.

Create or buy An API can be bought or implemented; the API allows for different implementations. This is the case for standardized APIs, where you can find different products on the market that provide different implementations for the same standard API.

A good example is Structured Query Language (SQL) for databases. SQL is a well-defined standard by International Organization for Standardization (ISO) and an American National Standards Institute (ANSI) approved committee that is implemented by many vendors. The implementation can be changed simply as long as only the standardized part of the SQL API is used [7, 8].

Easier maintenance and operations APIs are well-defined interfaces between components. These allow for simpler integration of infrastructure components, like proxying, load balancing, tracing, failovers in disaster cases, and so on. This simplifies maintenance.

Simpler adaptability An API can also be used to build adapters for several products. This allows for the creation of interoperability among multiple products with different APIs.

Time to market When using APIs to develop products, you may use some functionality only as Software as a Service (SaaS) offerings, so products can be brought to market faster.

Loose coupling and abstraction bring many benefits; entire books can be written on these topics alone. However, we will not go into more detail; we just wanted to discuss certain points that are relevant to our use cases.

Having considered all possible types of APIs and their benefits, we will now define communication over a network and the APIs used there.

Communication in General

Communication can be abstracted to the following high-level pattern: One component sends "information" over a channel to another component (Figure 2-3).

Communication itself never changes, but the pattern of implementation of this communication can vary heavily. Different mechanisms that are commonly used are discussed in the following sections.

Interaction patterns are defined by the interfaces the components use to communicate with each other and the objects they transfer (Figure 2-4).

Figure 2-3 Communication between two components

Figure 2-4 Communication between two components with interfaces

We will only focus on remote APIs; other mechanisms, like shared databases or file transfer over a network, will not be discussed, as they do not provide the abstraction of an API that we are looking for [4].

As a common example, we look at the interaction of a user with a web server when the user accesses a webpage: A personal computer (Component A) calls the web server (Component B) over the internet. There is a communication channel, here HTTP over TCP/IP with two HTTP interfaces, one for the sender and one for the receiver.

We dive into interfaces and the naming of the components of an interaction in more detail in section "Communication and APIs". However, first we need to look at the different kinds of communication and at the central part of it: the "message."

Messages as the Central Building Block

A message is a package of data that should be transmitted over a channel from sender to receiver at once [4]. The message is normally structured in metadata and a message body containing data in a custom schema.

Using HTTP, this is the request (or even the response) with header and message body. In an event streaming system like Apache Kafka, this is a message in a topic.

Communication Styles and Mechanisms

We defined communication on an abstract level. Let us look more closely at the different types of communication and how they differ from each other. Figure 2-5 provides an overview of the communication types: communication mechanisms, architectural patterns for integration architectures, and protocols that can be inherited from them. This list is far from exhaustive, but it offers an overview of common approaches. The diagram in Figure 2-5 shows the approaches and protocols (white boxes) covered in this book, as well as those (gray boxes) that are close to them, so they can be differentiated.

In the following pages, we take a look at each layer.

Communication Styles and Mechanisms 31

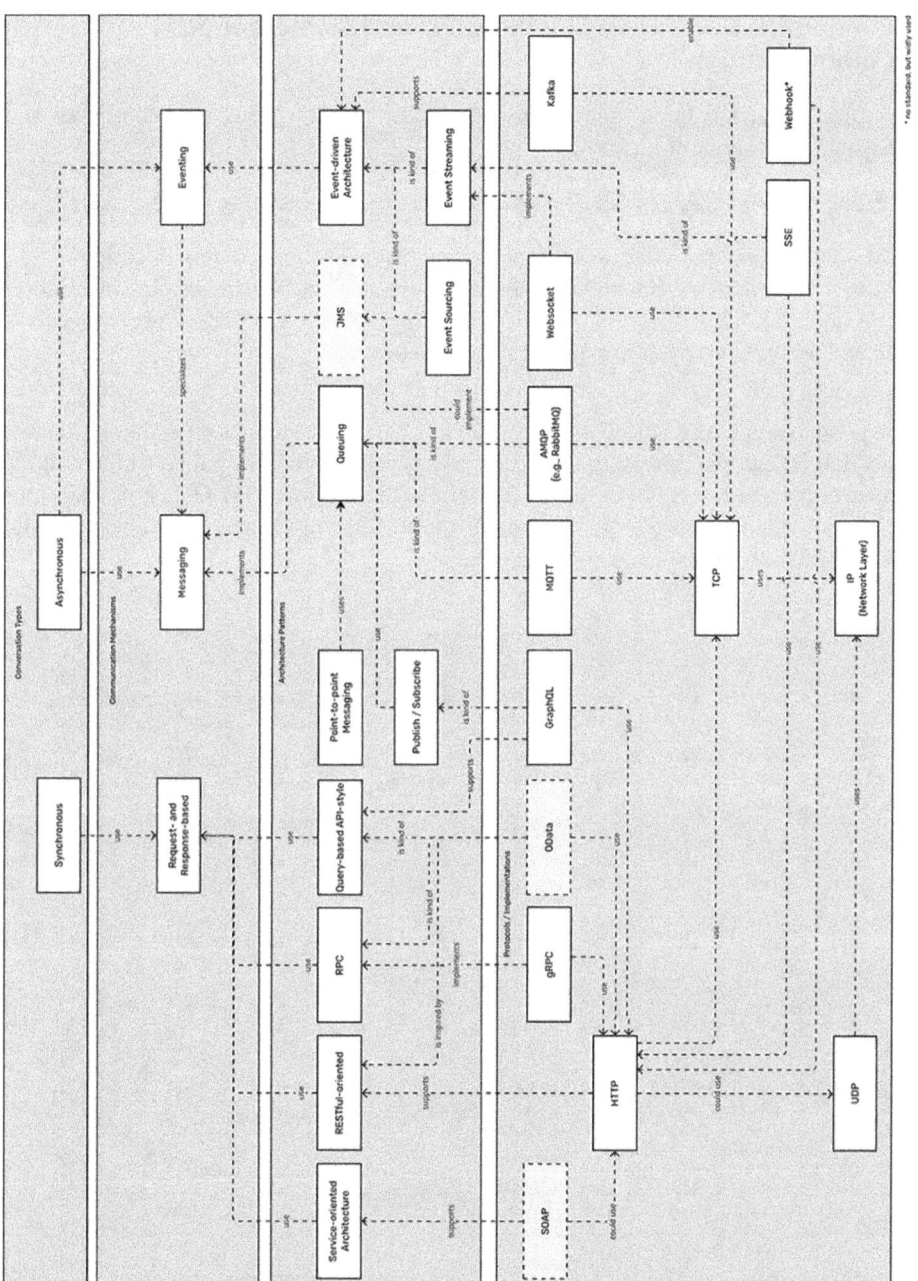

Figure 2-5 Overview of communication types, communication mechanisms, architectural patterns, protocols, and implementations. Asterisks ("*") indicate a commonly used approach, not standards. The darker gray boxes with dotted borders are protocols that the authors do not, in general, recommend

Communication Types: Synchronous and Asynchronous Communication

A core aspect this book focuses on is the differentiation between synchronous and asynchronous communication.

Distinguishing between synchronous and asynchronous communication

We distinguish between synchronous and asynchronous communication types. Please note that we are talking about the interaction principles of two systems (Figure 2-3). This does not automatically imply that each underlying connection or call will also be synchronous or asynchronous.

Example
A good example is a RESTful API over HTTP, which provides an endpoint to store a value. The communication type is clearly synchronous. The callee waits until the answer is stored, and it receives a result back (Figure 2-6a). HTTP is a synchronous protocol (mostly). However, looking at the underlying protocol, which is usually TCP,[9] we can see asynchronous patterns:

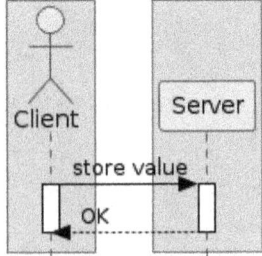
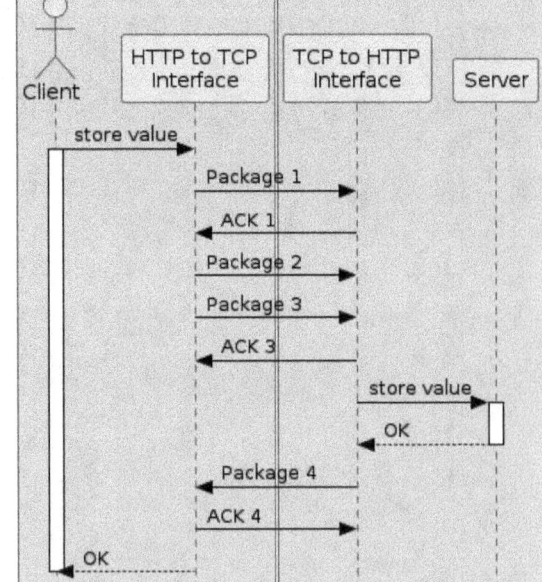

(a) HTTP store value (simple)　　　　(b) HTTP store value (with TCP)

Figure 2-6 Storage of values over HTTP is synchronous. However, the underlying TCP connection has asynchronous components

---

[9] Since HTTP/2 with HTTP over QUIC and the proposed standard HTTP/3, next to TCP, user datagram protocol (UDP) can be used, too [9].

In TCP, packages are sent out in an asynchronous fashion. The sender transmits (after a ramp-up phase) multiple packages to the sender without waiting. Later, it expects an acknowledgment (ACK). If it does not get the ACK, it will resend the data (Figure 2-6b) [10].

Thus, the TCP transmission is asynchronous; however, in this case, HTTP is synchronous. Recall that we must look at the overall intent and architecture to decide whether communication is synchronous or asynchronous.

At any rate, we can abstract beautifully and do not need to understand the underlying details.

Synchronous Communication

In synchronous communication, the callee gives a message to the other component (request) and waits until it receives an answer (response). This is visualized in Figure 2-7. Component A needs to wait until Component B has processed the data and sends a response.

This is like a face-to-face conversation. You talk face to face with someone who is focused on you, and you wait for an answer from that person.

Benefits

The synchronous approach has the benefit that Component A knows if the processing was successful. It might receive a valuable answer in addition (e.g., if you want to create an order, you can receive an order number, which you can refer to later).

Drawbacks

Waiting for an answer is the drawback. Actively waiting for an answer is necessary. Of course, if Component A is a service, it can wait in an additional thread and may not need to block everything; however, it needs resources. Additionally, Component B must be available at the moment the message is received and provide resources to process the request.

Asynchronous Communication

In asynchronous communication, actively waiting for an answer by the recipient is not necessary. Instead, Component A sends a message but does not wait for

Figure 2-7 Synchronous communication between two components

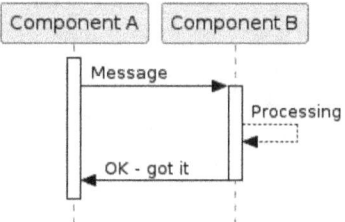

Figure 2-8 Asynchronous communication between two components

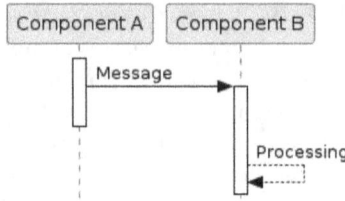

an answer. Depending on the implementation and protocol, Component A does not even wait for the recipient to receive the message. The principle is shown in Figure 2-8. Component A sends a message but does not wait for Component B to receive or acknowledge it.

This is exactly how a letter works. The letter is sent, but the sender does not wait for the letter to be read or a response from the recipient.

Benefits
Using asynchronous communication, waiting for the other component to process the message is not necessary. The message can be sent, and then the application can proceed with the subsequent steps. With asynchronous flow, the design is more resilient at dealing with latencies [11].

Drawbacks
The drawback, however, is that there is no knowledge if the processing was successful. The message is given to the other component. However, depending on the use case, it might be necessary to check afterward whether it was successfully processed or if the sender needed the result. In addition, the application still needs to retry if the other component does not respond to the message to ensure that the message was, in fact, received. At least an acknowledgment that the message was received is necessary. (However, there can be use cases where a response does not matter.)

Asynchronous Communication with Brokers
At any rate, we usually don't hand letters directly to recipients but send them through the mail. The post office batches letters, sends them together, and stores them in the receiver's mailbox. Letters can be sent with great flexibility, and the sender can know that the letters arrive safely thanks to the guarantee given by the postal service, even if the recipient is not available at the time to receive the letters directly.

A similar approach is taken in integration architectures when using asynchronous communication. Messages are sent to a broker, and the broker sends the messages to the recipient.

Communication Styles and Mechanisms

Figure 2-9 Asynchronous communication between two components

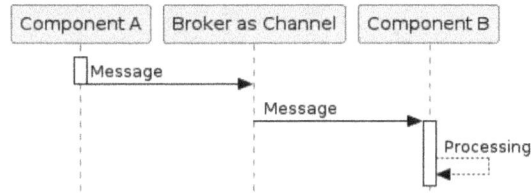

This is visualized in Figure 2-9: Component A sends a message to the broker. The broker is responsible for delivering the message to Component B (or Component B needs to retrieve the messages from the broker, depending on the protocol and the architecture).

As shown in Figure 2-5, communication mechanisms, messaging, and eventing are often used with a broker. The architectural patterns event-driven architecture (EDA) and Publish / Subscribe are almost exclusively implemented with a broker.

Benefits

The benefit to brokers is evident: A message can be easily sent to a broker, and then the task is complete. The message is at the broker, and Component A did its duty. There is no need to wait for Component B to become available or ready to receive the message. The processing of messages is outsourced to the broker, who takes on the responsibility of storing messages until they can be delivered.

In addition, the broker represents an additional layer of decoupling. The knowledge where Component B is deployed is no longer needed by Component A. Direct access to the application is not necessary, which allows for network isolation.

When the messages are modeled as a unidirectional event, Component A does not need to know that there is a Component B that is interested in the event. It just sends a state change for all interested parties (subscribers to the messages).

We will take a closer look at this in section "Messaging vs. Eventing".

Drawbacks

This architecture is really great, as it decouples components. However, it produces additional complexity. New infrastructural components are introduced that need to be released and maintained.

In addition, small teams might be too challenged to introduce a broker. The additional efforts for DevOps activities might overwhelm small teams.

Communication and APIs

As previously discussed in section "Communication in General", an interaction goes from one component to another. An interface was defined between them, or, to put a finer point on it, two interfaces are defined with a channel as intermediary.

Generally, a provider *provides* an API to a consumer, which *consumes* the API (Figure 2-10).

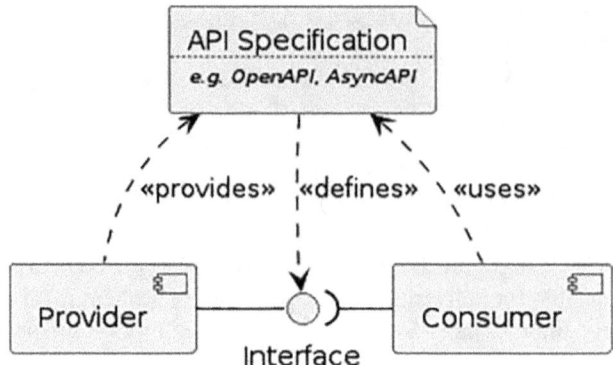

Figure 2-10 Communication and APIs

The API can be described in an API specification. An API specification describes the given API with all endpoints, how they behave, and what can be expected from them.

In theory, an API specification can be provided by any party. However, it should come from the API provider, who is most qualified to describe which endpoints exist, how they behave, and how they change over time.

Synchronous Communication
In synchronous communication (e.g., with HTTP), the provider is normally an HTTP server. The server provides an API that gets consumed by a client.

The client makes an outbound call to the server based on the API specification (the server gets an inbound connection from the client).

Synchronous Communication with Intermediates
Even in synchronous communication, "brokers" can be added between client and server. Normally, proxies, load balancers, or gateways are added. These are important infrastructure components; however, from a communication perspective, they are passive:

Proxies and load balancers are transparent, and the consumer does not even need to know that they are there.

Gateways can change the details of API calls. However, the provider can still be called, even though it is represented by the gateway (Figure 2-11).

> **More on that later**
>
> We will not yet go into detail about those components. In section "Publishing APIs", we will take a closer look at these components.

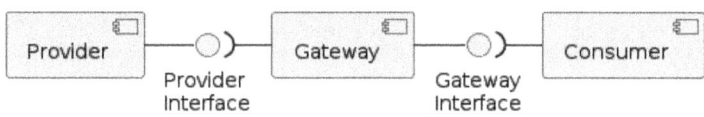

Figure 2-11 Synchronous communication with a gateway

Figure 2-12 Asynchronous communication with broker

Asynchronous Communication Without Broker
In asynchronous communication, a producer sends a message to a consumer.

Asynchronous Communication with Broker
In asynchronous communication with a broker, the broker provides the API interfaces (in general, of course, there are exceptions).

This is the same as in Figure 2-4, where there are technically two interfaces. Both are at the broker. One is for the producer and one for the consumer. One produces messages, the other consumes them. This is visualized in Figure 2-12.

The API specification is not visualized here. In section "Definition of Asynchronous Interfaces", we will introduce the API specification for asynchronous APIs; in this context, the specification can be defined in multiple places, which can have different benefits and drawbacks. However, in a broker setup, it is common for either the producer or the consumer to define the API specification depending on the architectural pattern.

Communication Mechanisms, Architectural Patterns, and Suitable Protocols

In the preceding section, we looked at communication types and mechanisms, as well as the definition of the term "API."

Here, we provide an overview of the communication mechanisms, architectural patterns, and protocols that support this approach, which were shown in Figure 2-5. In the following chapters, we will dive deeper into this. Therefore, we will provide many references containing more detailed descriptions.

Of course, these distinctions are not strictly defined in practice; they will constitute more of a mental model in deciding which approach, pattern, or protocol to start with.

> **! Warning: Protocols are sometimes called after products**

Some protocols are sometimes named after a product.

Let us take the example of Apache Kafka in Figure 2-5. The protocol that Apache Kafka implements is also called Apache Kafka (and sometimes Kafka Protocol or Kafka API).[10]

However, we see again the beauty of APIs. Initially, the product Apache Kafka defined the Kafka Protocol, but as the product became more popular, alternative implementations were created using the same protocol. This allows a seamless (or sometimes almost seamless) switch of the product without a change on the other side, here the producer and consumers that interact with Apache Kafka. Some examples: Apache Kafka was reimplemented by Redpanda as a replacement for Apache Kafka [12]. Microsoft made Azure Event Hubs a "data streaming platform with native Apache Kafka support" by implementing the Apache Kafka protocol as well as Advanced Message Queuing Protocol (AMQP) [13].

Request- and Response-Based Mechanisms

When we engage in synchronous communication, we mainly use a request- and response-based approach. This is natural as you wait for a request to be processed, and then you can send back an answer, as shown in Figure 2-7.

However, the type of request and, therefore, the answer can be quite different:

RESTful-oriented   A RESTful-oriented service is a service that is implemented using the REST architecture. This resource-oriented architecture style is the most common on the web. Richardson calls them "services that look like the Web" [14]. Implementations mainly use HTTP. The style has different endpoints modeled as resources that can be accessed, changed, created, and deleted with different operations; however, to make an API really RESTful and stable, many rules need to be followed [15].

We will look at RESTful APIs in section "Definition as REST with OpenAPI." In section "Introduction to REST," we will go into more detail on RESTful's definition, the separation from REST, and how we implement RESTful-oriented APIs.

RPC-style   Remote procedure call (RPC) is a style where you call a function or procedure on your conversation partner. It is similar to the design of local procedure calls [15].

---

[10] https://kafka.apache.org/protocol.html

We will look at gRPC in particular in this book; more details will follow in section "Definition as gRPC."[11] We also discuss Open Data Protocol (OData) in the following background box.

Service-Oriented Architecture In service-oriented architecture (SOA), protocols like Simple Object Access Protocol (SOAP) are used widely. They exchange messages that allow them to call functions with a certain payload.

It is different from EDA. EDA uses a "dumb pipe," whereas in SOA, a "smart pipe" is used. The "smart pipe" is mainly an Enterprise Service Bus (ESB), which contains the central routing logic with transformation and business logic. The pipe performs an orchestration of the other component.

A "dumb pipe" has smart endpoints. The pipe should be simple and not have any business logic. The endpoints, therefore, the components that integrate over the dumb pipe, are smart and contain logic. They are performing a choreography where each component knows its part and contains its part of the business logic. However, SOA with "smart pipes" lost popularity because of the complexity of the protocol and the central complex building blocks [15, 16].

Query-based API styles Query-based APIs are quite different from the others presented here. By design, the client decides which fields it wants. The server needs to deliver them [15].

We will take a closer look at the GraphQL protocol in this book (section "Definition as GraphQL"). We will not discuss OData here; a short summary can be found in the following background discussion.

OData

OData is a protocol specified by Organization for the Advancement of Structured Information Standards (OASIS), which is sometimes used as an example of a query-based API style [2]. However, it also has RPC styles in it, and they claim to be RESTful [17, 18]. The standard is pushed by Microsoft and SAP [18]. However, for the authors and other critics, the protocol is too complex for most cases and tries to achieve too many things at once [2]. The protocol benefits from an extensive specification, many operations, and capabilities already well defined on the specification level (e.g., sorting and filtering).

A good overview of the protocol can be found in the article "Understand OData in 6 steps" [17].

Now, with our overview of the request- and response-based styles, let us look at the others: messaging and eventing, but before going into the details, we will look at

---

[11] The "g" in "gRPC" does not always stand for the recursive definition "gRPC." The official definition of the "g" changes in every release of gRPC (see https://github.com/grpc/grpc/blob/master/doc/g_stands_for.md); however, it was created by Google, so there is, of course, also another interpretation possible as to what the "g" stands for.

the difference between messaging and eventing from a communication mechanism perspective.

Messaging vs. Eventing

The authors have had many discussions about events vs. messages and eventing vs. messaging. This is why we are trying to distinguish the terms cleanly here.

Messaging In messaging, messages get transferred from a producer to a consumer asynchronously. The messages are transferred over a channel, also called a queue. The consumer reads them, normally by subscribing to a channel up front, and usually deletes them by reading them [4].

Message Messages can be quite specific to a consumer, as they are normally produced for a specific consumer, containing exactly the fields the consumer needs.

Eventing In eventing, messages also get transferred from a producer to a consumer asynchronously. In eventing, events or event messages get transferred.[12] Events themselves are generally handled in a message envelope. The consumer can read the events that happened before if interested; it is not necessary to subscribe to them. Events are not deleted in eventing when they are read, as they may be interesting to other consumers, or even to the same one at a later phase.

Event Events are produced with or without less focus on consumers. In general, events should only represent a state change from a past action and, therefore, not be specific to a given consumer.

After defining the border between messaging and eventing, we will go into messaging and, subsequently, eventing and look at some architectural patterns in communication and some exemplary protocols.

Messaging

Normally, when we use messaging, we use queuing:

Queuing In queueing, messages are sent and held in a queue until they are consumed. Not all messaging mechanisms need to use queuing; however, as the producer just sends messages and does not wait for them to be consumed, it makes sense to store the messages before they are consumed. Otherwise, the sender may need to wait until the consumer is ready to send each message. When the sender transmits a message (and receives an ACK), he can then

---

[12] Events are sometimes also called event messages, as a specific form of messages (see also section "Messages as the Central Building Block").

immediately do something else, as the processing on the other side should be asynchronous. Thus, queuing can be considered a standard for messaging in integration architecture [4].

We can distinguish two types of messaging:

Point-to-point messaging   In point-to-point messaging, one sender sends messages to one destination. It is usually clear which component the receiver is, and the format can be highly specific to the communication between the two components. Sometimes there are competing consumers on a point-to-point channel. These are multiple instances that compete to consume messages, but this is still a point-to-point messaging style, as the consumers are all instances of the same application [4].
Point-to-point messaging can also just be used as an internal caching mechanism inside a single component.
Publish/subscribe   In publish/subscribe, the producer sends a message to a channel. Multiple consumers can consume the message if they are already subscribed to the channel [4].

To implement messaging, we will take a look at GraphQL, AMQP, and MQTT (see section "Definition of Asynchronous Interfaces").

Eventing

In eventing, we normally use EDA as an integration architectural pattern:

Event-driven architecture   In messaging, a message disappears after it is read; in EDA, it is not consumed but just read. Events are normally kept on an event broker and can be reread as often as the consumer wants to [16]. The focus is usually on the data and their relation, not on a single event. This leads to a more generic "message," which can be used more easily by other consumers for whom the event was not initially intended. Even services that were not known at the moment the message was produced can subsequently read the message at the broker.

There are two common subforms in EDA (which can be used together):

Event sourcing   In event sourcing, you keep all events from a point t_0 and then add all updates to the object as deltas. To obtain the current state, you need to read all changes from t_0 to now.
A good example of event sourcing is a code version control system (e.g., git, however, git does not use this approach [19]). In a version control system, you can only store the changes for each commit. When you process all changes from the beginning, you get the current state. This approach allows you to go back in time and get every state, just by the changes in between.
Event streaming   In event streaming, you have a stream of events that you can follow. Normally, newer changes are more interesting than older ones.

Message broker vs. event broker

Messaging normally uses a message brokers, and eventing uses an event brokers. What is the difference between these two?

The line between the products is rather vague, and many use cases can be covered by both types of products. However, some differences make one product beneficial, whereas others benefit in different use cases.

Message Brokers

Message brokers send messages to consumers that subscribe to them. The channel queues messages for a consumer; if they are read, the messages are deleted. Normally, messages need to be acknowledged before they are deleted [4, 20].

Event Brokers

An event broker, on the other hand, stores all messages for a channel or a topic, which is more producer-oriented. In general, they only get deleted after a period or never (e.g., for event sourcing). The messages are stored in an append-only log. The messages can be read by a consumer. However, there is usually no acknowledgment and no active deletion on the broker. The consumer can reread old messages as often as it wants. An offset is used to store a reading position [20].

Implementation

Some differences become clear when implementing these types of brokers using a specific product. In creating a queue for a message broker, you normally add the consumer first (or immediately after the producer starts publishing messages), so that the queue does not get filled up if the producer sends messages that are not consumed.[13]

In an event broker setup, you can start sending events long before the consumer consumes these messages. This allows simpler decoupling between the teams in the implementation phase, and the consumer team can already start testing and see what the events look like.

Points to Remember

APIs are everywehere. In the area of APIs, we can draw a line of protocols that go over the network. We distinguish between synchronous and asynchronous conversation types; however, the line is not strict. Table 2-1 shows a simple usage guide on when to prefer a particular communication type. However, this is just a first distinction; at the end of the day, you can use all protocols for all types of communication (see background, "Distinguishing between synchronous and

---

[13] Some message broker products like RabbitMQ have a concept of topics and queues to alleviate that risk and allow produced messages to be immediately dropped. Such approaches mitigate this problem for products applying the idea.

Table 2-1 When to use which communication type

Communication type	Usage
Synchronous communication	Caller needs a response, caller wants to know if something went wrong
Asynchronous communication	Caller does not need to wait for a response
Asynchronous communication with broker	Decoupling needed, responsibility for delivery should be handled centrally

asynchronous communication," in section "Communication Types: Synchronous and Asynchronous Communication").

Review Questions

2.1 What stands the term API for?

(a) Application Probability Integration
(b) Application Programming Interface
(c) Asterix and Paul in India
(d) Application Project Integration

2.2 What main communication strategies do you know?

(a) Synchronous and free communication
(b) Synchronous and asynchronous communication
(c) Synchronous and HTML communication
(d) HTML and RPC

2.3 How do you distinguish messages from messaging and events from eventing from each other?

(a) A message can only be read by the intended consumer.
(b) An event can only be used in an event-driven architecture.
(c) A message is deleted after reading, whereas an event can be stored forever.
(d) A message is sent outside of the network, whereas an event can only be consumed in one network zone.

References

1. Spichale K (2019) API-Design, Praxishandbuch für Java- und Webservice-Entwickler, 2. überarbeitete und erweiterte Auflage. Heidelberg: dpunkt.verlag, 1381 pp. ISBN: 978-39-60886-02-0
2. Higginbotham J (2021) Principles of Web API Design: Delivering Value with APIs and Microservices. Pearson Education, London. ISBN: 978-01-37355-63-1

3. Lauret A (2019) The Design of Web APIs. Manning Publications Co. LLC, New York, 1324 pp. Description based on publisher supplied metadata and other sources. ISBN: 978-16-38351-19-1
4. Hohpe G, Woolf B (2013) Enterprise Integration Patterns: Designing, Building, and Deploying Messaging Solutions. The Addison-Wesley Signature Series, 17. print. Addison-Wesley, Boston, 683 pp. ISBN: 032-1200-68-3
5. Fowler M (2010) Richardson maturity model. [Online] Available: https://martinfowler.com/articles/richardsonMaturityModel.html. Visited on 25 Jun 2024
6. Masse M (2011) REST API Design Rulebook Designing Consistent RESTful Web Service Interfaces, Designing Consistent RESTful Web Service Interfaces. O'Reilly Media, Sebastopol. ISBN: 978-14-49319-90-8
7. "ISO/IEC 9075-1:2023," International Organization for Standardization, Standard (2023). [Online] Available: https://www.iso.org/standard/76583.html. Visited on 10 Aug 2024
8. Kelechava B (2018) The SQL Standard - ISO/IEC 9075:2023 (ANSI X3.135). [Online] Available: https://blog.ansi.org/sql-standard-iso-iec-9075-2023-ansi-x3-135/. Visited on 10 Aug 2024
9. Bishop M (2022) HTTP/3, RFC 9114. https://doi.org/10.17487/rfc9114. [Online] Available: https://www.rfc-editor.org/info/rfc9114. Visited on 29 July 2024
10. Peterson LL (2022) TCP Congestion Control: A Systems Approach, Brakmo L, Davie B (eds). Systems Approach LLC, Wroclaw, 138 pp. . ISBN: 978-17-36472-14-9. See https://tcpcc.systemsapproach.org/
11. Vernon V, Tomasz J (2022) Strategic Monoliths and Microservices, Driving Innovation Using Purposeful Architecture. Pearson Addison-Wesley Signature Series, Jaskuła T, Poppendieck M (eds). Addison-Wesley, Boston, 313 pp. Literaturangaben. ISBN: 01-3735-546-7
12. Redpanda Data Inc. (2024) Introduction to Redpanda. Redpanda Docs. [Online] Available: https://docs.redpanda.com/current/get-started/intro-to-events/. Visited on 10 Aug 2024
13. Microsoft (2023) Introduction to apache kafka in event hubs on azure cloud - azure event hubs, Microsoft Learn. [Online]. Available: https://learn.microsoft.com/en-us/azure/event-hubs/azure-event-hubs-kafka-overview. Visited on 26 July 2024
14. Richardson L (2015) RESTful Web APIs, Amundsen M (ed), 1st edn, Second Release. O'Reilly, Beijing, 373 pp. ISBN: 978-14-49358-06-8
15. Indrasiri K (2020) gRPC: Up and Running, Building Cloud Native Applications with Go and Java for Docker and Kubernetes, Kuruppu D (Ed), 1st edn. O'Reilly, Beijing, 188 pp. Index: Seite 183–188. ISBN: 14-9205-833-5
16. Rocha O (2022) Practical Event-Driven Microservices Architecture, Building Sustainable and Highly Scalable Event-Driven Microservices, Filipe H (ed). Apress L. P., Berkeley, 1449 pp. Description based on publisher supplied metadata and other sources. ISBN: 978-14-84274-68-2
17. OData (2024) Understand OData in 6 steps. Source is at https://github.com/OData/odataorg.github.io/blob/b1a7da9bd77cd55684bacdd006b7184d097b5808/pages/gettinstarted/understand-odata-in-6-steps.html. [Online] Available: https://www.odata.org/getting-started/understand-odata-in-6-steps/. Visited on 10 Aug 2024
18. Pizzo M, Handl R, Zurmuehl M (2020) Odata version 4.01. part 1: Protocol, OASIS. [Online] Available: https://docs.oasis-open.org/odata/odata/v4.01/os/part1-protocol/odata-v4.01-os-part1-protocol.html. Visited on 10 Aug 2024
19. Vinkler AM (2023) How Snapshot and Delta Storage Differs. [Online] Available: https://blog.git-init.com/snapshot-vs-delta-storage/. Visited on 10 Aug 2024
20. Bellemare A (2020) Building Event-Driven Microservices, Leveraging Organizational Data at Scale, 1st edn. O'Reilly, Beijing, 1304 pp. ISBN: 978-14-92057-86-4

Quality Requirements for APIs 3

Having gained an understanding of APIs, recognized the characteristics of poorly designed APIs, and setting out a strategy to rectify these issues, we shall delve into the quality criteria that guide the development of superior software products, with an emphasis on great APIs.

A public API offers an external perspective on a company and the value it places on design quality [1].

Applying quality to a product is always a tradeoff, as high quality might be at odds with cost-efficiency and development speed [1]. However, low quality also dramatically increases costs over time. The goal is to apply the "appropriate" level of quality. However, this means quality needs to be measurable. It is hard, but we need to assume it is possible [2].

Quality, once lost, is hard to get back. Old systems degenerate over time and lose quality in doing so. A suitable degree of quality can only be achieved at high cost. At any rate, quality depends on customer requirements. One needs to know what is enough [3]. In this chapter, we will look at quality characteristics when it comes to API design and development.

The quality model of ISO 25010 is the cornerstone of product quality evaluation [4]. The quality model determines which quality characteristics will be taken into account when evaluating the properties of a software product and how they are structured.

Most of these quality attributes, or subsets of the main ones, are applicable to complete software products or projects. However, they can be applied to API messages (e.g., responses) as well [5].

The product quality model defined in ISO/IEC 25010 comprises nine quality characteristics with an overall 40 subcharacteristics (Figure 3-1). Even though further standards are available, the characteristics are widely used [6].

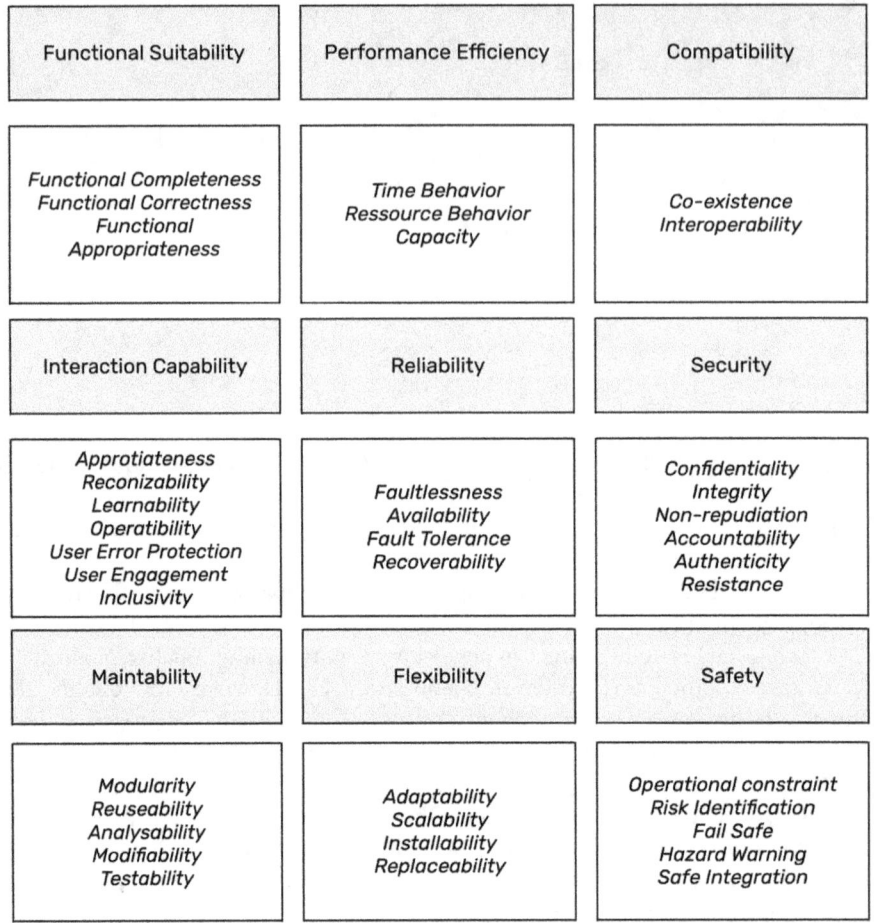

Figure 3-1 Quality requirements following ISO 25010 [4]

In the following sections, we will discuss quality characteristics in particular and further discuss how they are relevant for API design and implementation.

Functional Suitability

Functionality suitability means that a software fulfills the requirements of its users.

Functional suitability as a quality characteristic contains the following points [4]:

- *Functional completeness*
- *Functional correctness*
- *Functional appropriateness*

Functional Completeness

Let us first discuss functional completeness. The question after completeness is a challenging one. When will a software component be *functionally complete*? The answer might be when all customer requirements are implemented. However, it is widely known that customer requirements are incomplete because they change over time; new requirements emerge, others disappear. So completeness cannot be achieved. However, it is a valid requirement and must be fulfilled. It helps to think of *completeness* as *good enough*. What is *good enough* can be documented in domain stories and Bounded Context Canvas. API design means that business experts and stakeholders are tightly involved in the design process.

Functional Correctness

The same can be said of *functional correctness*. A function is correct when the customer says it is correct. A function is incorrect when it does not behave as the customer expects it to behave. A developer using the API specification has a confident expectation of how the API should behave (see example in section "APIs Without Documentation and Unexpected Behavior"). When it comes to differences between specification and behavior, the specification must be adapted accordingly.

Functional Appropriateness

Functional appropriateness goes in the same direction. It might be appropriate to have springing balls in a user interface in a math learning tool for preschool kids. However, it is inappropriate in an administrator interface of an API gateway. The example can be applied to API design. API specifications that are overloaded with functionalities no one asked for are inappropriate. An example of such an API is given in section "Overloaded World-Domination API". Implementing a functionally appropriate API requires balancing implementation time with customer requirements. We can support functionally appropriate APIs using various concepts of the workshops presented in the following chapters.

Performance Efficiency

This characteristic represents the degree to which a product performs its functions within a specified time and throughput the given parameters.

It composes the following subcharacteristics [4]:

- *Time behavior*
- *Resource utilization*
- *Capacity*

Time Behavior

Time behavior covers the response time an application needs to respond to a request. Fast responses are not free; lousy design can lead to unnecessary complexity and disturbing latency. It can be measured with a 95th percentile (p95) or a 99th percentile (p99) metric, where the 95th percentile means that the speed requirement is reached in 95% of cases and the 99th percentile means the requirement is reached in 99% of cases [7].

API designers can consider *time behavior* right at the beginning. So when considering *time behavior* at the API level, it is crucial to know the characteristics of the use case (e.g., how often, how big). However, API designers must have the whole infrastructure picture in mind and not only a single component, for example, is there additional routing latency because of geographical distribution and are there web application firewalls (WAFs) in between.

Resource Utilization

Resource utilization is the quality of how well the provided resources are used. To use more resources than necessary for the required function is a characteristic of bad quality. These are usually contradictory requirements, as more resources are often allocated up front, allowing faster execution on load spikes later. Ideally, the system would dynamically allocate resources depending on requests and effectively manage scenarios where resource limits are exceeded. However, it is a balancing act [4].

Capacity

Capacity is the degree to which a product can be scaled. An example is to use an unsigned 32-bit integer as a key: It can go from 0 to 4,294,967,295 [8, 9]. For most use cases (e.g., for most primary record keys), this is fine. However, if you aim to store log entries that are being written thousands of times per second, this approach may soon prove ineffective. Always using bigger numbers is, again, ineffective resource utilization as it requires more storage space and takes longer to transmit – again, a tradeoff [4].

The performance efficiency characteristic and its subcharacteristics have one excellent feature: They can generally be monitored and tracked quite well.

Compatibility

Compatibility is the ability of machines, especially computers or computer programs, to work successfully with other machines or programs [10].

We have established that an API is the crucial "glue" that binds software components together (Chapter 2). It is important to note that APIs are responsible for ensuring software components' compatibility.

Let us look at the subcharacteristics of *compatibility* in more detail [4]:

- *Coexistence*
- *Interoperability*

Coexistence

Coexistence requires different versions of an API to be supported at the same time. A common practice is to use two versions in production at the same time, called "two in production" [5]. This allows a high level of compatibility [11]. We will go into more details in sections "Versioning of APIs" and Publishing APIs."

Interoperability

Looking at standardized APIs like SQL and OpenID Connect, the implementation of the APIs needs to guarantee that their implementation is correct and complete [12–14]. Otherwise, interoperability is not given because implementations using those standards cannot work together.

Other examples are APIs of Android. The API remains stable despite different implementation and user interface changes. APIs guarantee *interoperability* because the contract can be held stable, whereas the implementation of the API specification can be changed.

Interaction Capability

This section will look at some of the most relevant quality objectives for an API design and its implementation. Primarily we will point out the qualities to be fulfilled to support developer experience; see also section "Developer Experience When Integrating an API". Interaction capability reflects all functionalities of an application to *interact* with the outside world. We will discuss interaction capability and the following subcharacteristics [4]:

- *Appropriateness recognizability*
- *Learnability*
- *Operability*
- *User error protection*
- *User engagement*
- *Inclusivity*
- *User assistance*
- *Self-descriptiveness*

Appropriateness Recognizability

Users can *recognize* whether a product or system is *appropriate* for their needs. Developers using an API *recognize* the necessary requests and the expected response from the API specification. Even more business experts can *recognize* the *appropriateness* of an API, because the API uses the commonly defined ubiquitous language. An anti-pattern is a tool that tries to show itself as an all-solving miracle, that you try for ages to get out what you want but fail in the end because it is impossible.

Learnability

Learnability is the degree to which the functions of a product or system can be learned to be used by specified users within a specified amount of time [4]. Using the ubiquitous language allows business users to learn faster and to decide about the suitability of a business function in an API.

Operability

Operability is the degree to which a product or system has attributes that make it easy to operate and control [4]. Generally, APIs allow systems better *operability* because services can be operated and deployed separately.

User Error Protection

User error protection means the degree to which a system protects users against operation errors [4]. The subcharacteristic *user error protection* can be achieved by a good API design. It must be clean and unambiguous and contain reduced parameters (if we look at our early example, there is no tenant parameter if it is not required, see section "Overloaded World-Domination API").

Those are good practices from the developer experience perspective and improve *user error protection*.

Another aspect is that a good API specification contains descriptions and listings of possible error situations, for example, the error responses in an OpenAPI specification (OAS).

User Engagement

User engagement is the degree to which a user interface presents functions and information in an inviting and motivating manner, encouraging continued interaction [4]. *User engagement* on a technical object like an API is certainly not as important as on a graphical UI. However, the specification can be well formatted, described with good examples, and supported by interaction diagrams showing possible usage scenarios from the developer experience perspective. In addition, using best practices will increase *user engagement*.

Inclusivity

Inclusivity is the degree to which a product or system can be used by people of various backgrounds (such as people of multiple ages, abilities, cultures, ethnicities, languages, genders, and economic situations, for example) [4]. From the authors' perspective, that means that API specifications should be written in English as English is used as a ubiquitous language in software development. To write API specifications in a local language might increase the understanding of the specification for local business experts, but it would exclude developers who do not speak the local language. However, software development is international these days. Therefore, the API specification should be in English. Local terms can stay when they are explained thoroughly.

User Assistance

User assistance describes a tool that can be used by people with a wide range of characteristics and capabilities [4]. Such assistance to developers using an APIs can be well-defined and well-described mockups. Here, mockup means an API implementation that provides example responses [15]. Another possibility might be the availability of a sandbox implementation, where developers can play with an API [16].

Self-Descriptiveness

Self-descriptiveness is the degree to which a product presents appropriate information without excessive interactions with a product or other resources [4]. *Self-descriptiveness* is the goal for every API specification. An API specification must be written so that a developer can understand what awaits in request and response, whether data are provided or received. Moreover, finding a common language between business experts and IT specialist (known as ubiquitous language) helps the understanding on both sides (Chapter 5). The goal is to look at the code while at the time keeping the whole process and the business goal in mind [1].

Reliability

In 2011, entrepreneur and venture capitalist Marc Andreessen published an essay titled "Why software is eating the world" [17]. He explained that software was crucial for every aspect of life. We perceived this when, for example, flights were cancelled and traffic lights were on the blink in July 2024. A flawed software update sent out by a cybersecurity company called CrowdStrike caused computer crashes all over the world - affecting everything from small personal computers to multinational corporations, including airlines and retailers. The integration of CrowdStrike into the Microsoft Windows operating system is so deep that computers crash when the update containing the flaw was automatically installed [18]. Because software is eating the world, we need to make the software reliable, robust, and resilient so that it becomes faultless, available, fault-tolerant, and recoverable.

The characteristic *reliability* contains the following subcharacteristics [4]:

- *Faultlessness*
- *Availability*
- *Fault tolerance*
- *Recoverability*

Faultlessness

Faultless software is a dream. As each ideal comes with costs, one must determine whether it is worth doing. Depending on the context described by business experts, the answers might differ. They are different when it comes to clinical diagnostic systems, retail systems, or airplane security systems. But we need to strive for it. Software needs to be designed and built for quality [19]. Quality does not happen by chance. *Flawless* APIs emerge when access to the API delivers what is expected. Faulty APIs output is something unexpected, as explained in section "APIs Without Documentation and Unexpected Behavior".

Faultlessness in operation is achievable when automating as much as possible. Removing the human factor from operational processes can reduce failures, as humans are imperfect when it comes to performing repetitive tasks [2]. We will look at this in section "Continuous Integration."

Availability

The *availability* of software and APIs is apparent to us almost every day. Most of us likely rely on the availability of our favorite streaming provider or retailer for our daily existence.

Availabilty is the degree to which a system, product, or component is operational and accessible when required for use [4]. Systems that rely on APIs are usually better suited to fulfill *availability* requirements because the services can be tailored to business requirements for availability. This includes the fact that availability is not a general quality requirement. It to a great extent depends on the business to fulfill.

Fault Tolerance

Fault tolerance is the degree to which a system, product, or component operates as intended despite the presence of hardware or software faults [4]. It detects and recovers from failures quickly and automatically. The metric mean time between failures (MTBF) measures robustness and is essential, and its value should be high. However, if the mean time to recovery (MTTR), which measures resilience, is small, the MTBF can be lower and still result in better availability over time [20]. A robust system experiences less failure; therefore, such a system is simpler and cheaper to operate[1] [20].

Recoverability

Recoverability is the degree to which, in the event of an interruption or failure, a product or system can recover the data that were directly affected and reestablish the system's desired state [4]. Smaller systems that communicate via APIs help to increase the *recoverability* because smaller systems can recover faster than large ones. However, the dependency between the systems needs to be taken into account. Those can be explained in good API documentation.

Security

Security is an essential topic in general. In the context of remote APIs, it is essential that these APIs go over networks or even the public internet and can be exposed to the public. Even if they are not exposed to the public but just to business partners, perhaps over a virtual private network (VPN), published APIs can dramatically increase an attack vector [21].

The subcharacteristics of security are as follows [4]:

- *Confidentiality*
- *Integrity*
- *Nonrepudiation*
- *Accountability*

---

[1] The most robust and failure-free application is one without code: https://github.com/kelseyhightower/nocode.

- *Authenticity*
- *Resistance*

The topic of security issues from the perspective of APIs is a vast one and would require a separate book. Please refer to the relevant literature [22–26].

A starting point can be the Open Worldwide Application Security Project (OWASP) top ten list. Also, one should take a look at the application risk profile. It estimates the likelihood of a specific event and assesses the possible impact. This results in a list of risks classified by severity (e.g., low, medium, high), which can help communicate and agree on necessary actions. It is important to classify the severity and decide on mitigation and what should be done from a business point of view. Again, security quality often contradicts development speed and increases costs [25, 27, 28]. These should always be considered in an application. Security quality is quickly lost by a wrong configuration or design flaw. It should be an integral part of the design process from the beginning, not an afterthought.

This book will cover authentication and, to some extent, authorization when specifying an API (Chapter 6, "Interface Definitions"). As a reduction of the attack vector, we will also look at API management and service meshes (Chapter 8).

Maintainability

Maintainability is not as prominent a quality characteristic as availability or faultlessness, but it is crucial for evolving software. Maintenance is the most costly aspect of a software system [29].

It contains the following subcharacteristics [4]:

- *Modularity*
- *Reusability*
- *Analyzability*
- *Modifiability*
- *Testability*

Modularity

Modularity is the degree to which a system or computer program is composed of discrete components such that a change to one component has minimal impact on the other components. Obviously, APIs by their very nature support this characteristic. Large systems can be tailored in lower systems when using APIs. This is valid for internal APIs and APIs communicating over a network we focus on in this book.

For API design and implementation, it means designing APIs with meaning because those can be structured along a business process and have meaning. This is the foundation of the key to successful software development.

Using the online library example, we will look at DDD in the next chapter, Chapter 5, "API Design Supported by Domain-Driven Design." In Chapter 11, "Brownfield Project," we will look at how to improve the maintainability of legacy systems by new and adapted APIs.

Reusability

Reusability is the degree to which a product can be used as an asset in more than one system or in building other assets [4]. APIs that expose a business function of a bounded context support *reusability* because using APIs can create new or other business models using available business functions via APIs. The exposed business function can be used multiple times in different contexts.

Analyzability

Analyzability is the degree of effectiveness and efficiency with which it is possible to assess the impact on a product or system of an intended change to one or more of its parts, to diagnose a product for deficiencies or causes of failures, or to identify parts to be modified.

APIs can be analyzed when the implementation provides figures. However, using infrastructure components like an API gateway allows even deeper insights into the efficiency of an API; see section "Publishing APIs".

Modifiability

Modifiability is the degree to which a product or system can be effectively and efficiently modified without introducing defects or degrading existing product quality [4]. Usually, it can be measured by the degree of encapsulation and polymorphism [30]. Modern software needs to be modifiable to enable reactions to technology and business changes; see section "Complexity in a Volatility, Uncertainty, Complexity, and Ambiguity (VUCA) World," for more details.

Testability

Testability means the degree of effectiveness and efficiency with which test criteria can be established for a system, product, or component, and tests can be performed to determine whether those criteria have been met [4].

In the case of APIs, this means not only that the system exposing the API can be tested. It includes the ability on the part of the APIs to be tested even in a very early phase of development using mockups and sandboxes, as discussed earlier in user experience, as testing and user experience are very closely connected with respect to APIs.

Achieving good analyzability and testability will be covered in Chapter 8, "Developer Experience and API Implementation."

Flexibility

Flexibility contains the following subcharacteristics [4]:

- *Adaptability*
- *Scalability*
- *Installability*
- *Replaceability*

Adaptability

Adaptability is the degree to which a product or system can effectively and efficiently be adapted for or transferred to different hardware, software, or other operational or use environments [4].

Systems using APIs can be better adapted to other components when an API is reused, as only the route to the new product needs to be changed.

Scalability

Scalability is the degree to which a product can handle growing or shrinking workloads or adapt its capacity to handle variability [4]. Using well-tailored systems allows the scaling of business functions along a business process where single functions are used to different degrees. They support cost savings by efficiently utilizing resources and managing peak loads, ensuring smooth operations even during high-demand periods.

Scalability is becoming increasingly important due to increasing real-time requirements, globally distributed systems, and clouds that allow the usage of resources only when needed. Fast, scalable systems are a boon in the modern tech landscape. Well-tailored APIs can help to scale systems.

Installability

Installability is the degree of effectiveness and efficiency with which a product or system can be installed and/or uninstalled in a specified environment. APIs support the installability of software systems because separate parts can be installed independently to a certain degree (section "Publishing APIs").

Installability of APIs is supported by software development kits. The provider of an API delivers a SDK, which allows integrators to create a client easily and quickly. It leads to a good developer experience and quick time to market. Another example might be plugins installed by one click into an integrated development environment (IDE). We will dive into more detail on this topic in section "Developer Experience When Integrating an API."

Replaceability

Replaceability is essential to overcome vendor lock-ins and should be handled using API standards that are widely applied, for example, OpenID Connect for authentication or SQL for relational databases (as described in section "The Term API"). We can use APIs to replace legacy software (Chapter 11). Using common standards for API specifications can support the replacement of APIs themselves. We will look at common standards in section "Introduction."

Safety

Safety is a topic that will not be covered extensively in this book. It represents the degree to which a product under defined conditions avoids a state where human life, health, property, or the environment is endangered [4]. It contains the following subcharacteristics [4]:

- *Operational constraint*
- *Risk identification*
- *Fail safe*
- *Hazard warning*
- *Safe integration*

For further information, please consult [31]. We will not discuss those specific topics in this book.

Summary

Quality requirements are not cut and dried. They need to be quantified by business requirements and balanced in close collaboration between technical and business experts. Sometimes, quality requirements even contradict each other, for example, simplicity vs. reusability and modularity. Those contradictions and costs of quality need to be balanced out carefully [5].

Some quality requirements can be measured, while others can only be governed [5].

Software quality needs to be taken into consideration during all stages in an API lifecycle, as shown in the development cycle in Figure 1-3. We will show the corresponding perspectives in upcoming chapters, starting with ideation up to operation and sundown.

Review Questions

3.1 What are the quality characteristics of ISO 25010? (multiple selections possible)

(a) Functional suitability
(b) Performance efficiency
(c) Maintainability
(d) Cost efficiency
(e) Effectivity
(f) Reliability
(g) Security
(h) Safety
(i) Beauty
(j) Complexity
(k) Flexibility
(l) Simplicity
(m) Interaction capability
(n) Compatibility

3.2 Can quality requirements contradict each other?

(a) Never
(b) Always
(c) Some are contradictory and need to be balanced out
(d) Only when they are not discussed with stakeholders

3.3 Are quality requirements important for APIs

(a) No, they are only important for implementation.
(b) Yes, they are important for specification, implementation, operation, and sundown.
(c) Yes, but only for operations.

References

1. Higginbotham J (2021) Principles of Web API Design: Delivering Value with APIs and Microservices. Pearson Education, London. ISBN: 978-01-37355-63-1
2. Duvall PM, Matyas S, Glover A (2013) Continuous Integration: Improving Software Quality and Reducing Risk. A @Martin Fowler Signature Book, 8. print. Addison-Wesley, Upper Saddle River, 283 pp. Literaturverz. S. 273–274. ISBN: 03-2133-638-0
3. Beyer B, Jones C, Petoff J, Murphy NR (eds) (2016) Site Reliability Engineering: How Google Runs Production Systems, 1st edn. O'Reilly, Beijing, 1 p. Description based on publisher supplied metadata and other sources. ISBN: 978-14-91951-18-7
4. ISO IEC25010 (2022). [Online] Available: https://iso25000.com/index.php/en/iso-25000-standards/iso-25010. Visited on 12 Aug 2024
5. Zimmermann O (2023) Patterns for API Design: Simplifying Integration with Loosely Coupled Message Exchanges. Addison-Wesley Signature Series, Stocker M, Lübke D, Zdun U, Pautasso C (eds). Addison Wesley, Boston, 508 pp. ISBN: 978-01-37670-10-9
6. What is software quality? (2024). [Online] Available: https://asq.org/quality-resources/software-quality. Visited on 15 Aug 2024
7. Novy L (2023) Performance testing metric - percentile. [Online] Available: https://jtlreporter.site/blog/2023/07/07/performance-testing-metric-percentiles. Visited on 15 Aug 2024
8. Integral numeric types (c# reference) (2022). [Online] Available: https://learn.microsoft.com/en-us/dotnet/csharp/language-reference/builtin-types/integral-numeric-types. Visited on 14 Aug 2024
9. Fundamental types (2024). [Online] Available: https://en.cppreference.com/w/cpp/language/types. Visited on 14 Aug 2024
10. Compatibility (2024). [Online] Available: https://dictionary.cambridge.org/dictionary/english/compatibility. Visited on 15 Aug 2024
11. ISO 20022 - to foster interoperability, independently of the underlying technology; universal financial industry message scheme (2022). [Online] Available: https://www.iso20022.org/development-new-api-resources. Visited on 15 Aug 2024
12. Kelechava B (2018) The SQL Standard - ISO/IEC 9075:2023 (ANSI X3.135). [Online] Available: https://blog.ansi.org/sql-standard-iso-iec-9075-2023-ansi-x3-135/. Visited on 10 Aug 2024
13. Sakimura N, Bradley J, Jones M, de Madeiros B, Mortimore C (2023) Openid connect core 1.0 integrating errata set 2. [Online] Available: https://openid.net/specs/openid-connect-core-1_0.html. Visited on 15 Aug 2024
14. What is openid connect (2024). [Online] Available: https://openid.net/developers/how-connect-works/. Visited on 15 Aug 2024
15. Helton A (2024) What is API mocking? . [Online] Available: https://blog.postman.com/what-is-api-mocking/. Visited on 12 April 2025
16. What is a sandbox? (2025). [Online] Available: https://smartbear.com/learn/api-design/what-is-an-api-sandbox/. Visited on 12 April 2025
17. Andreessen M (2011) Why software is eating the world. [Online] Available: https://a16z.com/why-software-is-eating-the-world/. Visited on 13 April 2025
18. Satariano A, Mozur P, Conger K, Frenkel S (2024) Chaos and confusion: Tech outage causes disruptions worldwide. [Online] Available: https://www.nytimes.com/2024/07/19/business/microsoft-outage-cause-azure-crowdstrike.html. Visited on 15 Aug 2024
19. Abrial J-R (2009) Faultless systems - yes we can! . [Online] Available: https://www.research-collection.ethz.ch/bitstream/handle/20.500.11850/69796/eth-5023-01.pdf. Visited on 15 Aug 2024
20. Hohpe G, Danieli M, Landreau J-F, Hashmi T (2021) Cloud Strategy: A Decision-Based Approach to Successful Cloud Migration. Architect Elevator Book Series. leanpub.com

21. Vernon V, Tomasz J (2022) Strategic Monoliths and Microservices: Driving Innovation Using Purposeful Architecture. Pearson Addison-Wesley Signature Series, Jaskuła T, Poppendieck M (eds). Addison-Wesley, Boston, 313 pp. Literaturangaben. ISBN: 01-3735-546-7
22. Hoffmann A (2024) Web Application Security: Exploitation and Countermeasures for Modern Web Applications. O'Reilly, Boston. ISBN: 978-10-98143-93-0
23. Siriwardena P (2019) Advanced API Security: The Definitive Guide to API Security. Apress, New York. ISBN: 978-14-84220-49-8
24. Madden N (2020) API Security in Action. Manning, Shelter Island. ISBN: 978-16-38356-64-6
25. Owasp Developer Guide, version 4.1.3, OWASP Foundation (2024). [Online] Available: https://owasp.org/www-project-developer-guide/assets/exports/OWASP_Developer_Guide.pdf. Visited on 14 Aug 2024
26. Owasp Top Ten, OWASP Foundation (2021). [Online] Available: https://owasp.org/www-project-top-ten/. Visited on 14 Aug 2024
27. Owasp Risk Rating Methodology, OWASP Foundation (2023). [Online] Available: https://owasp.org/www-community/OWASP_Risk_Rating_Methodology. Visited on 14 Aug 2024
28. Guide for Conducting Risk Assessments, National Institute of Standards and Technology (2012). https://doi.org/10.6028/nist.sp.800-30r1. Visited on 14 Aug 2024
29. Martin RC (2018) Clean Architecture. Pearson Education, London
30. Nwe N, Thu E (2018) Measuring modifiability in model driven development using object oriented metrics. Adv Sci Technol Eng Syst J 3(1):244–251. https://doi.org/10.25046/aj030130
31. Hammer R (2007) Patterns Fault Tolerant Software. Wiley, Hoboken. ISBN: 978-04-70319-79-6

Part II

Domain-Driven API Design

How to apply Domain-Driven Design and API design to get beautiful APIs?

Online Library

4

Throughout the book, we will use an example of an online library application. This chapter introduces the principal requirements for the online library. The example is entirely fictional; however, it should be close to reality. The example will be used not only on an application basis but also to look at the business side of this case.

The online library should provide a broad selection of books in all genres. The library will provide a platform for users to search books in the library and read them online. A primary focus will be on children's books, with the aim of providing a supportive environment for children to learn and develop their reading skills. The distinguishing factor of this example is that the library should be a community library financed mainly by the community. Additionally, the library raises donations from organizations, companies, schools, and private donors. The donors will be announced on the library application pages, for example, when specific book licenses can be purchased. This makes the example more straightforward, as a billing model will not be necessary.

First Requirements Gathering

As discussed earlier, the application journey should start with workshops to solicit a common idea of the goals and requirements for the business. These first workshops to formulate an idea should include developers, entrepreneurs, community representatives, and technical consultants. The workshop format can be supported by creative methods such as brainstorming [1]. The following features are found in connection with the online library – listed here alphabetically – without any weighing:

- Community financed
  The library will be financed primarily by the community. The financing must cover the library's technical costs, such as cloud infrastructure (e.g., servers) and their maintenance, as well as the costs of book licensing. Community

citizens will be recruited as volunteers to carry out the tasks of librarians and teachers. Those volunteers will need to receive round sum allowances for their expenses. The community will establish the library as a nonprofit organization. A member-elected board will decide on its further development and make budgetary allocations.

- Raising donations

  The other source of revenue streams will be donations. The library will raise donations from organizations and schools encouraging children to read.

  However, it can also generate donations from companies and local businesses. Private individuals interested in reading and books will also donate to the library.

- Making abstracts

  Library members and volunteers should be able to write abstracts of books that can be shared with library members.

- Marking text in books

  Readers can mark text they find enjoyable or memorable.

- Making notes at marked text

  Moreover, when marking a text, a reader can make notes on the marked text.

- Recommending books based on reading behavior

  Based on their reading behavior, members will receive recommendations about books they might be interested in. Fixed rules and an artificial intelligence (AI) module will generate book recommendations.

- Proposing books for purchase

  Members can suggest that certain books be purchased. Librarians and the library's board of trustees will decide which book licenses to purchase.

- Recommending books to others

  Members can recommend books to other members.

- Rating books

  Members can rate books.

- Reading books

  Members can read books (obviously, that's the purpose of a library).

- Reading books to children

  Volunteers can read books to preschool- and school-age children. The children will learn to love books and how to read.

- Searching for books in the library

  A member can search for books they would like to read in the library's catalog.

- Sharing notes with others

  Notes taken by a member can be shared with other members who might be interested.

- Sharing reading lists

  Members can share their reading lists with other members. Specific sublists can be created by members to share with others.

- Teaching reading
  Volunteers can teach kids how to read at parents' request. Related appointments are organized by parents.
- Writing and sharing reviews
  Members can write reviews of books and share their reviews with other members.

Establishing requirements always generates implementation ideas among workshop attendees. We will discuss them in the next section.

First Implementation Ideas

As often happens when establishing requirements, the first implementation ideas to arise have mostly to do with the user interface. There is no problem with this as long as the implementation ideas are just that: ideas; they should not become requirements. As long as they are used to understand the library requirements better, it is completely fine.

A first idea might be the user flow shown in Figure 4-1. The user flow of the graphic is described in the following points:

1. An anonymous user can **register** to become a library member. Once they do so, they will be directed to the login page.
2. When a user is already registered, she can start with the **login**. The login can be done using a username and password or via social media accounts.
3. Once logged in, the member will first see the landing page. The **landing page** contains library news, newly purchased licenses, and funded donations. Donors are named or anonymized in accordance with their wishes.
   From the landing page, members can navigate to their reading lists or a search. Users can return to the landing page from each application page – like a home button.
4. Members **search for books** in the library.
5. Members see the books they want to read, are reading, or have finished but not yet deleted from their **reading lists**. Each member has a general reading list. They can add additional reading lists using tags. Tagging a book will cause it to appear in the corresponding list.
6. Members can **read** a book. The book to be read can be selected from a result list from a search or via a reading list. The member can mark text in the book.
7. When members mark text in books, they can **add notes** to the marked text.
8. Members can **share** notes on a specific book or reading list with other library members by searching for members with whom they want to share the notes or a list.

User experience experts do not create the shown user flow. It helps to understand the requirements and can be used as a discussion base. First drafts can contain even user interface sketches, as shown in Figure 4-2.

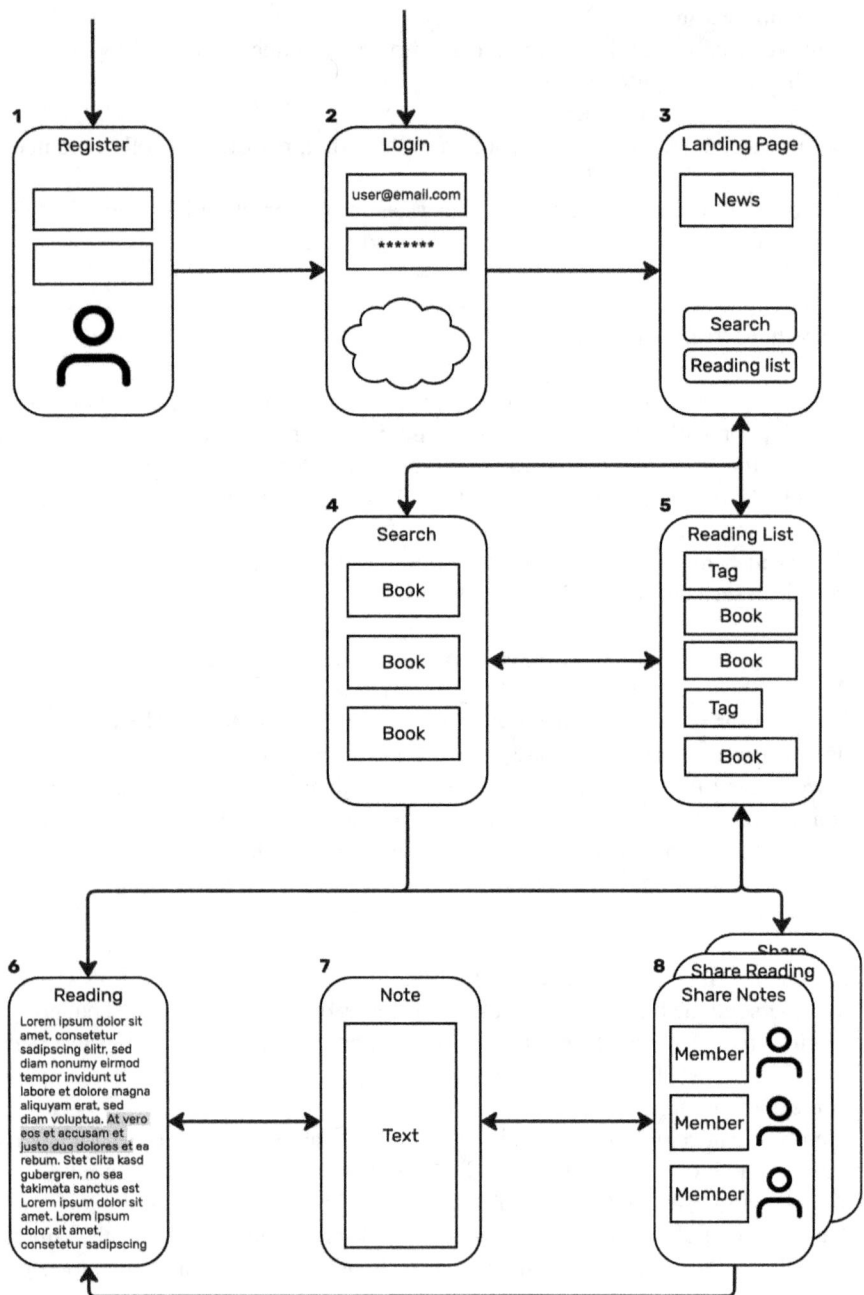

Figure 4-1 First ideas of a user flow

First Implementation Ideas

Figure 4-2 Ideas for a possible user interface

The considered user interface should not be understood as a graphical design. It helps to understand what data in what combinations are necessary and how to obtain the data.

The display of search results should contain not just the title and authors as text but also thumbnails of the book cover and a miniature portrait of the author. Additionally, book ratings should be shown, for example, by stars or hearts.

Member profiles should contain not just names and email addresses but avatar pictures as well. Additionally, profiles can show indicate whether a member is, for example, a volunteer or donor at the library.

This user flow and first mockups should help give a shared understanding of what should be achieved. Taking the conversation back to a discussion about requirements, we can now sort the found capabilities into different categories, as discussed in the next section.

Start of Prioritization

Before we build a list of requirements, the requirements and features that were found need to be prioritized next. They can be handled as the business capabilities of the organization discussed. So early in the ideation process, only a rough categorization is meaningful, so capabilities are sorted into "core," "supportive," and "generic." These terms are based on the work of *Evans* and will be introduced first [2]:

domain   A domain is everything an enterprise – in this example a library – does [2].

core subdomain   A core subdomain or short core domain covers all capabilities without which the business of the organization would not be possible [2]. In the case of the library, it would be everything having to do with collecting and reading books.

generic subdomain   A generic subdomain or subdomain contains capabilities that do not pertain directly to the business [2]. This is something that functions in almost all organizations in a similar way. For example, identity and access management systems are typical examples of capabilities of a generic subdomain.

supportive subdomain   A supportive subdomain or short supportive domain covers capabilities that have core characteristics as well as generic characteristics [2]. A typical example of such a supportive domain would be a records management system, where records can be archived quite generically. However, the handling of records with regard to essential metadata and retention periods depends largely on the particular business. Retailers usually need to retain orders from customers no longer than 5 years, whereas a life insurance company needs to store contracts as long as the insurant lives and even longer when it comes to specific inheritance regulations. However, those retention periods even depend on legal regulations in the location where the corresponding organization is active.

With these definitions, we can focus on using the categories to sort out our requirements.

Obviously, to prioritize the capabilities to be implemented, the core capabilities get the highest priority and the generic ones the lowest. The characteristics of the collected capabilities are shown in Figure 4-3.

Those capabilities will be used to dive deeper into the online library and find out what APIs are needed.

We will work on the listed capabilities; however, we will exclude digital rights management of the books because it is not of interest to the general reader and

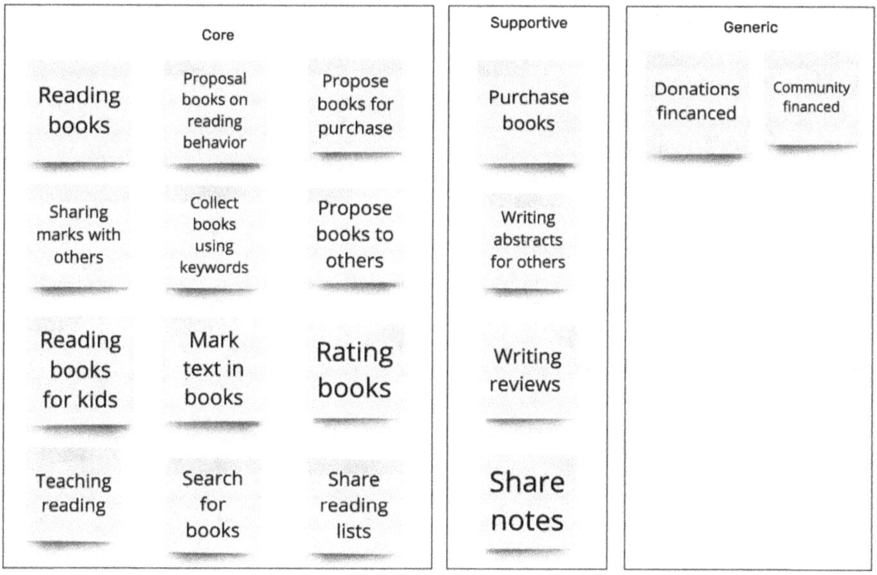

Core			Supportive	Generic	
Reading books	Proposal books on reading behavior	Propose books for purchase	Purchase books	Donations financed	Community financed
Sharing marks with others	Collect books using keywords	Propose books to others	Writing abstracts for others		
Reading books for kids	Mark text in books	Rating books	Writing reviews		
Teaching reading	Search for books	Share reading lists	Share notes		

Figure 4-3 Prioritized capabilities

complicates the matter considerably. The goal is to look at capabilities that will be easily recognizable for people experienced in using a library.

Points to Remember

- The online library is an example that will be used throughout the following chapters in the book.
- A domain is everything an enterprise does.
- A core subdomain is a subdomain of an organization containing activities essential to the business of the organization.
- A generic subdomain contains capabilities that do not directly pertain to the business.
- A supportive subdomain covers capabilities that have core characteristics as well as generic characteristics.
- We can prioritize requirements by "core," "supportive," and "generic."

Review Questions

4.1 How are first ideas implemented?

(a) Pixel perfect design
(b) Design as paperwork with measurements
(c) Sketch out with mockups
(d) Click dummy with perfect design

4.2 What are different kinds of subdomains?

(a) Core, supportive, generic
(b) Technical, business, both
(c) Insurance, manufacturing, retail

References

1. van Kelle E, Verschatse G, Baas-Schwegler K (2025) Collaborative Software Design. Manning, Shelter Island. ISBN: 978-16-33439-25-2
2. Evans E (2004) Domain-Driven Design: Tackling Complexity in the Heart of Software. Addison-Wesley, Reading

API Design Supported by Domain-Driven Design

In Chapter 4, "Online Library," we collected first ideas for the online library. In this chapter we want to structure and prioritize them and find first approaches to bounded contexts and necessary APIs. We use methodologies of Domain-Driven Strategic Design to determine what APIs are necessary to create the online library. We start by qualifying the business ideas collected in Chapter 4. Then we structure and detail the business requirements using Domain Storytelling and Event Storming. Creating a context map, we obtain a first logical design of the library, including necessary APIs.

Qualifying Business Ideas

To create a successful business, it makes sense to create a business plan first. The business plan should help the founders and investors arrive at a common understanding of what the business will do and how it will generate revenue. The main part should be a description of the business model. It should describe how the organization will generate, provide, and retain value. Business plans are, generally speaking, long, boring texts that provide detailed descriptions of all aspects of a business [1].

A shorter and sufficient start is a business plan canvas, which will focus on the business model. In this book, we will focus on the Business Model Canvas and leave the rest to the specialist literature.

Business Model Canvas

To describe a business to be established or an already established business, one can use a Business Model Canvas, which is a tool for managers and entrepreneurs to describe the business model and the underlying ideas and discuss and challenge them [1, 2]:

- Key partners

  This section contains the key business partners.

  For the online library, the key partners are the community served by the library and the local schools and preschools.
- Key activities

  This section contains the key business activities.

  In our case, the key activities are reading books, reading books to children, recommending books to others, marking text in books, making notes, and searching for books.
- Key resources

  The key resources are resources that are necessary for the core activity of the business. They include materials, machines, and people. Books are needed for an online library. Volunteers take on the tasks of teachers and librarians. Members read books and are prospective donors. Besides the community, donors will provide financial support for the library.
- Key propositions

  The fundamental propositions contain differences from those of other comparable businesses. The online library offers easy access to books and easy access to resources for learning how to read.
- Customer relationships

  The section discusses the frequency with which customers will have contact with the business. In the case of the online library, a member can read books daily.
- Channels

  The channels section contains all channels a customer can use to come in contact with the business. In the case of the library, a mobile app and a web app should be provided.
- Customer segments

  The section lists the customer segments in which the business will be active. The online library's customer segments are parents, preschool and schoolchildren, teens, and adults.
- Cost structure

  This section contains all items that incur costs. The following cost drivers relate to the online library: licenses, server or cloud environment, development and maintenance, and volunteer allowances.
- Revenue streams

  This section discusses the sources of revenue that need to cover costs. Because the online library does not rely on member fees and is organized as a nonprofit organization, it does not actually earn revenue. Donations and municipal assistance by the community will provide financial support for the library.

The Business Model Canvas of the library is shown in Figure 5-1.

Qualifying Business Ideas 73

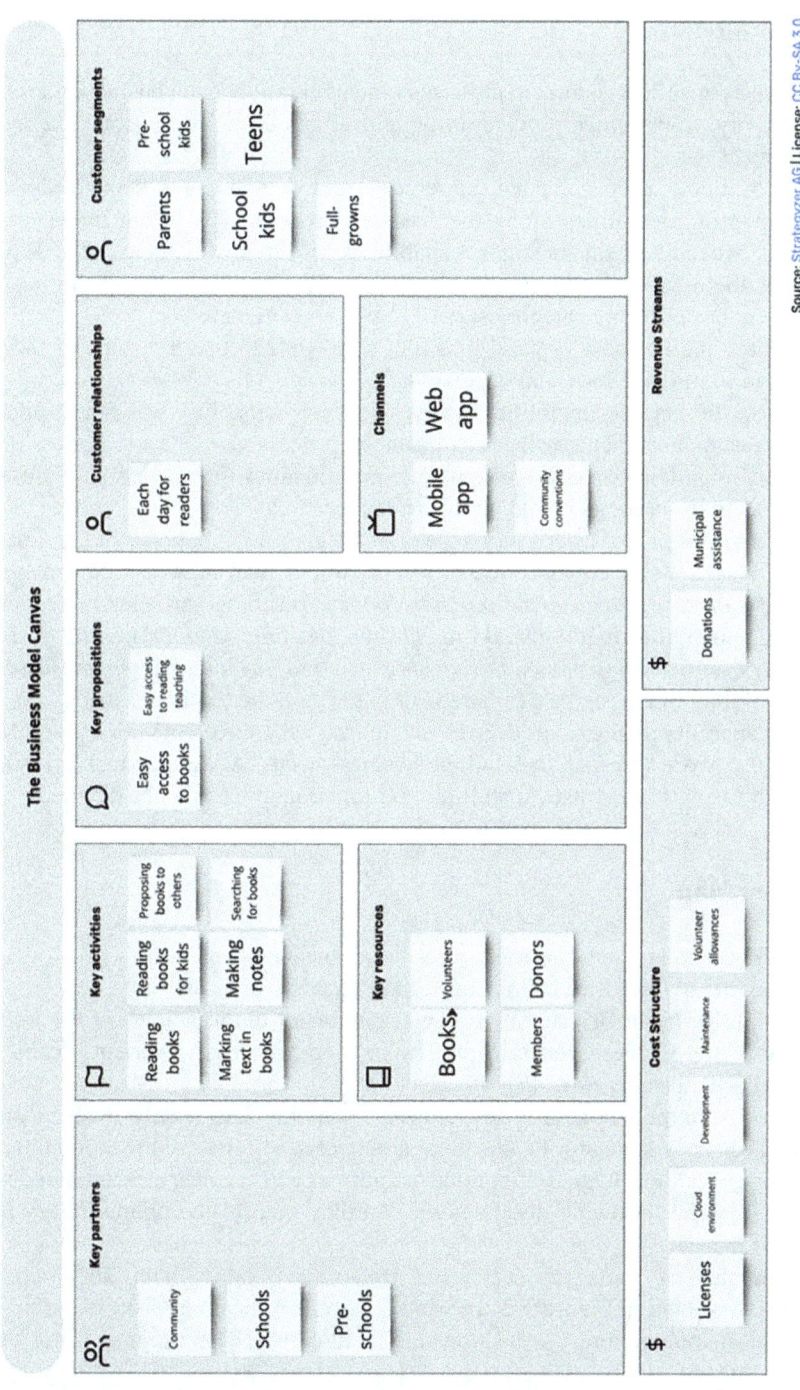

Figure 5-1 Business Model Canvas of online library; see Strategyzer [2]

Capability Map

A business capability document abstracts fundamentally what a business does as its core activity. Capabilities provide an organization's capacity to achieve a desired outcome [3].

Capabilities can be collected, structured, and detailed in a map to show the importance of capabilities along the business process [4]. During brainstorming sessions, we collect the business capabilities in an unstructured way. We can structure them further using a capability map. We map them along the business process in the following categories: core, supportive, generic [4].

To show the business process in a first step, we can use the steps of the user journey in Figure 4-1 shown in Chapter 4.

Next to the known capabilities, a new necessary capability was found: identity and access management, using the business process as a new criterion in the capability map. It is necessary to manage users and allow them to log in by different means, such as via social media platforms (Figure 5-2).

The business process steps *purchase* and *teaching* do not appear in the first user journey idea. *Purchase* was added in the capability map to cover the capabilities *propose books for purchase* and *purchase books*. Teaching was added to cover the essential capabilities *reading books to children*, *teaching reading*, *writing abstracts for others*, and *writing reviews*. *Teaching* and *reading to children* are essential because public funds can be guaranteed only through the teaching aspect.

The capability map gave us deeper insight into what the online library should do. We can use a Wardley map to detail the business needs. A Wardley map gives more profound insights into what capabilities and functionalities must be detailed.

Wardley Map

A Wardley map is a solid tool for prioritizing business capabilities [5]. It enables organizations to formulate and navigate their strategic choices.

Simon Wardley, a British entrepreneur, combined the principles of the military strategy of the Chinese general *Sun Tzu* [6] and the observe, orient, decide, act (OODA) principles by *Boyd* [7].

Since its introduction in 2005, Wardley mapping has widely evolved and is recognized as a standard in business architecture. It has a broad community. Creating a Wardley map is supported using a specific canvas, recommended by *Mosior* [8]. The following discusses the Wardley map of the online library using a *Mosior* canvas.

It contains the parts purpose, users, user needs, value chain, and map. The structure is shown in Figure 5-3. In what follows, application of the Wardley map canvas to the online library will show individual parts. The entire map is not shown for readability.

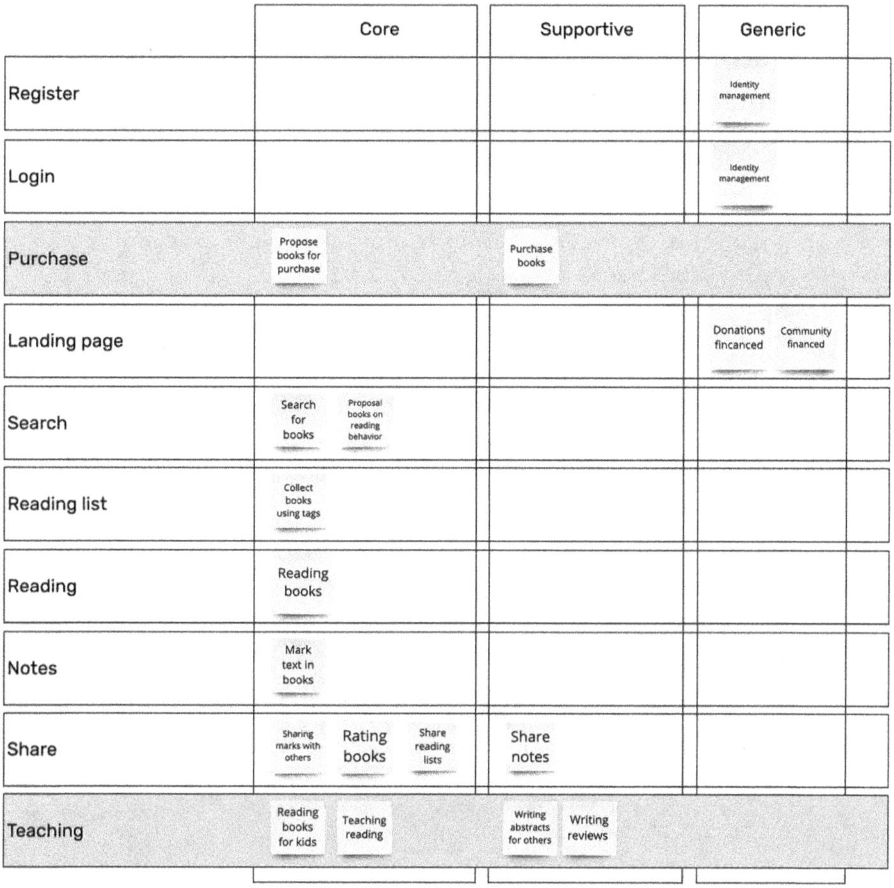

Qualifying Business Ideas

	Core	Supportive	Generic
Register			Identity management
Login			Identity management
Purchase	Propose books for purchase	Purchase books	
Landing page			Donations fincanced / Community financed
Search	Search for books / Proposal books on reading behavior		
Reading list	Collect books using tags		
Reading	Reading books		
Notes	Mark text in books		
Share	Sharing marks with others / Rating books / Share reading lists	Share notes	
Teaching	Reading books for kids / Teaching reading	Writing abstracts for others / Writing reviews	

Figure 5-2 Capability map of online library

1. Purpose
   The purpose area gives information about the map's purpose. The purpose is specific and should be from the customer's perspective (Figure 5-4).
2. Scope
   The scope area gives information about the scope of the map – it can be specific, for example, for a specific department of a company, or it can be, as in the case of the online library, high-level (Figure 5-5).
3. Users
   The third step of Wardley mapping details the users of the business. Users are the key customers identified in section "Business Model Canvas." Parents, preschool and schoolchildren, teens, and adults are the customer segments of the online library. For the Wardley map, we use more precise users: readers, nonreaders, and donors (Figure 5-6).

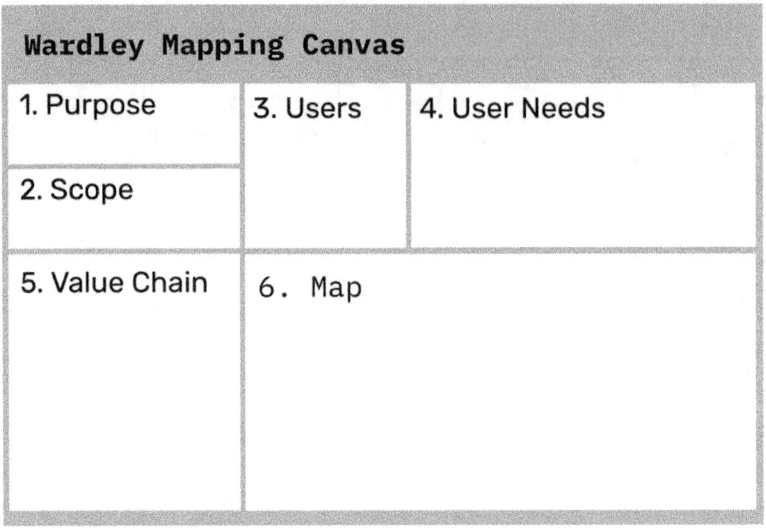

Wardley Mapping Canvas		
1. Purpose	3. Users	4. User Needs
2. Scope		
5. Value Chain	6. Map	

Figure 5-3 Overview of parts of Wardley map canvas

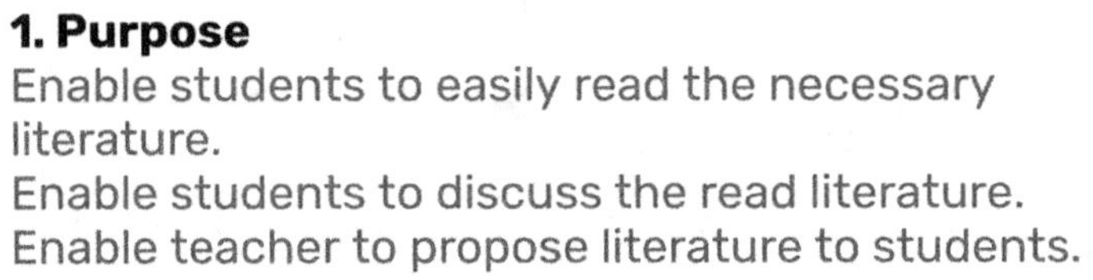

1. Purpose
Enable students to easily read the necessary literature.
Enable students to discuss the read literature.
Enable teacher to propose literature to students.

Figure 5-4 Purpose section of Wardley map

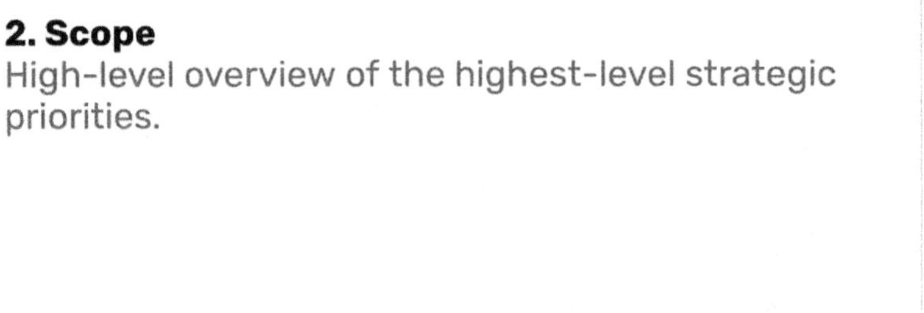

2. Scope
High-level overview of the highest-level strategic priorities.

Figure 5-5 Scope of Wardley map

Figure 5-6 User section of Wardley map

3. Users

Figure 5-7 User needs section of Wardley map

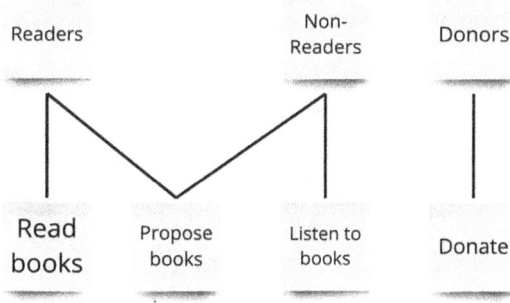

4. User Needs

4. User needs

   The fourth section of the Wardley map contains user needs, which are connected to the users of the user section. Readers want to read and recommend books. Nonreaders want to recommend books as well, and they want to listen to books. Donors want to support the library with donations (Figure 5-7).

5. Value chain

   The value chain section contains the capabilities in order of their visibility to the users. The most visible capabilities and user needs are in the upper part of the diagram. The less visible capabilities are at the bottom of the area. Invisible capabilities include identity management or community finance. Users and their needs are copied into the area to create an overview. Then a channel is added over which customers can access the business. The Business Model Canvas (Figure 5-1) shows the channels of a web and a mobile app. In the case of the online library, both apps are equal, so differentiation of both is not necessary. In what follows, capabilities are ordered depending on their visibility to customers. The arrows indicate the dependencies of the capabilities. For example, rating books depends on recommending books for others. All capabilities depend on identity management, which is almost invisible to customers. All sorted capabilities are shown in Figure 5-8.

6. Map

   The Wardley map is created in the last step. The capabilities sorted along the value chain are also sorted in sections of market evolution. The evolution states are discussed in the following items.

5. Value Chain

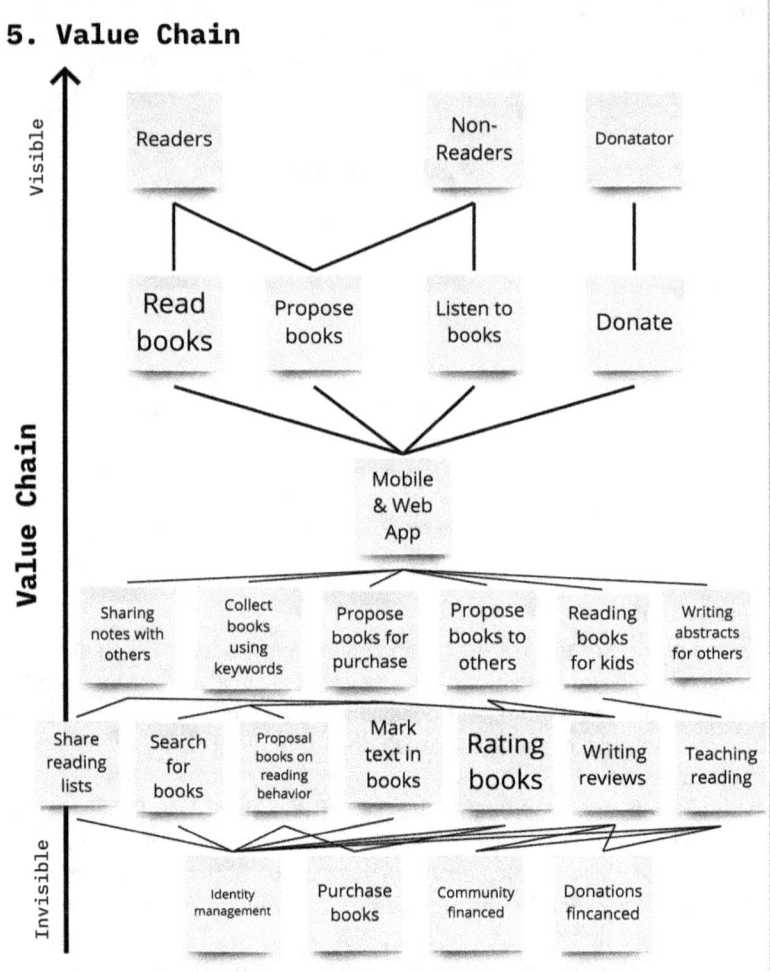

Figure 5-8 Value chain section of Wardley map

- Genesis
  Capabilities in market evolution appear in the Genesis section of the Wardley map. In that phase, it is unclear whether the corresponding capability will reach market maturity [9]. In the case of the online library, no capability is sorted in that evolutionary stage.
- Custom
  Custom means custom-built. Capabilities in this section create the unique selling point of the business [9]. The online library is established as a nonprofit organization. Therefore, the capabilities of sharing and proposing belong to

the custom part. Only those capabilities can guarantee that public financing is possible.
- Product
  Product means capabilities in this section can be obtained on the market but need heavy customization. In the case of the online library, the capabilities of writing reviews, rating books, and marking text in books are sorted into that section. Even the capabilities of teaching reading and reading books to children are sorted into that section, even though they are among the library's core competencies. However, they can be easily technically implemented. Books for purchase are sorted into the product area. Products for purchase are available on the market with quite different characteristics. Even so, managing digital licenses of the books offered in the library needs specific integrations and adaptations.
- Commodity
  Commodities are capabilities that can be easily purchased on the market and need only slight or no adaptations to meet the needs of the business. Typical examples are identity and access management systems available at each cloud provider. Even integrations with social platforms are available. So it is not surprising that the corresponding capability for the online library is also sorted into this section. However, this section also discusses the capability of reading. Reading is one of the core activities in a library, but a standard web or mobile reader can easily be obtained.

The map can be found in Figure 5-9.

The prioritization is given using the Wardley map. Capabilities in the sections Custom and Genesis can be mapped to core capabilities. Capabilities in the Product section can be mapped to supportive capabilities, and capabilities in the Commodity section can be mapped to generic ones.

The outcome will be discussed in the following subsection.

Prioritization of Capabilities

Based on the determination of the Wardley map, the capabilities in the capability map (Figure 5-2) can be restructured as shown in Figure 5-10.

Multiple capabilities were regrouped (the regrouped capabilities are marked with an information sign in Figure 5-10):

- Search for books
  This capability was initially sorted into the core domain. However, it can be handled as generic because search functionalities are technically widely supported. Specific adaptations for the online library are not necessary.
- Collect books using tags
  This capability was called initially *collect books using keywords*. It belongs to the core domain in the first approach. It is now in the supportive domain because

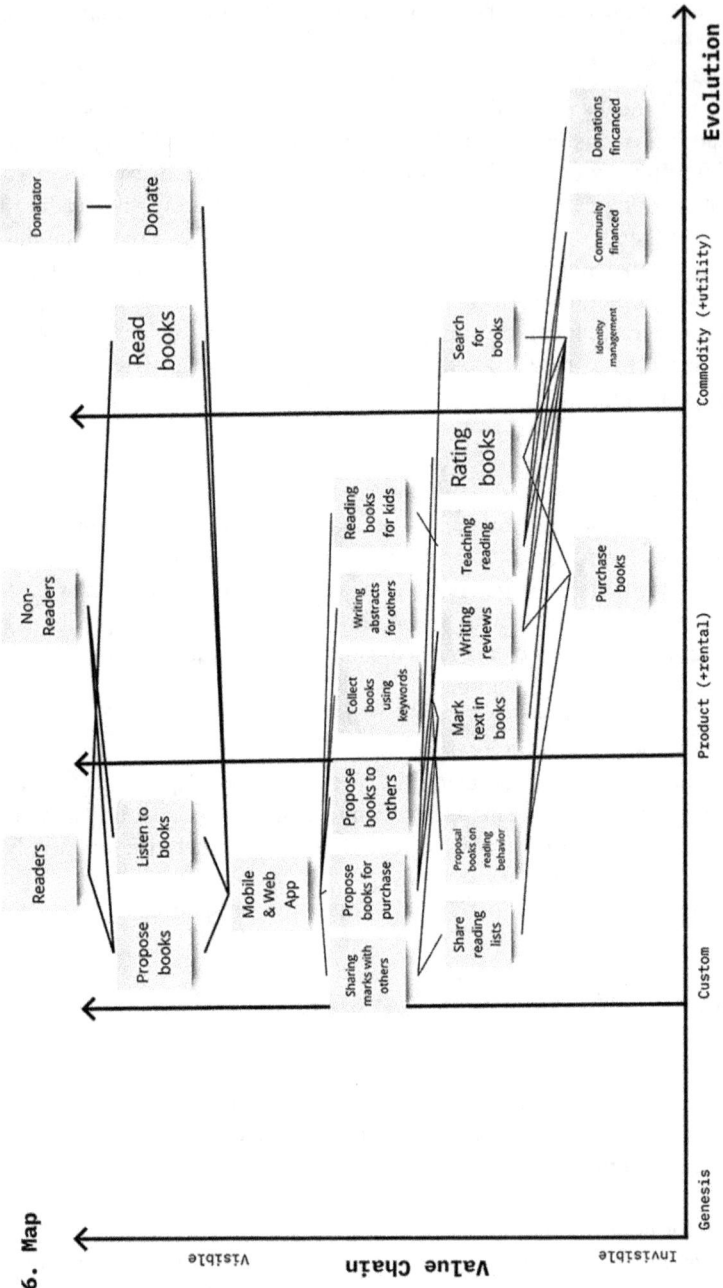

Figure 5-9 Wardley map of online library

Qualifying Business Ideas

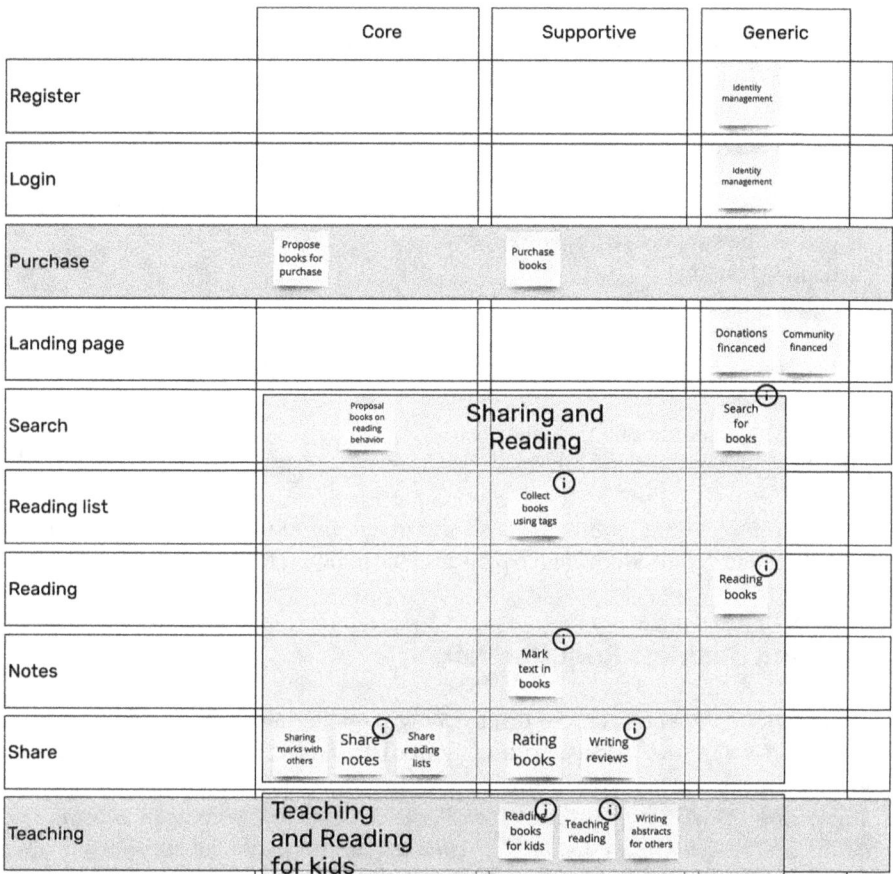

Figure 5-10 Capability map restructured

tagging books is a standard capability, but members' individual tag creation needs to be customized. The capability was renamed because of a slightly different technical meaning between tag and keyword.
- Reading books
  From a poor business point of view, *reading books* belongs to the core domain of an online library. From a more technical perspective, it can be handled as generic because online readers are easily obtained on the market. Specific adaptations are not necessary.
- Mark text in books
  Similar to the capabilities before, this capability can be easily adapted.
- Share notes
  Initially, *share notes* was seen as a supportive capability. However, due to the library's community nature, it needs to be handled as a core capability.

- Writing reviews
  Writing reviews was first seen as belonging to the *teaching* step. However, analysis revealed that it belonged to the *share* step but was still part of the supportive domain.
- Read books to children
  This capability is an essential part of the library. But it can be easily provided using standard software. However, certain adaptations are needed. Therefore, it is part of the supportive domain.
- Teaching reading
  Similar to the preceding point.

When we look at the core points of the library, two major areas appear:

- Sharing and reading and
- Teaching and reading to children.

Both of these core points will be discussed in more detail using Domain Storytelling and Event Storming. We discuss those in the following section.

Gathering Business Requirements

In this section, we would like to discuss how to gather and detail requirements in a structured way, how to develop a ubiquitous language, and how to prepare the logical architecture of an application.

In section "Qualifying Business Ideas", we discovered the areas of *sharing and reading* and *reaching and reading to children* as the major areas. We will dive deeper into the areas using Domain Storytelling and Event Storming to determine the bounded contexts and the associated ubiquitous language.

First, we determine the areas' business requirements using Domain Storytelling.

Domain Storytelling

Domain Storytelling is a collaborative workshop method to gather requirements. The workshops are done in cooperation between business experts and IT specialists. It was developed by *Hofer* and *Schwentner* [10].

Introduction to Domain Storytelling
Business experts tell their stories, and moderators capture the story graphically. The business experts can tell whether they are being understood correctly on the screen or a whiteboard. Business experts use their vocabulary as they normally do. It is captured in the Domain Story.

Experts tell their story sentence by sentence so that it appears step by step. Each sentence is checked to see if the moderator is correctly capturing the experts'

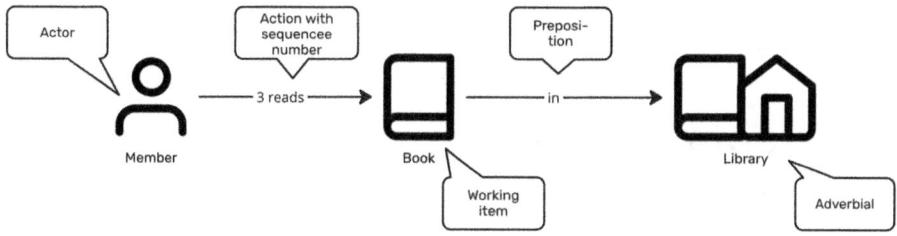

Figure 5-11 Example of a domain story sentence

meaning. Workshop attendees listen to the experts and ask questions to determine the precise meaning of a sentence or the entire story.

Sentences are marked with numbers in such a way that the sequence of the story can be followed. A sentence contains:

- An actor
  An actor is a role in the process, not an actual human. Technical actors are allowed, but the story must be told from the perspective of a business expert. An actor should be described as precisely as possible. The user is too generic; the word member better describes an actor in a library. An actor is symbolized by an icon, typically a user or a server. Actors should only appear once in a graphic to center the story around them.
- An action
  An action is a strong verb. Strong verbs describe actions perfectly. The story should not contain weak verbs such as have or be, which do not describe actions comprehensively. An actor performs the action. An action is symbolized by a numbered arrow between the actor and the object.
- A work item
  An action is done on an item. The work item can be a document, a book, a phone, or something else. An icon symbolizes the item. Each sentence should have its item icon.
- Adverbials
  Adverbials can enhance the sentence, for example, in time or place. Adverbials are symbolized by icons as well. Adverbials are connected to the work object by arrows in the direction of the adverbial and containing a necessary preposition like in, of, or to.

Figure 5-11 shows an example of a domain story.

Multiple business experts can tell their stories to provide a comprehensive picture, which can be adapted from the first story and the relevant language.

The workshop attendees listen to the experts and can ask questions to elicit more details on the story.

When the story is ready, the first bounded contexts can be marked. A boundary of bounded contexts can be indicated by a shift in actor or by a work item change.

The bounded context can be indicated by boxes around the accompanying sentences and acquire a meaningful name.

In what follows, we want to introduce the domain stories of the prioritized areas.

Domain Story for Searching and Lending

First, we want to examine how a member can lend a book. Figure 5-12 shows the Domain Story.

In Figure 5-12, we have the bounded contexts: inventory management, catalog management, and lending. Let us look at the sentences in detail.

- Inventory management

    1. The librarian secures the needed funding.
    2. A member recommends a book for purchase.
    3. The librarian purchases the book.

- Catalog management

    4. The librarian creates a catalog entry with tags in the catalog.
    5. A member searches for a book in the catalog.

- Lending

    6. The member checks his reading list containing individual tags to see if they have read the book already.
    7. The member lends the book.
    8. The reading list stores the book to be read.
    9. The member returns the book.
    10. The reading list proposes a book to the member.

The story is expressive. From bounded context *inventory management* to *catalog management*, the actor changes from librarian to member, which indicates a boundary. From catalog management to lending, the book changes from a general catalog entry to an individual entry and then to a reading list.

Domain Story for Reading and Sharing

The following story contains reading and sharing activities (Figure 5-13).

The story covers the reading and sharing parts of the library. It also contains some optional parts. The optional parts are demarcated by boxes.

- Reading

    1. A member reads a book.
    2. Optional: The member marks text in the book.
    3. Optional: The member creates a note for the marked text.

Gathering Business Requirements

Searching and Lending

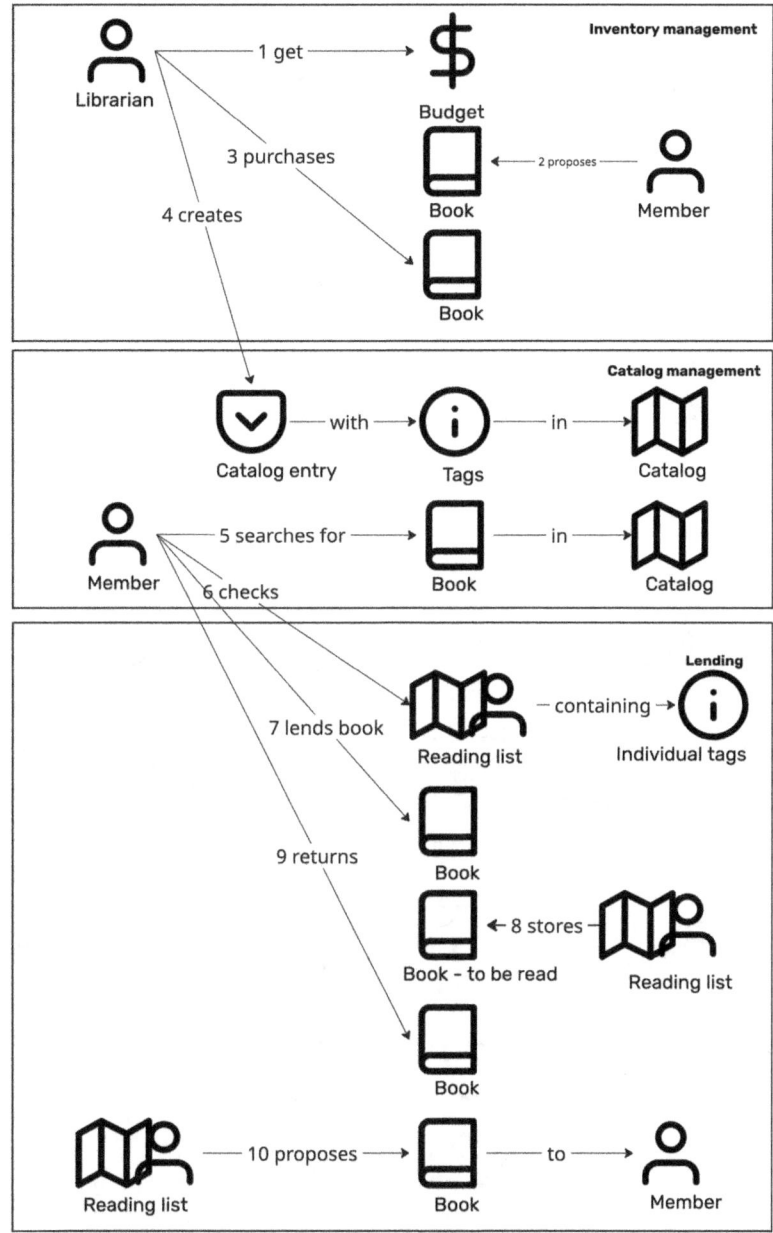

Figure 5-12 Domain story for searching and lending

Reading and Sharing

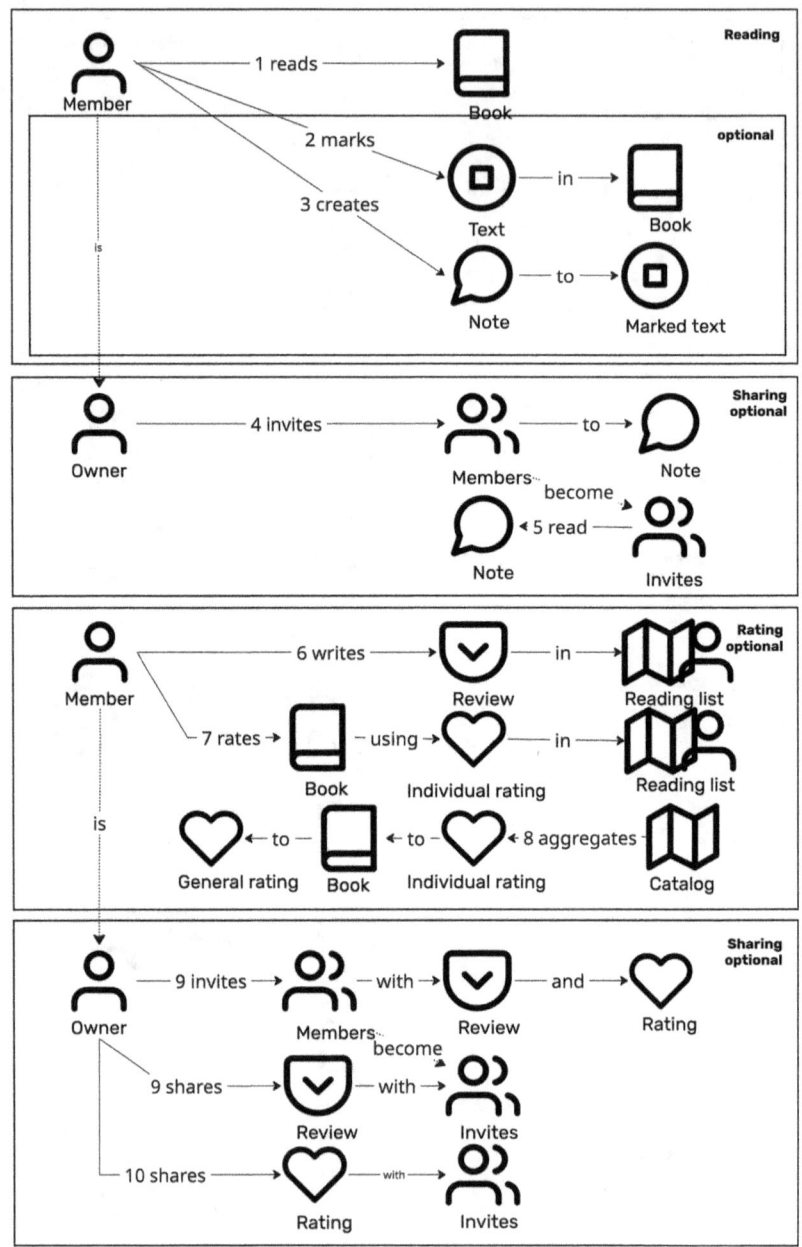

Figure 5-13 Domain story for reading and sharing

- Sharing

    4. Optional: The owner invites other members to read the note.
    5. Optional: The invited members read the note.

- Rating

    6. Optional: A member writes a review in the reading list.
    7. Optional: The member rates the book using individual ratings in the reading list.
    8. Optional: The catalog aggregates individual ratings of the book to come up with a general rating shown in the corresponding catalog entry.

- Sharing

    9. Optional: The owner invites members to read the review and see the rating.
    10. Optional: The owner shares the review with invitees.
    Similar to invitations to read notes, invitations to read reviews and reading lists are members invited by the review or rating owner.
    11. Optional: The owner shares the rating with invitees.

The domain story is expressive. Members can read books and share their opinion about them. They can make notes and share them as well. When a member rates a book, the rating can be shared with other members. Ratings are aggregated to the catalog as a general rating.

Members want not only to have one reading list; they want to organize their books by keywords or tags. The following domain story explains that requirement.

Individual Reading List

A member of the library wants to arrange the books they have borrowed. The arrangement will be made using certain tags, such as *Domain-Driven Design* or *Urban Fantasy*.

The corresponding Domain Story is shown in Figure 5-14.

The entire story belongs to the bounded context individual reading list. Let us dive into the story:

- Individual reading list

    1. A member creates a tagged list with a tag.
    A tagged list means a reading list that contains a specialized tag. A tag is a string like "*DDD*" or "*Urban Fantasy.*"
    2. The member assigns a tag to a book. Assigning a tag to a book allows the member to find the book quickly in their lists. The book can have multiple tags so it will appear in various lists arranged by the member.

Individual reading list

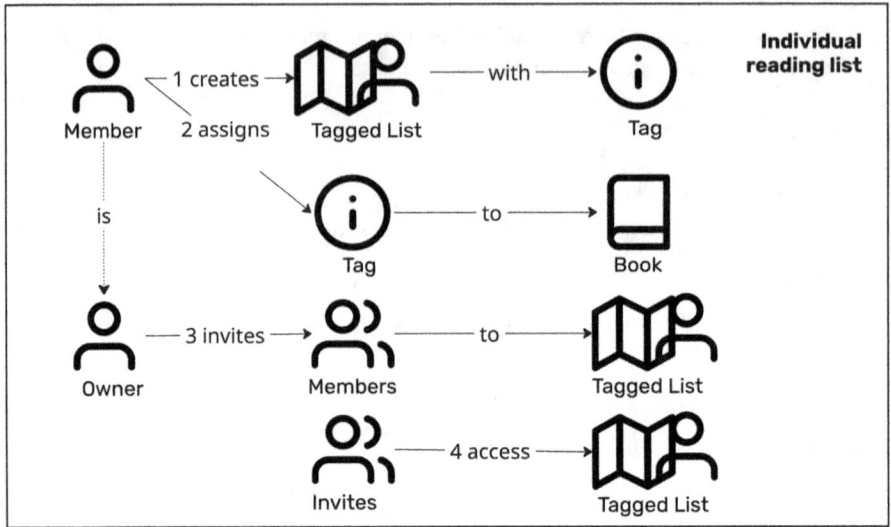

Figure 5-14 Domain story reading list

3. The owner invites members to the tagged list.
4. Invitees access the tagged list.

The respective explanations of the terms "Reading list," "Tagged list," and "Tag" will be discussed in section "Visual Glossary."

The domain stories around reading and sharing are complete. According to the capability map, we need to discuss the *teaching and reading to children* parts. We will do so in the following domain story.

Teaching and Reading to Children

The community aspect of the online library requires bringing children closer to books and reading. Children should love reading rather than loathing it. Volunteers at the library can read books to kids. The following domain story expresses it (Figure 5-15):

- Identity and access management

    1. The parents register their children at the online library.

- Appointment management

    2. A volunteer offers time slots to parents.
    3. The parents set an appointment with the volunteer for their children.

Gathering Business Requirements

Teaching and Reading for kids

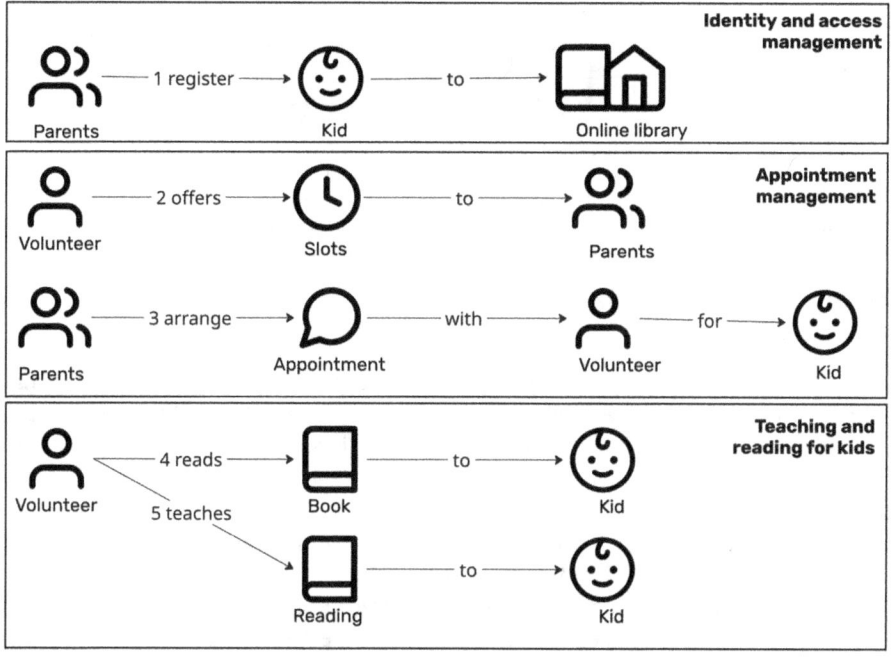

Figure 5-15 Domain story on teaching and reading to children

- Teaching and reading to children

    4. The volunteer reads the book to children.
    5. The volunteer teaches reading to children.

The domain story contains a bounded context or capability that has not yet come up: appointment management. It is necessary to arrange appointments between volunteers and parents to read books to their kids.

The last domain story is about catalog management. Let us look into it.

Catalog Management

Besides the standard catalog entries and a blurb provided by the publisher, the catalog should contain abstracts. Library volunteers write the abstracts (Figure 5-16).

- Catalog management

    1. The volunteer writes an abstract of a book.
    2. The member finds the abstract in the catalog.

Catalog management

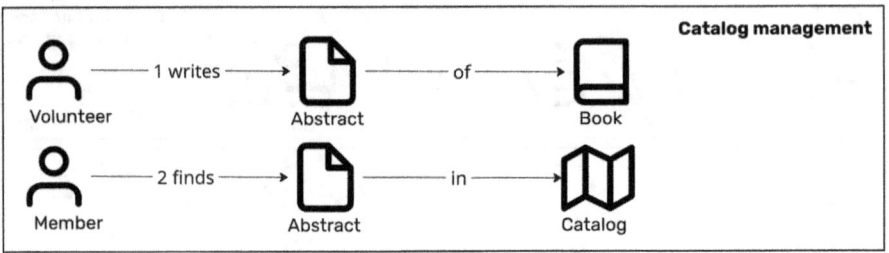

Figure 5-16 Domain story on catalog management

Writing book abstracts can be tiresome and time-consuming. Volunteers work on a charity basis; they are not paid. The abstract does not usually contain better information than the blurb provided by the publisher. Therefore, we decided to postpone the feature and rate it again after the first implementation of the necessary features.

Domain Storytelling is great for gathering business requirements and deciding what to do, as well as for deciding what not to do.

For the prioritized capabilities, we gathered all domain stories. During domain storytelling, we captured the terms and structured them in Visual Glossaries, which will be explained in the following section.

Visual Glossary

A Visual Glossary is a great tool to define terms and their quantities derived from a domain story.

Terms used during a Domain Storytelling workshop or within a development team are collected, for example, as sticky notes. Afterward, the terms are connected with arrows using business relations. Business relations even describe the quantity relations between terms, for example, "1..*" for a 1:n relation [11].

The relations between the terms get expressive verbs. For example "A member **creates** no or multiple invitations." In such a relation, the quantities are shown – like a classical Unified Modeling Language (UML) diagram. However, it contains the strong verb "creates," which makes the Visual Glossary much more expressive than a class or entity relationship diagram of UML.

Let us apply the principle to the domain stories of the online library. We start with the area of lending and searching.

Visual Glossary Lending and Searching

Figure 5-17 shows the Visual Glossary of the searching and lending story (Figure 5-12).

Let us look closer at the Visual Glossary.

Gathering Business Requirements

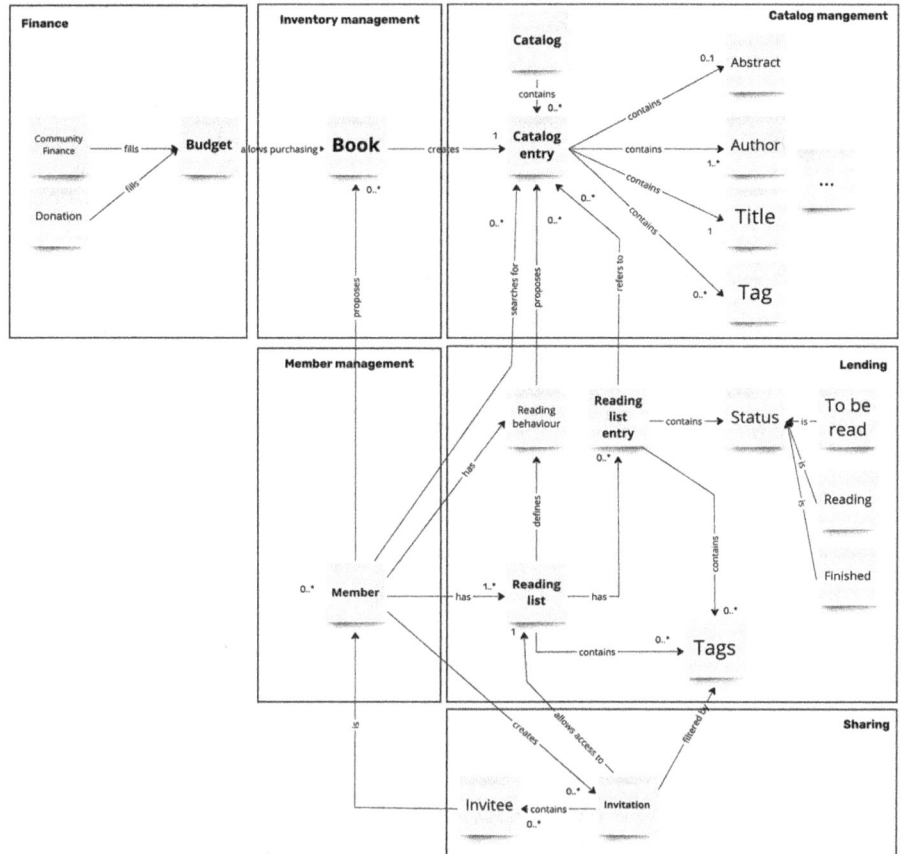

Figure 5-17 Visual Glossary: searching and lending

- **Finance**
  - Community finance
    Budget for the library delivered by the community.
  - Donation
    Donations by members, private individuals, parents, or organizations are used to meet the *budget* of the library.
  - **Budget**
    The library's budget is usually proposed on an annual basis; however, it can be done on a quarterly or monthly basis, too; it is met by *community finance* or by *donations*. The budget contains allocations for book purchases.
- **Inventory management**
  - **Book**
    A digital representation of a book that can be read by members. It can be purchased thanks to available budgetary allocations. *Members* can recommend books for purchase.

- **Catalog management**
  - **Catalog**
    The library catalog can be used by members to search for books. It contains various *catalog entries*.
  - **Catalog entry**
    An entry in the catalog representing a *book* that can be read by members. There are more properties in a catalog entry that are not listed in the Visual Glossary. A catalog entry contains at least a *title* and *authors*. Volunteers can write an *abstract*. The responsible librarian can create *tags* for the catalog entry so members can easily find the book. Catalog entries can be proposed based on the *reading behavior* of members. A member's *reading list* refers to a catalog entry.
  - Abstract
    A volunteer can write an abstract of a book that can be stored in a *catalog entry*.
  - Author
    A *catalog entry* contains one or multiple authors of a book.
  - Title
    A *catalog entry* contains the title of the book.
  - Tag
    A volunteer librarian can create one or more tags for a *catalog entry* so members can easily find the desired book.
- **Member management**
  - Member
    A registered user of the library. A member can recommend *books* to be purchased. Members can search for books using the catalog and invite other members to share *reading lists*. A member has at least one *reading list* and a *reading behavior*.
- **Lending**
  - Reading behavior
    A member engages in a certain reading behavior. It is defined by the member's *reading list*. Based on the member's reading behavior, the system can recommend special *catalog entries* for a member.
  - **Reading list entry**
    An entry in a member's *reading list*. It refers to a corresponding *catalog entry* and contains a given status.
  - Status
    The status of a certain *reading list entry*.
  - To be read
    The status of a *reading list entry* that means that the member has not yet started reading the entry.
  - Reading
    The status of a *reading list entry* that means the member has started reading the entry.
  - Finished

Gathering Business Requirements

The status of a *reading list entry* that means the member has finished the book.
- **Reading list**
A list owned by a member containing the borrowed books. The reading list includes various *reading list entries*.
- Tag
A tag means a word assigned to a list used for grouping *reading list entries*. A tag assigned to a list creates a new tagged list.
• **Sharing**
- **Invitation**
An invitation can be sent to members by an owner of a shareable item. Shareable items are notes, ratings, reviews, or *reading lists*. The invitation contains *invitees*.
- Invitee
An invitee is a *member* who has been invited to a shareable item.

Using Visual Glossary and Domain Storytelling allows first guesses of the aggregates of the corresponding bounded contexts. They are marked in bold in the list. We will discuss aggregates and their roles in a bounded context in the next section, section "Event Storming".

Now let us take a look at the Visual Glossary of *reading*.

Visual Glossary Reading

Figure 5-18 shows the Visual Glossary of the *reading* area.

The area is explained in detail in the following list. Already discussed bounded contexts as *sharing* and *member management* will not be addressed again.

• **Reading**
- Text marked
A reader can mark text in a book. The marked text will be stored so the book's reader can easily find the passage.
- **Note**
A reader who has marked text can create a note on the passage. A note is always created when a reader marks text. But the note can be empty.
- **Book**
A reader reads a book. The current text position is stored for the reader without further interaction. Additionally, the reader can create *notes* to find and comment on certain text passages.
• **Rating**
- **Review**
A *member* can write a review of a *book*. *Members* can share their reviews with other *members*.
- **Rating**
A *member* can rate a *book* by giving it a natural number between 1 and 5. *Members* can share their ratings with other *members*.

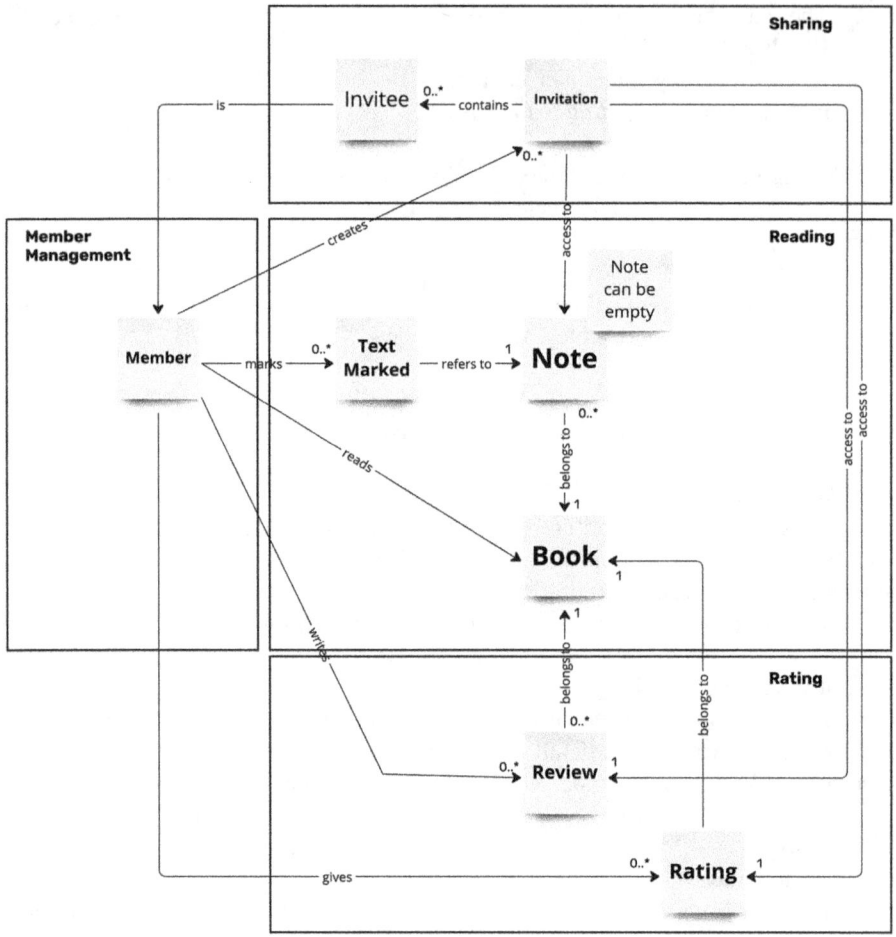

Figure 5-18 Visual Glossary: reading

The third area we need to discuss is the *reading to children* area.

Visual Glossary Reading to Children
Figure 5-19 shows the Visual Glossary of reading to children, where volunteers read books to kids.

Let us look into the details of objects:

- **Member management**
  - **Parents**
    Parents can register themselves at the library. Then they can register their *kids*.
  - **Kid**
    Parents register their kids at the library to be able to arrange *appointments* for them.

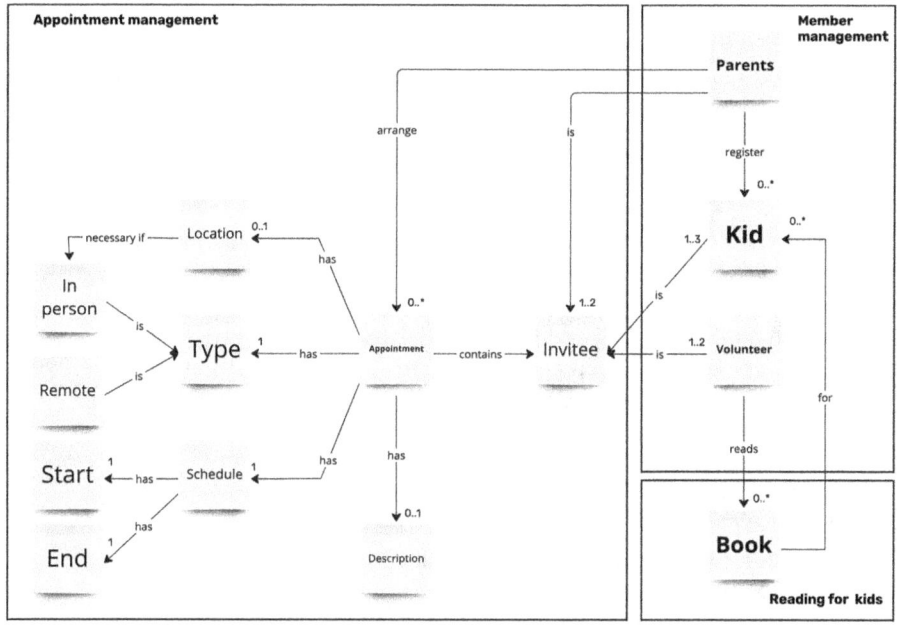

Figure 5-19 Visual Glossary: reading to children

- **Volunteer**
  A volunteer can read books to *kids* when an *appointment* is arranged.
- **Appointment management**
  - **Appointment**
    Parents can set appointments between their *kids* and *volunteers*. An *appointment* can be made *in person* or *remotely*. It will have a *schedule* and probably a *description*.
  - Location
    An *appointment* can have a location if it is not done remotely.
  - Type
    An *appointment* has a type. The type is either *in person* or *remote*.
  - In person
    In person is a type of appointment.
  - Remote
    Remote is a type of appointment.
  - Schedule
    An *appointment* has a *start* and an *end* as a schedule.
  - Start
    means the start of an *appointment*.
  - End
    means the end of an *appointment*.

- **Reading to children**
  - **Book**
    This is a book that volunteers read to children. Children can have a reading list in the same manner as members. The book can be rated and reviewed, and the text can be marked. Volunteers do it on behalf of children.

The Visual Glossary contains parts of member management to show that different kinds of users are necessary: members, volunteers, parents, and kids. Appointment management should be a commodity, but it needs to be customized for the specific requirements of volunteers, parents, and kids to avoid misuse.

The Domain Storytelling and combined Visual Glossary give a first impression of bounded contexts and accompanying aggregates. Those impressions need to be deepened by Event Storming, which we will discuss in the following section.

Event Storming

Event Storming is a workshop format invented by *Alberto Brandolini* to describe business flows via domain events [12]. It combines the idea of brainstorming and process modeling.

Brainstorming is an ideation technique. The term was first coined by *Osborn* in 1953 [13]. Event Storming helps in holding successful workshops following the rules of brainstorming workshops [14]:

- Focus on quantity and not on quality,
- Withhold evaluation in a first step,
- Encourage wild, outlandish ideas,
- Combine or build on ideas from others.

Introduction to Event Storming

The aim of an Event Storming session is deeper insight into the domain. Event Storming sessions can be done based on previous Domain Storytelling workshops, but they can also be done as a single workshop.

In what follows, we will describe Event Storming as a tool to obtain an overview of the entire domain. Such a workshop performs the following steps:

1. Collect events
   All workshop attendees collect domain events as the attendees believe they would fit in the process in scope, for example, the *reading* process of a library. The events are not judged; all input is welcome. Domain events are essential to the business process. They happen to be business objects. Usually, a business object's status changes, for example, the status of a book from `To be read` to Read.

Gathering Business Requirements

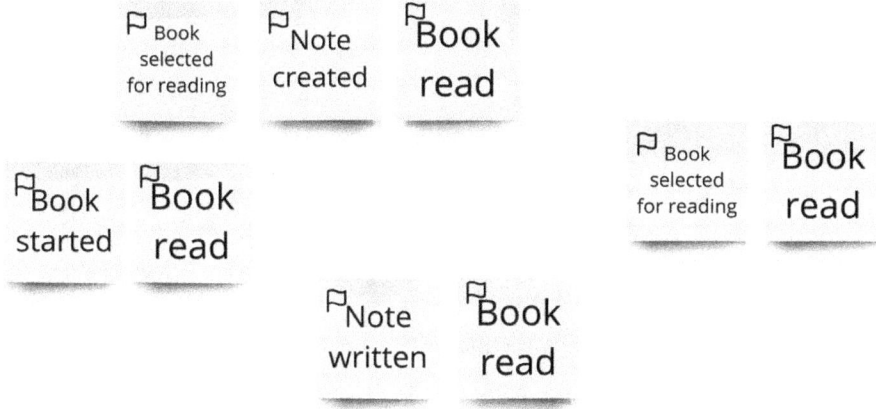

Figure 5-20 Event Storming: collection of events

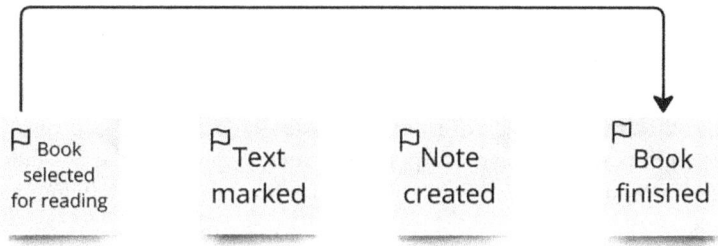

Figure 5-21 Event Storming: consolidated events

Events are formulated in the past tense, e.g., `Book finished`. Figure 5-20 shows an example of such a collection for the *reading* area.[1]

2. Consolidate events
   In the second step, workshop attendees sort and consolidate events. They discuss all the events found. Usually, the event group of one attendee and then the events found by the next attendee are discussed. Events are sorted by the time they occur, from left to right. Events that happen simultaneously are stacked on top of each other. The same applies to events that occur exclusively. Furthermore, attendees find common names for events. For example, they use `Book finished` instead of `Book read` because it is more expressive and better describes what happened. Arrows can be used to show that certain events happen optionally and can be skipped. Figure 5-21 show an example of the sorted events.
3. Find commands and roles and processes
   In the third step of the workshop, attendees discuss the command that triggers the event. For example, the event `Note created` is triggered by the command

---

[1] Usually events are shown as orange sticky notes. In grayscale print, color coding cannot be applied. Therefore, events are marked with a flag icon.

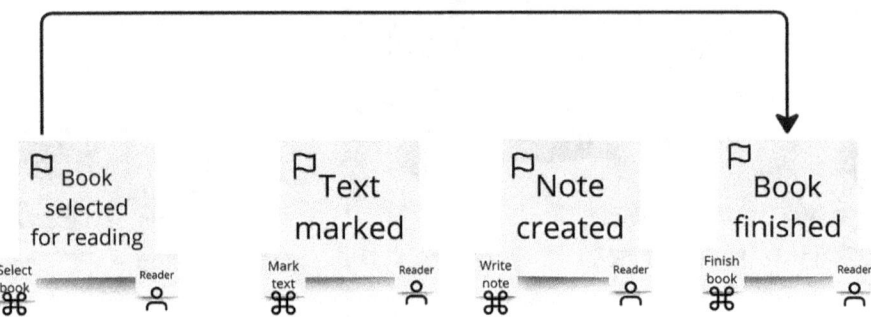

Figure 5-22 Event Storming: commands, roles, and processes

Create note. That double occurrence of words – past tense for the event and imperative for the command – is inherent in the methodology. However, there are occurrences where the words in command and event can differ. The workshop attendees need to discuss the commands carefully. Additionally, the attendees discuss who or which system ordered the command. People are presented by the roles they fulfill and systems by their names. Commands[2] are placed in the bottom left corner of an event, and roles or processes[3] are placed in the bottom right corner of the corresponding event. Figure 5-22 shows the result of the discussion about commands, roles, and processes for the *reading* area as an example.

4. Find aggregates and read models

In the last step, workshop attendees discuss the data that have changed or are needed for the event. Data changed by an event are usually collected in an aggregate. An aggregate is a referenceable object that usually represents the essence of a bounded context. For example, catalog entries are aggregates for catalog management. They can be referenced and contain an invariant. According to *Evans* [15], an invariant is a business rule that does not change when the aggregate changes. The invariant is usually a status model of the aggregate, but other business rules might also be applied. The data needed to perform the event but not changed are called read models. Read models[4] are placed in the upper left corner of an event. Aggregates[5] are placed in the upper right corner of an event.

---

[2] Commands are usually presented by blue sticky notes. Because of the print, the commands are marked with a command icon.

[3] Roles are usually marked with dark-yellow sticky notes, whereas processes are marked with pink sticky notes. Because of the print, roles are marked with a user icon and processes with a steering wheel.

[4] Usually, read models are shown by green sticky notes. But in grayscale print, they are represented by a screen icon.

[5] Usually, aggregates are shown by light-yellow sticky notes. Because of the grayscale print, they are marked with a document icon.

Gathering Business Requirements

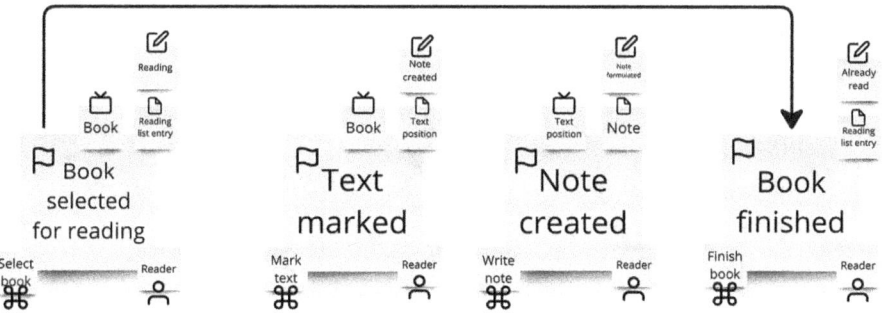

Figure 5-23 Event Storming: read models and aggregates

The aggregate of the *reading* area is a book. Over single events, it holds different states. Those states symbolize the invariant of Book. Business rules[6] or invariants are shown above the associated aggregate. Figure 5-23 shows an example of such a discussion.

During this discussion, the associated status model must be discussed as well. The status model can be created using sticky notes and arrows presenting the necessary transitions. The transitions contain the found commands. The status model for the discussed *reading* area is shown in Figure 5-24. When discussing the status model, additional commands might be found. In the example, the additional command Delete text and Unmark text were found for the Note status model. For the book, the additional command Reopen was found. Those additional commands do not falsify the event model. They can be used as additional requirements to be implemented in the corresponding service.

5. Find bounded context

   The last step of an Event Storming workshop is the finding of the bounded context. Bounded context are indicated by a change in role, for example, from Librarian to Member or by a change in aggregate, for example, from Appointment to Book. The *reading* example only contains one role, Reader. However, the aggregate changes from Reading list entry to Note and back to Reading list entry. However, attendees can decide that all events belong to one bounded context: *reading*. That corresponds to the result of the domain storytelling. This means that marking text and creating notes belongs to *reading*. The bounded context is shown as a rectangle with rounded corners or as an ellipse around the associated events. The form is labeled with the name of the bounded context. The result is shown in Figure 5-25.

Now we can apply the Event Storming format to areas of online library.

---

[6] Usually, business rules are shown by purple sticky notes. In this book, they are symbolized by an edit icon.

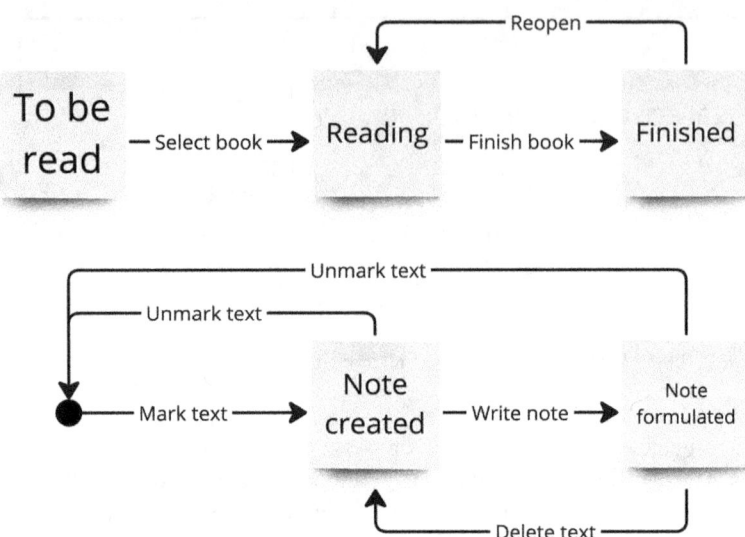

Figure 5-24 Event Storming: status model *reading* area

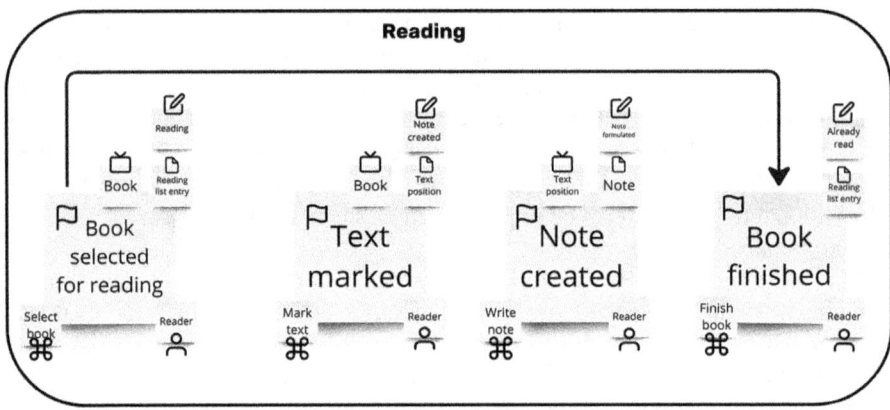

Figure 5-25 Bounded context *reading*

Event Storming of Online Library

We start with the *inventory management* and *catalog management*.

Figure 5-26 shows the results of the Event Storming of the bounded contexts.

A librarian can select a book from the list of recommendations and check whether the budget is sufficient. If it is, the book can be bought. After the book is purchased, a new catalog entry is created. A librarian can review the entry and probably add specific tags to it. The boundary between the bounded contexts is between the events `Book purchased` and `Catalog entry created`.

Gathering Business Requirements

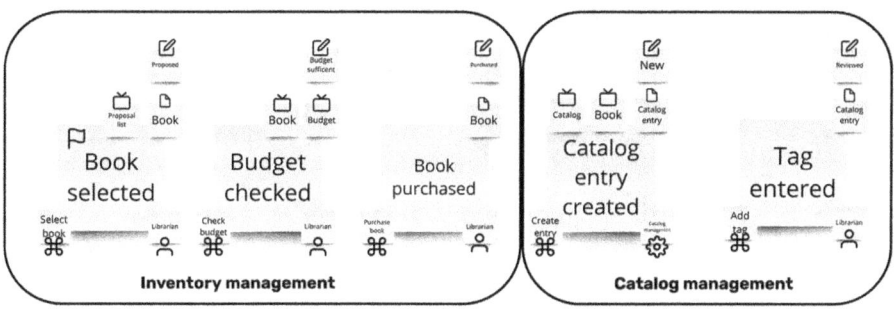

Figure 5-26 Event Storming results: inventory management and catalog management

When a catalog entry is created, a member can find the entry and read the corresponding book. The result of the Event Storming is shown in Figure 5-27.[7]

Library members can search for books using the catalog management catalog. When they find a suitable book, they borrow it and start reading. Lending a book to a library member means the `Reading list` is updated with a new entry in the status `To be read`. The boundary between *catalog management* and *lending* is marked by the events `Book searched` and `Book borrowed`. The boundary between *lending* and *reading* are marked by `Readling list updated` and `Book selected for reading`.

The next bounded contexts *rating* and *sharing* are shown in Figure 5-28.

When a book appears in a member's reading list independent of its state, the member can "tag" it. That means the member assigns a word as a kind of keyword to it. A book can contain several tags, as discussed in the corresponding visual glossary; see Figure 5-17. A member can rate a book or write a review. A rating, review, or tagged list can be shared with other library members. The owner invites the member to those "shareable items."

A closer look into the results for *sharing* is shown in Figure 5-29.

When members create an `item`, they become the owner of it. The owner can invite other members to the `item`. The invitees are able to review the `item` (Figure 5-30).

Besides members, the library needs to support parents and kids, as shown in Figure 5-31.

Volunteer librarians or even teachers for the kids do not appear in the event storming results even though they are part of the identity and access management. However, they are not part of member management. The onboarding process for volunteers is more complex and requires a couple of documents to ensure the library's and children's safety.

---

[7] Usually, the events in this sequence should be sorted from left to right; however, because of the limited printing locations, they are placed above each other. The sequence is indicated by an arrow.

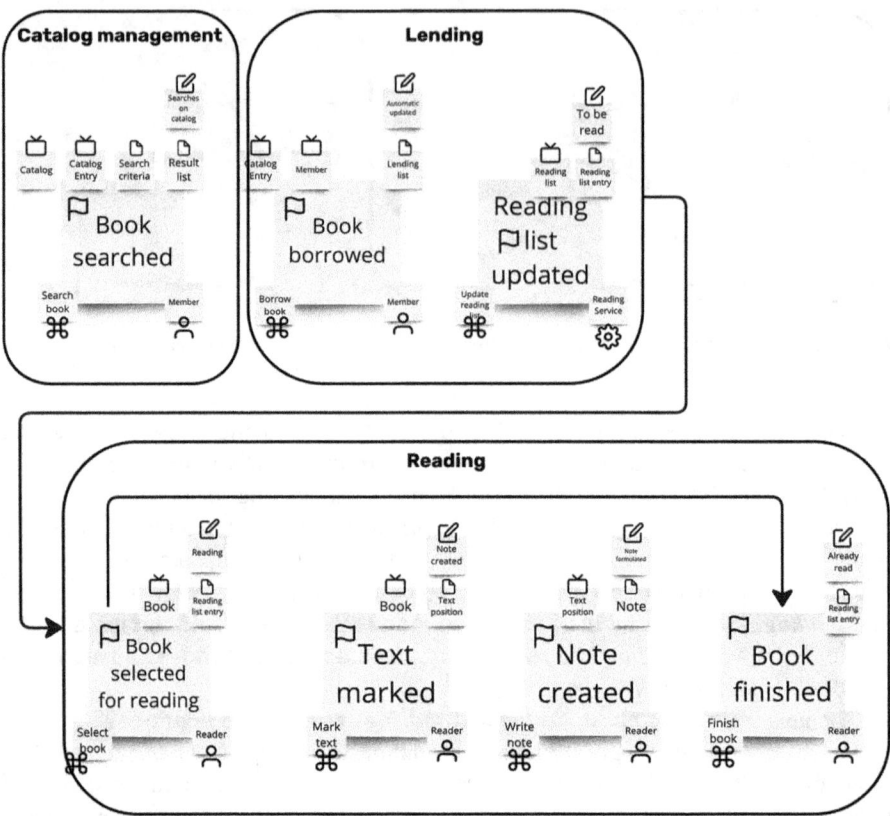

Figure 5-27 Event Storming results: catalog management, lending, and reading

As a last result, appointment management is shown here. Other areas, like *reading to children*, are not as complex and do not need a separate discussion.

A volunteer who wants to read to children can offer free slots. Parents who want to make an appointment for their kids can see all the open slots and book one. They select a slot and provide all the necessary information, such as their kids' names or which book they want to hear. When a slot is booked, the appointment is set. Parents or the volunteer can cancel an appointment.

Domain Storytelling and Event Storming, together with a Visual Glossary, gave us all the information needed to create a first architectural design. The architectural design sketches the team dependencies and the necessary data exchanges via interfaces. We use the technique of a context map to sketch the architecture. The context map of the online library will be discussed in the following section.

Context mapping is the method to find the right cut for systems and teams based on DDD. The context map was introduced by *Evans* [15].

A context map shows the bounded contexts of an application and the associated team and data dependencies. In the next subsection we will dive deeper into this topic.

Gathering Business Requirements

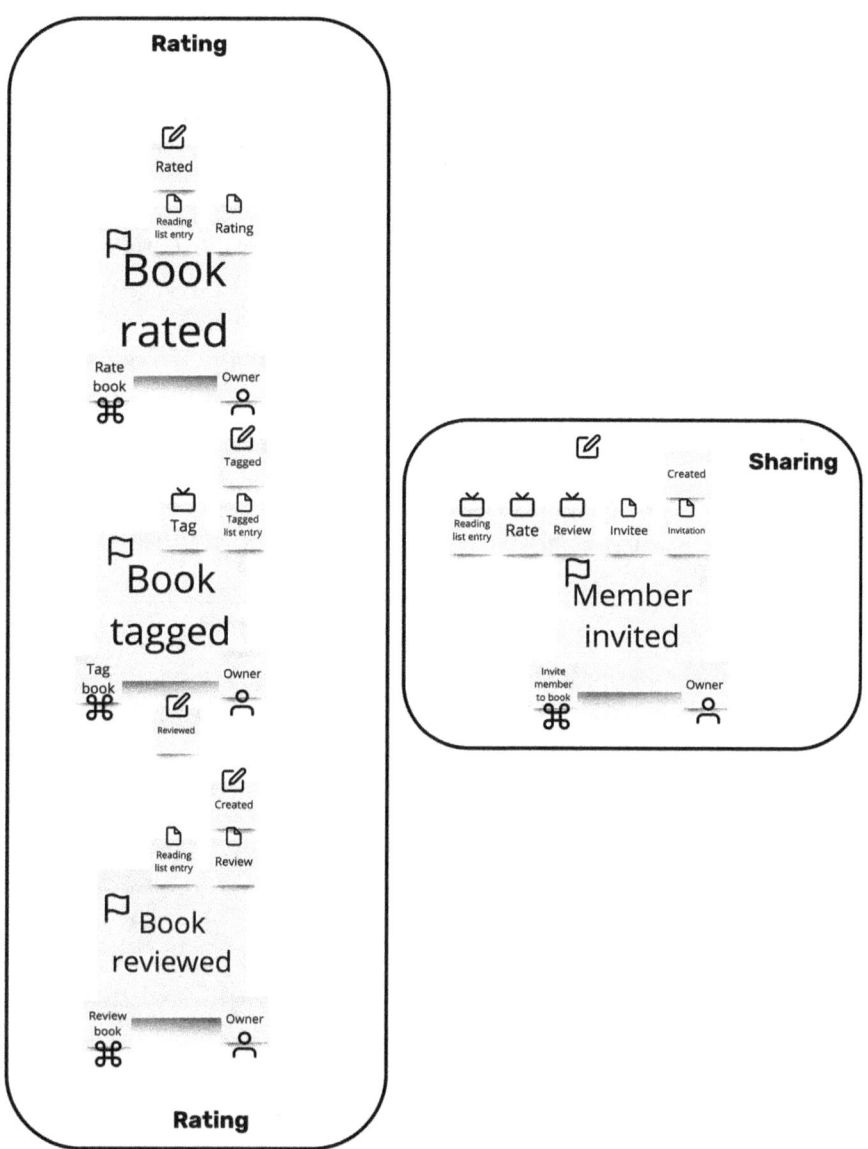

Figure 5-28 Event Storming results: rating and sharing

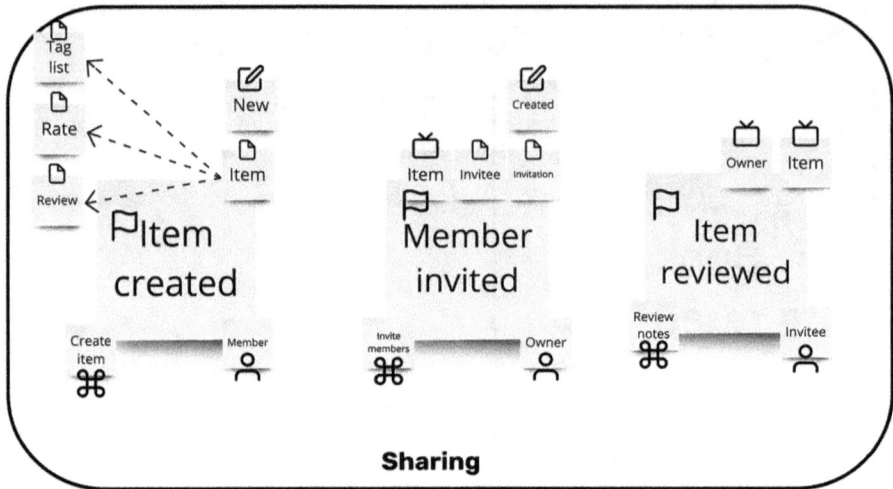

Figure 5-29 Event Storming results: sharing

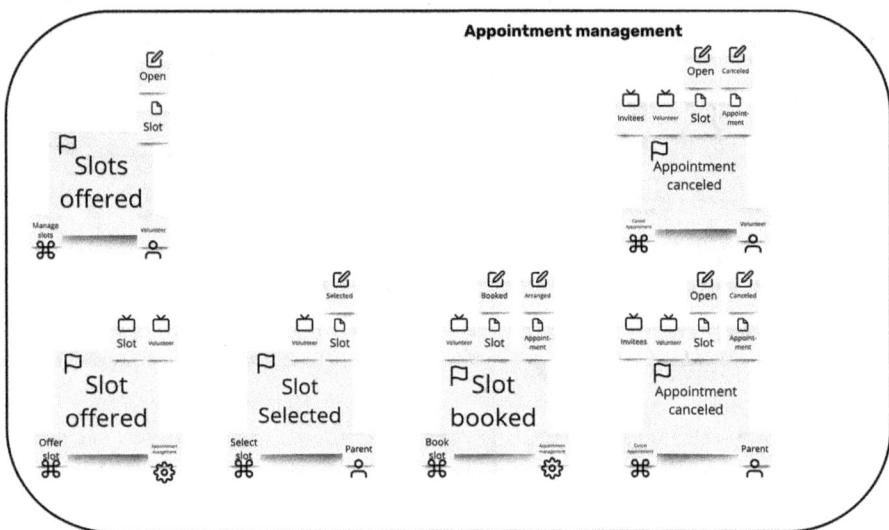

Figure 5-30 Event Storming results: sharing

Context Map

A context map can be created using the found bounded contexts in a Domain Storytelling or Event Storming session. A rectangle is drawn for each found bounded context.[8]

---

[8] Originally, ellipses are used for bounded context. We use rectangles here because the diagrams are more straightforward to draw using rectangles.

Gathering Business Requirements

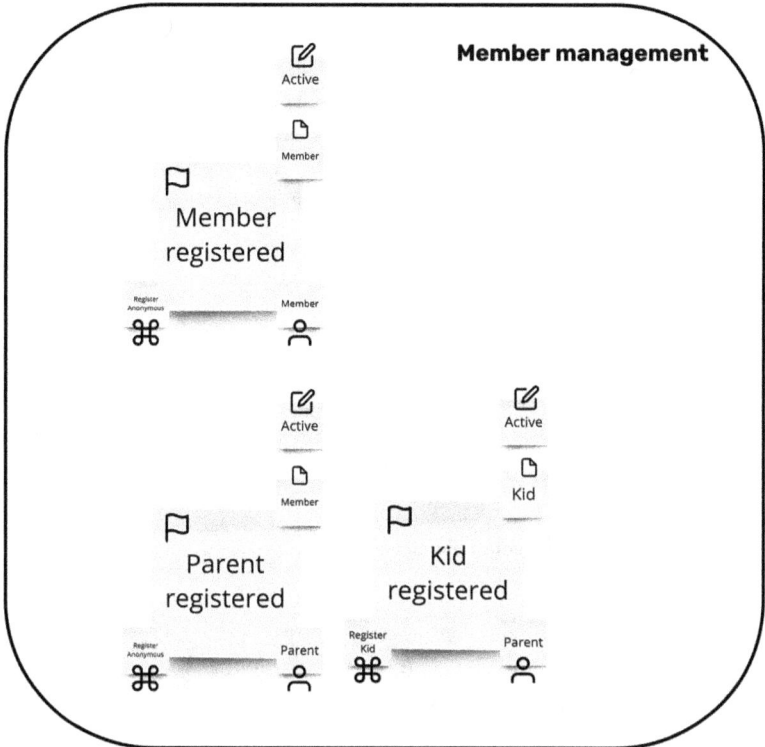

Figure 5-31 Event Storming results: sharing

Then team dependencies are determined. *Evans* determined six typical team dependencies:

- Shared kernel
- Customer/supplier development teams
- Conformist
- Anticorruption layer
- Separate ways
- Open host service

Later, the *DDD crew* adds a partnership and Big Ball of Mud (BBoM) to it [16].

The ubiquitous language can be externalized via a published language [15], for example, using an API.

We will look into the individual types of team dependencies in the following sections.

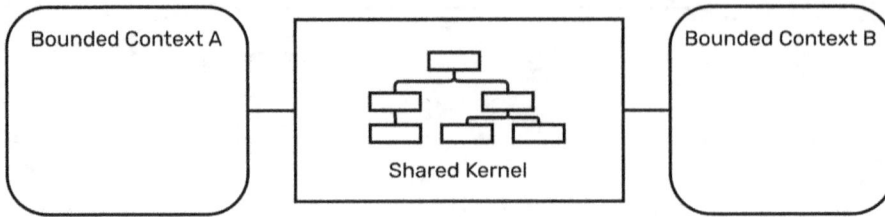

Figure 5-32 Shared kernel

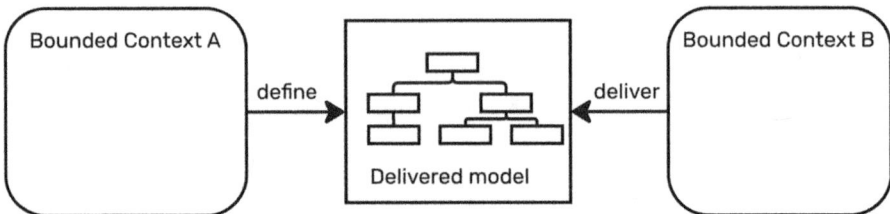

Figure 5-33 Customer/supplier development teams

Shared Kernel
A shared kernel exists when two teams use a shared model overarching two bounded contexts. That shared model might be problematic when the teams run together for a while, but the models will eventually diverge significantly. Then the teams must find a model like a conformist. The shared kernel is shown in Figure 5-32 [15, 16].

Customer/Supplier Development Teams
Customer and supplier teams are comparable to the shared kernel. They produce a commonly used model. However, the customer team defines the model, and the supplier team uses it. Typically, such a cooperation model is needed when the user interface team is separated from the business layer team. Then the business layer team defines the model as a customer and the supplier, and the user interface team needs to deliver it in the user interface. The corresponding dependency is displayed in Figure 5-33 [15, 16].

Conformist
A conformist is a team that uses the model defined by another team without changing it. Usually, that is necessary when the team uses standards where the model cannot be changed. For example, when one uses identity standards such as Open ID Connect [17], it simplifies the integration enormously [15]. The corresponding relations are shown in Figure 5-34 [15, 16].

Anticorruption Layer
An anticorruption layer is necessary to avoid one model leaking into a model using it. Usually, poorly modeled systems, as in historically mature systems, leak into

Gathering Business Requirements

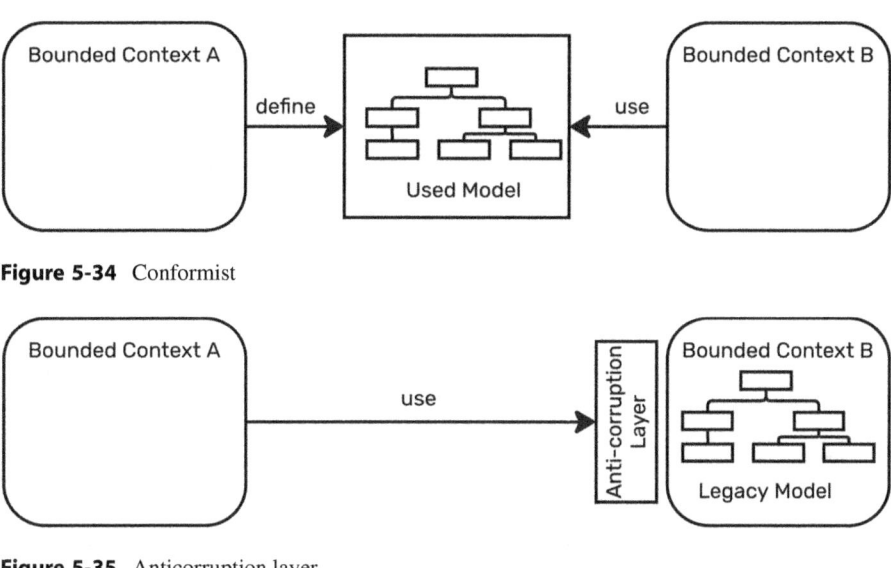

Figure 5-34 Conformist

Figure 5-35 Anticorruption layer

Figure 5-36 Separate ways

more suitable models developed later [15]. Another example are third-party systems purchased from the outside. When developing modern systems based on legacy systems, which should be replaced step by step, the modern systems need to be secured so that the legacy model does not leak into the contemporary model. We will discuss the modernization of legacy systems in greater detail in Chapter 11, "Brownfield Project." The model is shown in Figure 5-35 [15, 16].

Separate Ways
Integration is expensive. When possible, bounded contexts should be defined so that the teams can act entirely independently [15]. They go their "separate ways." The relation is shown in Figure 5-36 [18].

Open Host Service
A way to get independent teams is to implement open host services. An open host service gives consuming services a protocol that they can use. The open host service

Figure 5-37 Open host service

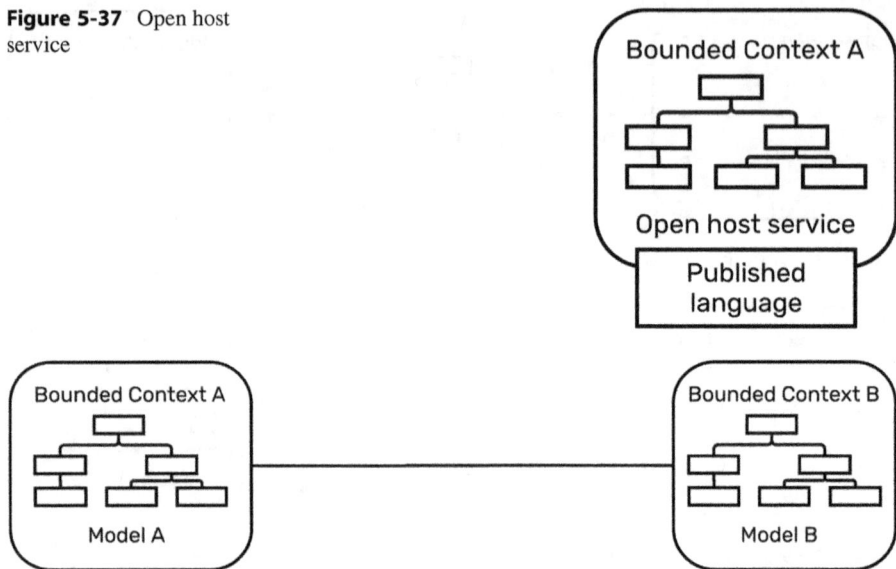

Figure 5-38 Partnership

acts independently of its consumers [15]. It should be the desired form of service in a modern software application.

The protocol can be externalized via published language [16].

For several services in a software application, the models can be externalized per bounded context in one published language in one API. Figure 5-37 shows the corresponding service [16].

Partnership

Two teams might depend on each other in a way that affects both teams when an error occurs. Those teams need to forge a partnership through a process that allows them to evolve their interfaces in collaboration [18]. The relation is shown in Figure 5-38 [16].

Big Ball of Mud (BBoM)

A BBoM is created when the relations we discussed earlier were not considered during the system's design. It has mixed models and inconsistent boundaries. Such a model needs to be surrounded by high walls – anticorruption layers – so that it does not leak into the models of well-designed bounded contexts.

Using the models explained above should avoid such a mess. The following section discusses how we can apply the models to the online library.

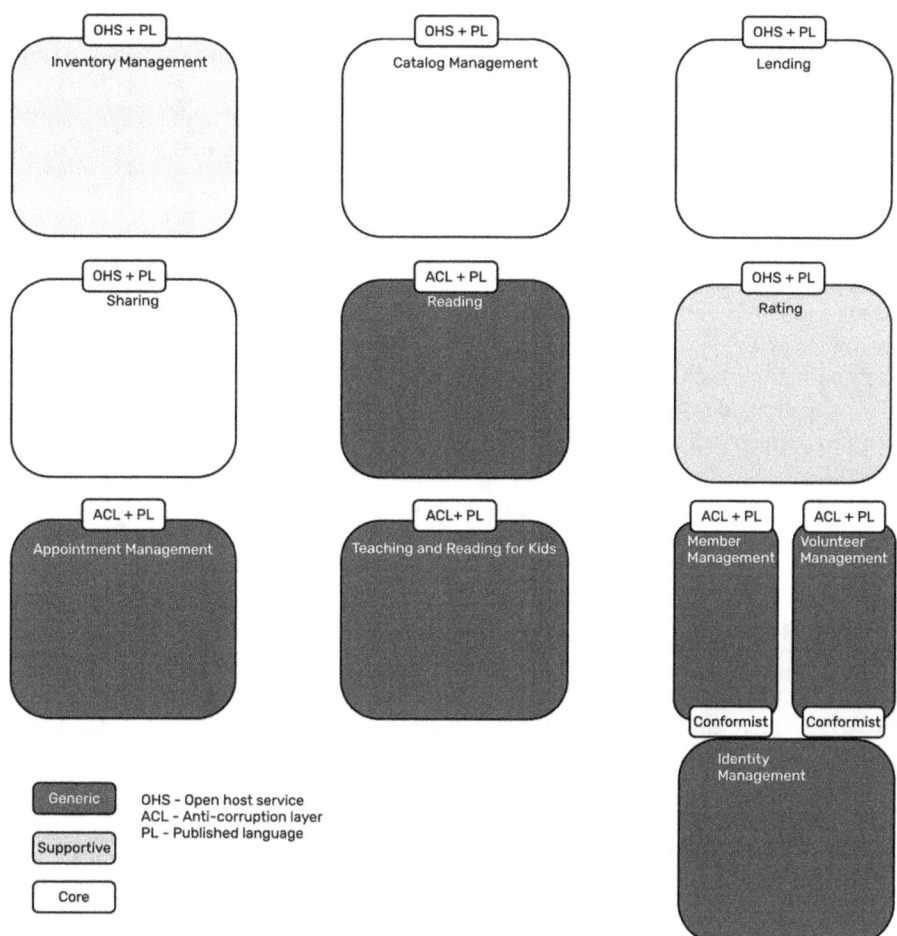

Figure 5-39 Context map of online library

Discussion of Data Exchange Using API-First Approach

To create a context map of the online library, we draw a rectangle for each bounded context found in section "Event Storming of Online Library." The bounded contexts are also marked as core, supportive, or generic, corresponding to Figure 5-10. The result is shown in Figure 5-39.

Catalog management, Lending, and Sharing are core bounded contexts of the online library. All of them can be implemented as independently as possible as open host services. Their ubiquitous language is exposed as a published language.

Inventory Management and Rating are supportive bounded contexts containing aspects of the core domain and specific points. They publish their ubiquitous language, too.

Reading, Appointment management, Teaching and Reading to children, Member management, Volunteer management, and Identity management are generic bounded contexts.

As commodities – see section "Wardley Map" – they need an anticorruption layer so that the ubiquitous language can be published corresponding to the requirements of the online library.

Moreover, Member management and Volunteer management are specifics of a completely generic Identity management. Both variations are necessary because volunteers and members have different onboarding processes. A volunteer needs a secured identity, whereas the onboarding threshold of members should be as low as possible. As conformists, they use the standard interfaces of a provided identity management.

Using that map, we can discuss the necessary data exchange. To do so, we use the corresponding borders between the bounded contexts from section "Event Storming."

Looking at the border between Inventory management and Catalog management, the event Book purchased needs to be published (Figure 5-26).

Going further, we see that we need a piece of information from Catalog management to Lending. To lend a book is usually a user interaction done by a member. So, technically, it is too expensive to route such an event via the server. Therefore, a client-side event informs Lending about a book to be lent.

The Sharing bounded context needs information about the shareable items Reading list and Reading list entry from Lending. The same holds for Rating and Review from Rating. A synchronous call is necessary when a member actively wants to share them with other members. The context map shows the data flows. Therefore, the arrow runs from Lending to Sharing.

When a member has borrowed a book, a corresponding event is broadcast and consumed by Reading. The same holds when a book is returned. When a member has finished a book, Reading produces an event. Lending consumes the event, and the member can return the book. The same event is consumed by Rating. Rating will prompt the member to rate the book and write a review.

When a member wants to rate a book or write a review of it, they need the reading list entry to refer correctly to the book. Again, it is a user interaction. Therefore, the communication is modeled as synchronous. The data flow goes from Lending to Rating.

When a member is newly registered, an event is produced. Lending consumes the event and creates a new Reading list for the new member.

Appointment management needs information about parents and their children from Member management to assure that only registered parents can arrange appointments for their kids.

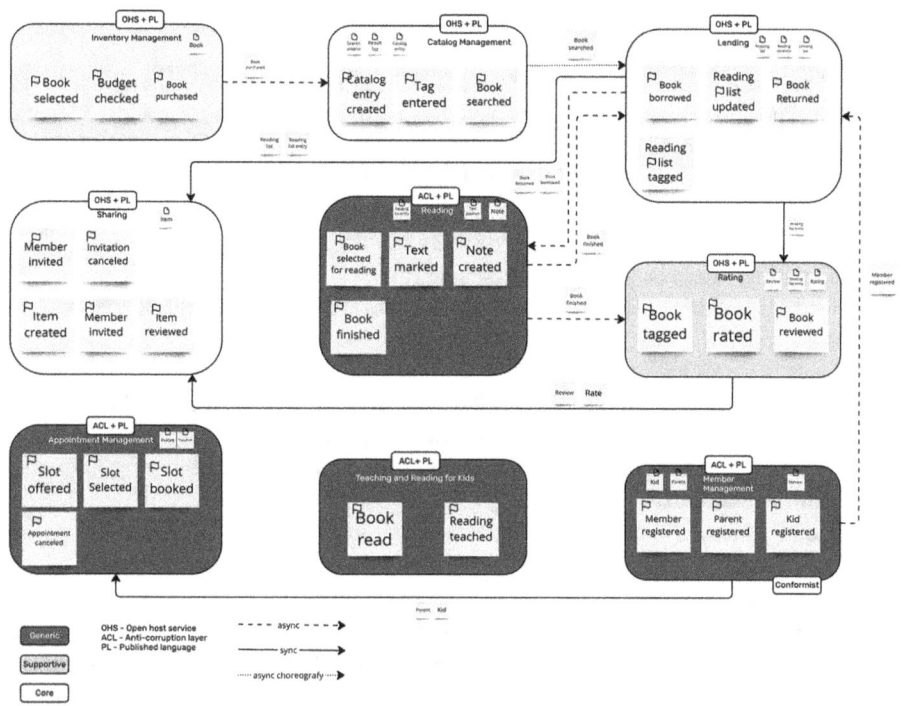

Figure 5-40 Context map of online library with data flows

Discussion of Synchronous APIs

The context map in Figure 5-40 shows two synchronous flows:

- Data flow from Lending to Sharing with Reading list and Reading list entry and
- Data flow from Rating to Sharing with Review and Rating.

Those data flows will be discussed in the following section.

Besides the synchronous data flows, synchronous Endpoints need to be provided to be accessed by the user interfaces of the associated bounded contexts.

Synchronous APIs for Data Exchange

The data flow from Lending to Sharing can be implemented as a GET call of Lending.

The Reading list can be accessed via the identifier of the corresponding library member. The library member's identifier does not need to be provided as a parameter because the call is done in a particular user context, and the identifier can be provided via an access token. We will discuss such a call in the following chapter, Chapter 6, "Interface Definitions."

The corresponding call looks like this:
GET ../lending/reading-lists

The response to this call will be a list of reading lists of the corresponding library member. The structure of the reading list entry is shown in the corresponding Visual Glossary in Figure 5-17 in section "Visual Glossary Lending and Searching".

The corresponding identifier of a reading list entry can be used to access a single reading list entry.
GET ../reading-lists/415d7d89-3e84-4ce1-b4ef-ebf4918e64f3

Comparably, access to the ratings and reviews of a library member can be organized. In addition to granting access by library member identifier, access to ratings and reviews must be filtered by the book identifier. A call of the associated getter access could look like this:
GET ../ratings/?book="2c0560ae-595c-41ae-b963-bdbc314f39ca"

Additionally, `Appointment managemnet` requires information about the relations between parents and their children to allow parents to arrange appointments on behalf of their kids.

In detail, we will discuss the interface definitions for the online library in Chapter 7, "Defining the Online Library Interfaces."

Synchronous APIs for User Interactions

We must combine the domain stories with the context map to define which synchronous APIs are necessary for user interactions. Actions such as creation, changing, and deletion are usually needed on the defined aggregates. Sometimes it is necessary to define actions on subresources or properties, for example, for changing states. In our case, a status change on a reading list entry would be necessary, for example, when a library member starts or finishes reading a book.

- `Member management:`
  - Create a new member (POST),
  - Register a parent (POST),
  - Register a child (POST).
- `Catalog management:`
  - Create a new catalog entry (POST),
  - Search for a catalog entry (GET) by author, title, and so forth,
  - Update a catalog entry (PUT).
  - Write an abstract of a book (POST),
  - Update an abstract of a book (PUT),
  - Create a tag for a catalog entry (PUT).
- `Lending:`
  - Create new reading list entry (POST),
  - Update reading list entry (PUT),
  - Change status of reading list entry (PUT),
  - Add keyword to reading list (PUT),
  - Add keyword to reading list entry (PUT).

Gathering Business Requirements 113

- Reading:
  - Store reading position in a book (PUT),
  - Create note (POST) at a text location,
  - Update note (PUT),
  - Delete note (DELETE).
- Rating:
  - Write review for a book (POST),
  - Update review (PUT),
  - Access review of a particular book (GET),
  - Rate a book (POST),
  - Update rating (PUT).
- Sharing:
  - Create a share for a shareable item and invitations (POST),
  - Add additional or delete invitations (PUT),
  - Cancel share (DELETE).
- Teaching and reading to children
  - Store reading position (PUT).
- Appointment management
  - Publish slot (POST),
  - Cancel slot (DELETE),
  - Book slot (PUT),
  - Cancel meeting (PUT).

Those individual APIs will be formulated in Chapter 7, "Defining the Online Library Interfaces." In great detail, we will discuss the requests and responses based on domain stories, Visual Glossaries, and context maps. First, we need to take a closer look at how to define APIs in the following chapter, Chapter 6, "Interface Definitions."

Discussion of Events

The context map in Figure 5-40 shows four asynchronous data flows.

The event `Member registered` is produced by `Member management` and consumed by `Lending`. `Lending` can react to the event by creating an empty reading list of the new member.

The event `Book finished` is produced by `Reading` and consumed by `Rating` and `Lending`. In `Lending`, the corresponding reading list entry can be updated. And in `Rating`, the user can rate the book and write a review.

The events `Book borrowed` and `Book returned` are produced by `Lending` and consumed by `Reading`. Thus, the service `Reading` can determine what books are and are not accessible to a given member.

The event `Book purchased` is produced by `Inventory management`. The event is consumed by `Catalog management`. The service can automatically create a new catalog entry based on the event.

We will discuss events and their payload in Chapter 7, "Defining the Online Library Interfaces."

The sections show how a context map can be created from the result of an Event Storming workshop. The context map gives a detailed overview of the bounded contexts and the necessary data exchange synchronously as well as asynchronously. APIs can be formulated based on a context map.

In the next section, we want to discuss how AI can support the formulation of APIs based on the workshop results of Domain Storytelling, Event Storming, and context map.

Generation of API Definitions with AI

To define APIs, certain definition languages are available. To use those languages manually might be tedious. Therefore, using generators to define APIs might be desirable.

We know generators of API specifications from service code. But, as described in section "API-First vs. Code-First," it is advisable to use a consequent API-First approach. However, when using API-First, the API cannot be generated from code.

Today we have better tools. We no longer need implemented controllers to generate APIs. They can be generated from the documented results of workshops, as discussed earlier in sections "Domain Storytelling, Visual Glossary, Event Storming, and Context Map".

Let us look at a short example using ChatGPT 4o.[9]

First, we let ChatGPT know it is an expert in OpenAPI and DDD, as shown in Figure 5-41.

In a second prompt, ChatGPT is asked to create an OpenAPI specification based on a Visual Glossary and a provided skeleton.[10] We use the reading list of a member as an example. The prompt is shown in Figure 5-42.

ChatGPT answers with its considerations and the generated OpenAPI (Figure 5-43).

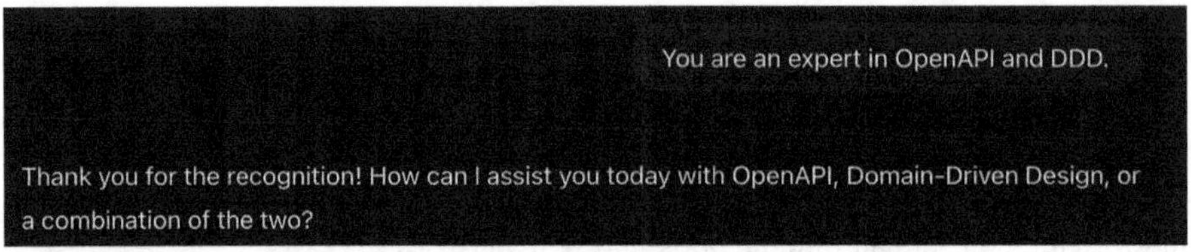

> You are an expert in OpenAPI and DDD.
>
> Thank you for the recognition! How can I assist you today with OpenAPI, Domain-Driven Design, or a combination of the two?

Figure 5-41 Start ChatGPT chat, done September 27, 2024

---

[9] ChatGPT 4o is a large-language model product of OpenAI. https://chatgpt.com/ .
[10] The used skeleton can be found under https://github.com/Grinseteddy/DomainDrivenApiDesign/blob/main/OnlineLibrary/RestSkeleton.yaml.

Conclusion

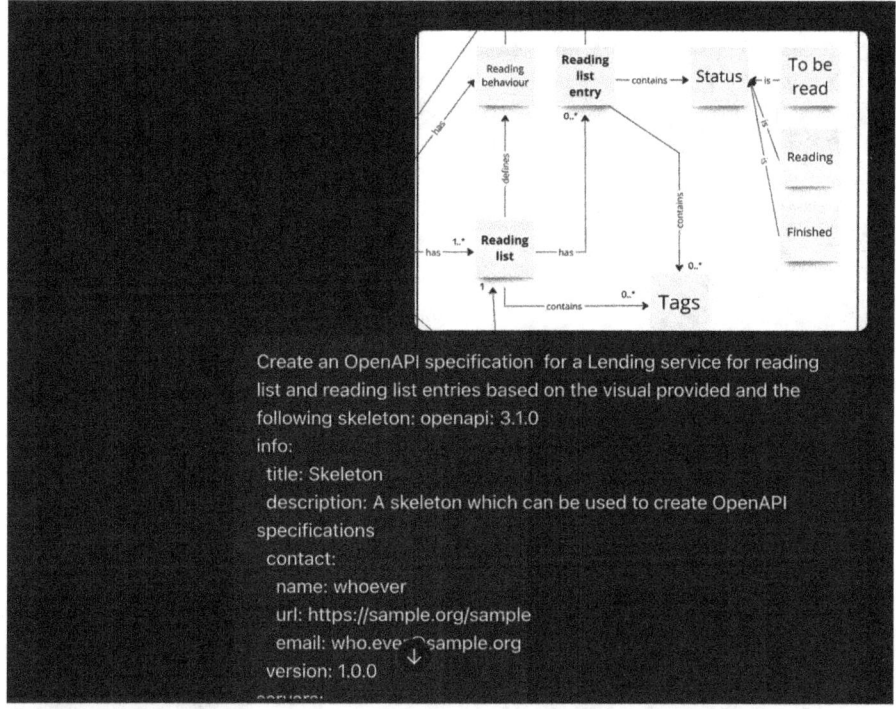

Figure 5-42 Prompt ChatGPT to create an OpenAPI, chat done on September 27, 2024

The generated result is acceptable. It can be used to tune the interface to the requirements of the consumers. But one needs to be critical.

Try using different large-language models, which will likely fit your requirements better. Doing the same chat in Claude[11] delivers slightly better results from the authors' point of view (Figure 5-44).

This can only be a starting point, but it helps to avoid tedious coding. And the responsible architect or developer can concentrate on the business requirements. However, using the visuals from the workshops guarantees that the API will use the defined ubiquitous language.

Conclusion

We saw in this chapter that the process as it is shown in section "Introduction to Modern Development Processes" was sufficient to obtain a reliable basis for defining APIs.

---

[11] Claude is a large-language model product of Anthropic. https://claude.ai/new.

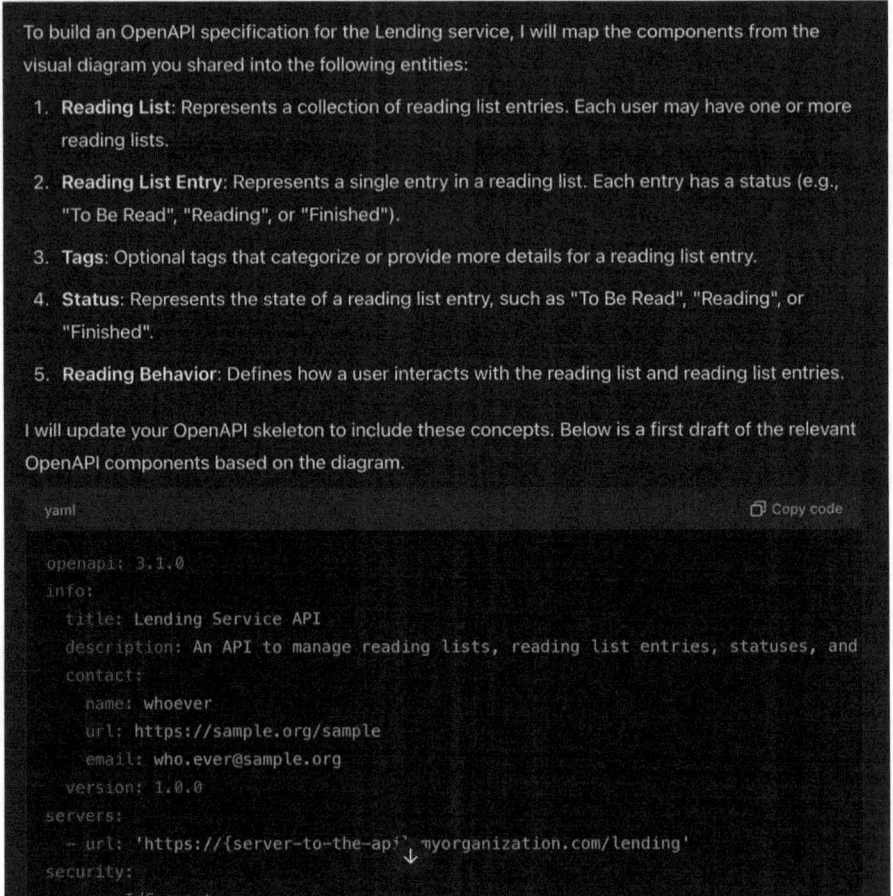

Figure 5-43 Generated OpenAPI by ChatGPT, chat done on September 27, 2024

Defining APIs depends largely on a reliable definition of a ubiquitous language. The language must be defined in cooperative workshops between business experts and IT specialists. Those workshops define the bounded contexts and the corresponding ubiquitous languages. As workshop formats, the methodologies support the cooperation of experts as well as the documentation of the results in a visual way.

Using those results will help to define elegant and well-crafted APIs.

In the following chapter, we will discuss the technical aspects of API formulation. In addition, we will discuss how to select the most suitable communication approach and the most suitable protocol for the problem at hand.

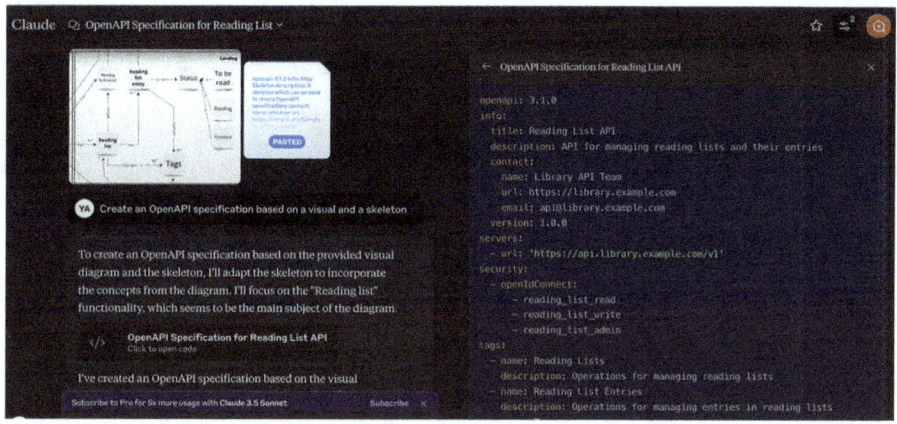

Figure 5-44 Chat with Claude to generate an API, September 28, 2024

Points to Remember

- Use a Business Model Canvas to sketch business ideas.
- Use a capability map to organize the business capabilities as core, supportive, and core capabilities.
- Use a Wardley map to prioritize business capabilities along the lifecycle of an application.
- Use Domain Storytelling to understand the business and to gather business requirements in a visual format. Use collaborative workshops to create a common understanding of business experts and IT specialists.

Review Questions

5.1 What are the main parts of a Business Model Canvas?

(a) Key partners, key activities, key resources, key propositions, Customer relationships, channels, customer segments, cost structure, revenue streams
(b) Unique selling points, value stream, employee skills
(c) Core capabilities, supportive capabilities, generic capabilities
(d) Technology requirements, business requirements, legal requirements

5.2 What are the evolutionary steps of an application in a Wardley map?

(a) Visionary, early adopters, main stream, legacy
(b) Genesis, custom, product, commodity
(c) Invented, superior, used
(d) Next generation, current, previous

5.3 What are the parts of a sentence in Domain Storytelling?

(a) Role, action
(b) Source, verb, target
(c) Actor, action, work item, adverbial
(d) Character, step, stage

5.4 Are quantities parts of a Visual Glossary?

(a) Yes
(b) No
(c) Only between aggregates
(d) Only between value objects

5.5 What are the steps in Event Storming?

(a) Event definition and bounded context definition
(b) Collection of events and sorting
(c) Sorting of events, definition of bounded contexts, finding aggregates
(d) Event collection, event consolidation, event enhancing, definition of bounded contexts

5.6 Can you use a context map to define data exchanges?

(a) No
(b) Only for synchronous data exchange
(c) Only for asynchronous data exchange
(d) Yes

References

1. Osterwalder A (2013) Business Model Generation: A Handbook for Visionaries, Game Changers, and Challengers, Pigneur Y (ed). Wiley, New York. ISBN: 978-04-70876-41-1
2. Strategyzer (2024) The business model canvas. [Online] Available: https://www.strategyzer.com/library/the-business-model-canvas. Visited on 21 July 2024
3. Rosen M (2011) Business capabilities - the rosetta stone of business/it alignment. [Online] Available: https://technologytransfer.it/business-capabilities-the-rosetta-stone-of-business-it-alignment/. Visited on 27 Aug 2024
4. Alex SM, Bryan Lail BJ (2022) Business capabilities, version 2, the open group. [Online] Available: https://pubs.opengroup.org/togaf-standard/business-architecture/business-capabilities.html. Visited on 27 Aug 2024
5. Wardley S (2022) Wardley Maps, 2. überarbeitete und erweiterte Auflage. : Simon Wardley, Heidelberg
6. Gagliardi G (2021) Sun tzu's five elements. [Online] Available: https://scienceofstrategy.org/main/content/sun-tzus-five-elements. Visited on 27 Aug 2024

References

7. Luft A (2020) The ooda loop and the half-beat. [Online] Available: https://thestrategybridge.org/the-bridge/2020/3/17/the-ooda-loop-and-the-half-beat. Visited on 27 Aug 2024
8. Mosior B (2018) Understand context and diminish risk: How to build your first Wardley map. [Online]. Available: https://miro.com/blog/wardley-maps-whiteboard-canvas/. Visited on 27 Aug 2024
9. Wardley S (2016) Exploring the map, chapter 3. [Online] Available: https://medium.com/wardleymaps/exploring-the-map-ad0266fad59b. Visited on 27 Aug 2024
10. Hofer S, Schwentner H (2022) Domain Storytelling: A Collaborative, Visual, and Agile Way to Build Domain-Driven Software. Pearson International, London. ISBN: 978-01-37458-91-2
11. Zörner S (2015) Softwarearchitekturen dokumentieren und kommunizieren, Entwürfe, Entscheidungen und Lösungen nachvollziehbar und wirkungsvoll festhalten, 2., überarbeitete und erweiterte Auflage. Hanser, München, 277 pp. Literaturverz. S. [269]-272. ISBN: 978-34-46443-48-8
12. Brandolini A (2024) Event storming. [Online] Available: https://www.eventstorming.com/. Visited on 30 Jun 2024
13. Osborn A (1953) Applied Imagination: Principles and Procedures of Creative Thinking. Charles Scribner's Sons, New York. ISBN: 978-06-84162-56-0
14. Miller BC (2012) Quick Brainstorming Activities for Busy Managers: 50 Excercises to Spark You Team's Creativity and Get Results Fast. American Management Association, New York. ISBN: 978-08-14417-92-8
15. Evans E (2004) Domain-Driven Design: Tackling Complexity in the Heart of Software. Addison-Wesley, Reading.
16. DDD Crew (2023) Context mapping. Visited on 08 Sep 2024
17. Sakimura N, Bradley J, Jones M, de Madeiros B, Mortimore C (2023) Openid connect core 1.0 integrating errata set 2. [Online] Available: https://openid.net/specs/openid-connect-core-1_0.html. Visited on 15 Aug 2024
18. Evans E (2015) Domain-driven design reference. [Online] Available: https://www.domainlanguage.com/wp-content/uploads/2016/05/DDD_Reference_2015-03.pdf. Visited on 08 Sep 2024

Interface Definitions   6

Having examined how API design can be supported by DDD, we take a look in this chapter at how interfaces need to be defined. Interfaces are essential to modularize the software so that individual modules can be developed independently by different teams or even a single developer, as we already discussed in Chapter 5, "API Design Supported by Domain-Driven Design." The mobile application example discussed in Chapter 2, "Communication Categories," shows different kinds of Application Programming Interfaces. As we saw, we want to concentrate in this book on interfaces using different network communication channels. Therefore, in this chapter we will look at how we can define these interfaces. We will first look at the different components of an interface definition (section "Components of an Interface Definition"), then go over the different data types and schemas that we will use, as well as their benefits and drawbacks (section "Data Formats and Their Schemas"). Then the API definition languages will be discussed for synchronous (section "Definition of Synchronous Interfaces") and asynchronous APIs (section "Definition of Asynchronous Interfaces"). We will conclude the chapter with a comparison of the different protocols in combination with the different data schemas (section "Advantages and Disadvantages of Different Schemas and Protocols"), antipatterns, and counter options (section "Typical Antipatterns") and how artificial intelligence (AI) can help in the process (section "AI Generation of API Definitions: Chances and Limitations").

Components of an Interface Definition

In this section, we will discuss the components of an interface definition. First we discuss, Web Services Description Language (WSDL), RESTful API Modeling Language (RAML), Swagger, and OpenAPI.

Before we discuss the different types of interfaces and their definitions, let us examine the history of interface specifications.

Short History of Interface Definitions

Interfaces hide the implementation of modules from the outside world. Other modules can access the module without any knowledge about its internal implementation. Because the implementation is hidden, changes do not disturb the accessing module. It is only interested in the specification of the interface.

The specification functions as a contract between the different modules as defined by *Bertrand Meyer* in his programming language *Eiffel* [1]. Such a contract can be described by an IDL. The definition language allows definitions independent of the programming language used by the interface provider and consumer. For example, a web application is usually written in a JavaScript framework, whereas the interface providers can be written in Java, Go, or again in JavaScript (e.g., with Node.js). The contract (e.g., defined by an IDL) promises that the request or call will behave as specified. In most type-safe languages, an IDL can guarantee at the syntax level that the contract is being fulfilled.

The interface description languages have evolved dramatically over time; a selection of such description languages is listed below, sorted by first publication date:

- **Sun's Open Network Computing Remote Procedure Call (ONC RPC), first published 1988 [2]** ONC RPC was introduced to allow clients to remotely access CISC applications (Customer Information Control System by IBM) and their services. CISC applications provide middleware to process several commands fast [3].
- **IBM's System Object Model (SOM) [4]**
  SOM is an object-oriented library developed by IBM. It makes it possible to share libraries across different programming languages. It was used in IBM's OS/2 operating system and in Workplace Shell. The active development of the technology ended in the mid-1990s, when Apple withdrew its support [5].
- **Open Software Foundation's Distributed Computing Environment (DCE) [6]** This framework includes remote procedure call (RPC) mechanisms known as DCE/RPC, a naming service, a time service, an authentication service, and a distributed file system (DCE/DFS) [6]. DCE never achieved commercial success.
- **Microsoft RPC, later developed to COM (Component Object Model) and DCOM [7]**
  RPC was developed to allow the creation of distributed client-server programs. RPC provides stubs and libraries that manage most network protocols and communication processes. RPC can be applied to Windows operating systems implemented with the C/C++ programming language [8].
  It was later developed further into Component Object Model / Distributed Component Object Model (COM/DCOM). DCOM (Distributed Component Object Model) is a proprietary Microsoft technology. It is an extension of COM.

COM defines a binary standard to create reusable software components. DCOM enhances COM by network capabilities [9].

COM/DCOM is proprietary to Microsoft's operating systems. A broader application across different operating systems is, therefore, not possible.

- **Object Management Group's Common Object Request Broker Architecture (CORBA) [10]**
  CORBA stands for Common Object Request Broker Architecture and is a standard defined by the Object Management Group. CORBA enables different systems developed on different platforms in different languages to communicate with each other, where a client uses an object reference pointing to an implementation on the server. It uses an IDL to specify the interfaces that objects present [11].
  The first version was published 1991. The latest version of CORBA was published in November 2012 [11].
- **WSDL, first published 2000 [12]**
  WSDL documents describe web services. It specifies the location of the service, the belonging methods of the service, and the associated data structures [13].
  We will discuss WSDL in more detail later in this section (section "WSDL, RAML, Swagger, and OpenAPI") because interfaces using XML like SOAP are still broadly used.
- **Mozilla's Component Object Model (XPCOM) [14]**
  XPCOM is a cross-platform component model from Mozilla and is comparable to Microsoft's COM/DCOM, CORBA, or even IBM's SOM. XPCOM was used by the browser Firefox but replaced by WebExtensions API in 2017.
- *Apache's* **Thrift [15]** Apache Thrift provides a framework to develop scalable, cross-language services.
  The method definitions look like a method in C, but Thrift supports several languages, for example, Java, Python, and PHP [16].

All those languages are mostly no longer used for new implementations. The provider-specific definitions like IBM's SOM or Microsoft's COM/DCOM lack independence from providers. Others lack independence from the data representation in different forms like WSDL, which is only valid for XML.

Modern IDLs support different platforms, programming languages, and data containers. We will look at the evolution of OpenAPI in the next section (section "WSDL, RAML, Swagger, and OpenAPI") to understand what is essential for an interface definition.

WSDL, RAML, Swagger, and OpenAPI

The modern interface definition language OpenAPI was developed based on the predecessors WSDL, RAML, and Swagger. Let us take a deeper look at them to understand the structure of modern IDLs.

WSDL

WSDL is an XML schema definition (XSD) document, defining a web service and its XML documents. It contains the following elements [17]:

- `definitions`
  All XML documents need one root-level document, and the WSDL specification contains `wsdl:definition`.
- `documentation`
  This element contains arbitrary text and elements that benefit human readers.
- `message`
  A message in a WSDL document is an abstract definition of the data sent or received by a service, for example, `SearchBookRequest` and `SearchBookResponse`.
- `portType`
  This tag contains a list of operations with corresponding input and output messages, for example, the operation `searchBook` contains the input message `SearchBookRequest` and the output message `SearchBookResponse`.
  - `operation`
    - `input`
    - `output`
    - `fault`
      The item defines error messages a SOAP system delivers in case of faults.
- `binding`
  The binding element defines the protocol and data formats for the operations and messages defined below `portType`. In the case of SOAP, it contains the bindings for a SOAP data exchange.
- `service`
  The service can be used to group the given ports and to give the associated addresses.
  - `port`
    The port tag defines the corresponding addresses, such as http://library.org/library-search.

The definition of the service can be enhanced by the definition of custom types, for example, `Book` or a list of books.

Even in those early stages, we can see the most important parts of an interface definition:

- Custom definitions of types,
- Definitions of input parameters and responses,
- Definitions of access points with defined behavior.

WSDL was used to define SOAP web services. However, during this time, RESTful services developed as a resulting application of the HTTP protocol. A first approach to defining those interfaces was RAML. We will take a deeper look at RAML in the next section.

RAML

The definition of SOAP interfaces became less critical with the rise of HTTP-based-APIs in the 2010s. RESTful interfaces rely on the HTTP protocol and, therefore, do not need a detailed operation definition and associated binding. With RAML first proposed in 2013, a first approach to defining interfaces based on HTTP was published [18]. It is a definition of APIs that uses Yet Another Markup Language (YAML), but without a resulting application of RESTful principles. An essential role was played by Mulesoft and, later, by Salesforce, which took over Mulesoft in 2018. Salesforce now owns the trademark [18]. At any rate, it was necessary to define an interface that could be used independently of the implementation behind it.

As other IDLs, RAML does not just document an API, it models it [19].

RAML defines resources and methods following the structure of paths [20].

It could not take hold against OpenAPI because of its close dependency on the API management tools of Mulesoft. Swagger was published shortly afterward and was more successful than RAML.

Swagger

Work on Swagger started in 2009 by *Tony Tam* at Wordnik (an open dictionary of the English language). Originally, it should have been called Web API Description Language (WADL) as a kind of WSDL. However, a friend asked Tam: "Why WADL when you can swagger?" And the first version of Swagger was published in August 2011 [21].

The second version was published in 2014 and contained the Swagger editor. Swagger supported at that time JSON as well as YAML. Also at that time it became very popular for the RESTful API design in an API-First approach [22].

The second edition of Swagger was donated to the OpenAPI initiative in December 2015 and became OpenAPI 2.0.0.

OpenAPI

OpenAPI became fast a de facto standard for API definition. It combines the elegance of good documentation and a resulting API-First approach. The specification can be rendered as an Hypertext Markup Language (HTML) document, so it is human readable. However, the specification as a JSON or YAML document can be used to generate either client or server code to implement the corresponding components.

What follows constitutes a brief timeline of the major milestones in the history of OpenAPI:

- 2009: First work on Swagger by the Wordnik team [21]
- August 2011: Publication of Swagger 1.0 [21]
- August 2012: Publication of Swagger 1.1 [21]
- March 2014: Publication of Swagger 1.2 as first formal specification [21]
- September 2014: Publication of Swagger 2.0 [21]
- March 2015: Takeover of Swagger by SmartBear [21]

- December 2015: Donation of Swagger to OpenAPI initiative [21]
- July 2017: Publication of OpenAPI 3.0.0 [23]
- February 2019: Publication of OpenAPI 3.1.0 [24]
- September 2022: Start of work on OpenAPI 4.0 under name "Moonwalk" [25]

OpenAPI defines a standard to describe and define the programming language-agnostic RESTful APIs. A programmer can understand the interface without any knowledge of the implementation behind it.

The OpenAPI initiative describes the function of the specification as follows:

> The OpenAPI Specification (OAS) defines a standard, programming language-agnostic interface description for HTTP APIs, which allows both humans and computers to discover and understand the capabilities of a service without requiring access to source code, additional documentation, or inspection of network traffic. When it is properly defined via OpenAPI, a consumer can understand and interact with the remote service with a minimal amount of implementation logic. Similar to what interface descriptions have done for lower-level programming, the OAS removes guesswork in calling a service [24].

Later in this section, we will discuss the structure of such a specification (section "Components of an Interface Definition for Synchronous and Asynchronous Interfaces").

AsyncAPI

With the rise of EDA and microservices [26], it became essential to define not only synchronous APIs but also, even more importantly, events.

Definitions of events should work in the same manner as the definitions of synchronous interfaces. Based on the ideas of OpenAPI, the original structure and ideas are used to define events [27].

An AsyncAPI document contains, besides the schema definition and access points – like an OAS, binding information like channels to make it possible to support different technologies and protocols [28].

Necessary Information for Interfaces

With the foregoing samples we can collect the necessary information an API specification needs to provide:

- Information about the provider and the author of the specification,
- Information about the binding of the data to be exchanged to the protocol used,
- Information about the data to be exchanged and properties under which they are exchanged,
- Information about how the data exchange should be implemented on the provider and consumer sides.

Especially for events, the protocol bindings and the data representations need to be defined carefully because there is no standard protocol for data exchange using events. On the other hand, when defining synchronous APIs, it is not necessary to define a deep data binding, because one can rely on HTTP.

Let us see how the necessary information is provided in the two specification formats for synchronous and asynchronous APIs.

Components of an Interface Definition for Synchronous and Asynchronous Interfaces

The structure of API specifications can be detailed in different blocks, as shown in Figure 6-1.

In Figure 6-1, the common parts of OpenAPI and AsyncAPI are marked in gray. Let us dive deeper into the different blocks.

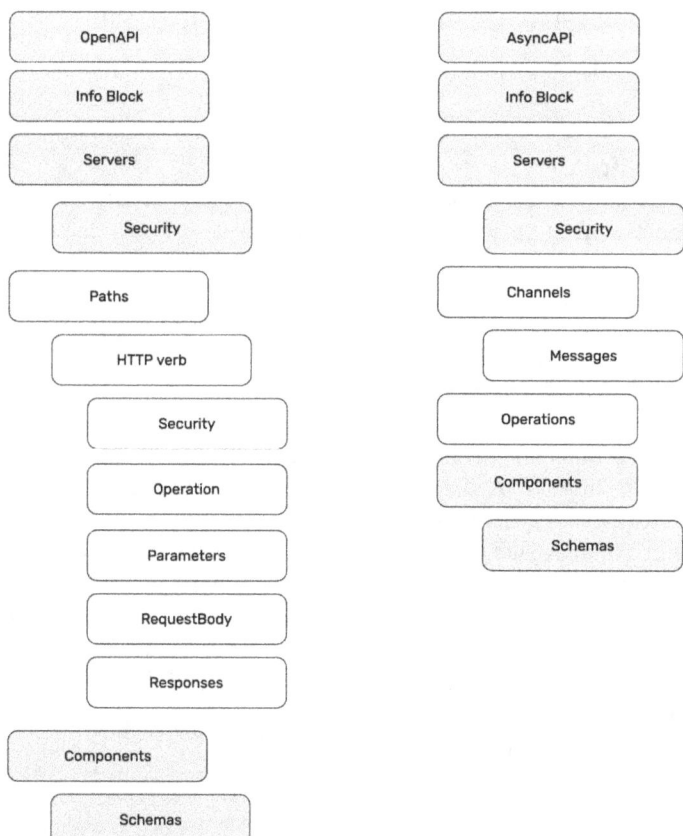

Figure 6-1 Structure of API specifications

```
info:
  title: Skeleton
  version: 1.0.0
  description: |
    Skeleton for an AsyncAPI definition
  contact:
    name: Fabrizio Lazaretti and Annegret Junker
    url: https://online-library.org/sample
    email: fabrizio@online-library.org
```

Listing 6-1 Example: Info block of an API

```
servers:
  production:
    # other messaging brokers bindings can be defined -- look in the
    documentation
    host: "events.myorganization.com:9092"
    protocol: "kafka"
    description: "Production broker"
    security:
      - $ref: '#/components/securitySchemes/user-password'
```

Listing 6-2 Example: Servers block of an asynchronous API

Information Block

The `info` contains information about the specification itself, for example, the title and the specification version. Additionally, a description can be given. The description is rendered as a markdown and can be nicely used to present the specification on an API homepage.

The function and form of the information block are equal in both specifications. An example of an `info` block is given in Listing 6-1.[1]

Servers

The `servers` section contains information about the message provider.

In the case of an AsyncAPI, it is the corresponding message broker, whereas the servers in the case of a synchronous API are given with their URL.

An example of an asynchronous `servers` block is given in Listing 6-2.

Additionally, the `servers` block contains information about the security protocol, as in the example with the user name and password. Other security protocols can be defined in the security schemes of the components block, for example, OpenIdConnect. Please consult the documentation [24, 28] for further information.

---

[1] See https://github.com/Apress/Crafting-Great-APIs-with-Domain-Driven-Design/blob/main/Chapter-6-8/AsyncSkeleton.yaml

Paths

The block `paths` is defined for synchronous APIs. It contains the path behind the basic paths given in the `servers` block.

The block contains information about:

- HTTP verb: This contains the verbs defined by the HTTP protocol [29]. Usually, the following verbs are used: GET for retrieving information, POST to create, PUT to change, and DELETE an object on the server side. The verb PATCH is also used widely but has certain disadvantages, which we will discuss later in this chapter. The verbs CONNECT, OPTIONS, and TRACE are defined by the HTTP protocol. However, they are uncommon for API specification.
  - Security as a specific application of the security scope for the particular path
  - Operation as an optional part as the name for the method that needs to be implemented on the provider side
  - Parameters as an optional part containing parameters that need to be applied to the corresponding request
  - RequestBody as an optional part to define the input parameters for a request as a complex structure – not allowed by the HTTP protocol for GET metods
  - Responses related to a corresponding HTTP request. They follow the defined response status codes of HTTP [29]
    - 100 group: Informational (usually not used in API specifications)
    - 200 group: Successful operation
    - 300 group: Redirection
    - 400 group Client error
    - 500 group: Server error

  Additionally it is allowed to define a default response.

Channels

Instead of paths, asynchronous APIs `channels` need to be defined. A channel defines the logical component of a broker as a transport channel for messages. The corresponding logical architecture is shown in Figure 6-2.

The channels are defined by the messages they transport and the address they are connected to.

An example of a `channels` definition is shown in Listing 6-3.

The content of the messages can be defined in the schemas, as for synchronous APIs, to make the whole specification more readable and to reuse certain definitions in one specification.

Operations

For asynchronous APIs, the operations need to be defined as for synchronous APIs. The operations for synchronous APIs can be defined in the `paths` block because it is defined exclusively from the provider's point of view. That is impossible for asynchronous APIs because the services can be providers and consumers simultaneously. Therefore, the operations are defined with the channel they point

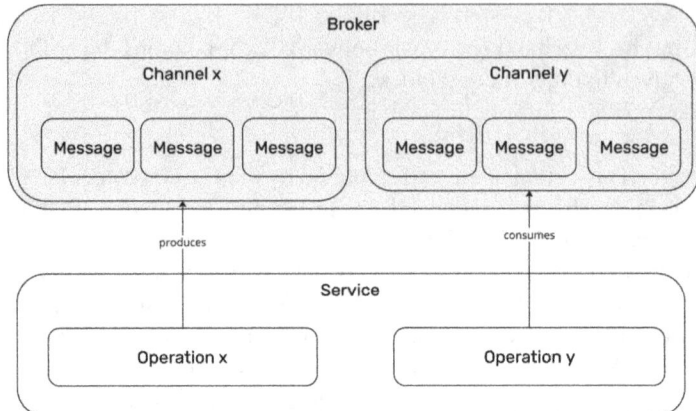

Figure 6-2 Logical architecture of a message broker

```
1  channels:
2    taskCreatedEvent:
3      address: "task-management/taskCreatedEvent"
4      messages:
5        taskEventOut:
6          $ref: '#/components/messages/taskCreatedMessage'
7    taskDoneEvent:
8      address: "task-management/taskDoneEvent"
9      messages:
10       taskEventOut:
11         $ref: '#/components/messages/taskDoneMessage'
```

Listing 6-3 Sample of a channel definition

to and the corresponding action as `send` or as `receive` depending on whether the corresponding service is a producer or a consumer of the corresponding message.

Schemas

The `paths` section is used to define the content of messages.

Depending on the format of the messages, the schemas can contain different formats, for example, a JSON or an XML schema.

We will discuss how to define schemas and the corresponding data types in the following section.

Data Formats and Their Schemas

Here, we will briefly introduce different data formats and their schemas that are commonly used in web-based APIs. Data formats are formats that are used to transfer the data of an API call. In HTTP, you can send data in a format of your choice with a request and get a response with data in a particular format. With a message or event-based asynchronous API, you send messages to a message or event bus or receive them in a particular data format.

However, to know exactly how the data looks, for example, which properties are required or exist in a JSON data format, a schema is needed. A schema describes which fields exist, which of them are required, and which data type is required. Some schemas also allow us to describe each field and add examples, which is unfortunately not the case in all formats, as we will see.

Unfortunately, not all APIs have a formally defined data schema. APIs lacking this feature can lead to poor quality as they are not self-descriptive, and the user will not know what to provide or what to expect as a result. Without clarity, users will be unsure as to what input is needed or what output to expect, forcing developers to rely on guesswork and trial and error.

There is no user error protection. A request may work with a field hundreds of times but then fail once, as the value is not in the expected range because a wrong data type was assumed (e.g., the API was providing an integer and not a long data type). This, again, does not support the quality objective functional correctness/completeness.

For these reasons, it is crucial to start with a data schema. We will look at two types: binary serialized and text-based formats [30]. Both types have benefits and drawbacks, which we will discuss next.

Text-Based Formats
Text-based formats are formats that are based on encoded text (e.g., UTF-8) (Figure 6-3). This makes them human-readable and editable. Examples of text-based formats are JSON, XML, and YAML.

They have the benefit of easy editing using standard editors, as they are always readable [30]. They can be edited without the knowledge of a schema, as they are in plain text; however, this may lead to a parsing error when using the data. A common problem is the ambiguity of numbers; in some formats, it is not clear if it is a number or a string; further, often, the number type itself (float or decimal), as well as the precision, is not clear

These formats also often lack the ability to pass binary data without an encoding. A workaround is often to use Base64 encoding on top. With this approach, binary

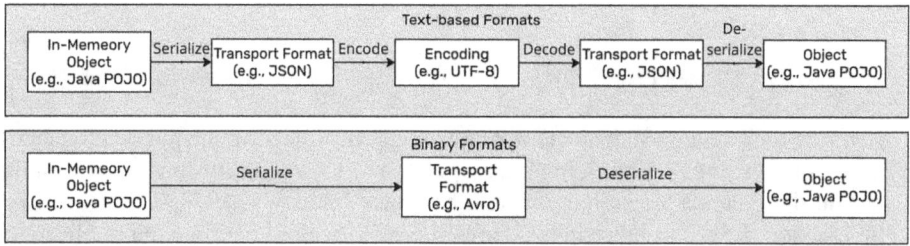

Figure 6-3 Serializing and deserializing data formats

information, for example, an image, first gets encoded in Base64 and then passed as a string inside the format. This works well; however, it adds an additional encoding/decoding step and increases the size by 33% (as Base64 only uses 6 bits per byte; it uses a subset of the ASCII table, starting with A and ending with /) [31, 32].

The text-based examples used here will have better-established standards, which have broader library and tooling support overall compared to binary serialized formats [30].

Binary Serialized Formats

Serializing and deserializing are more straightforward with binary formats. These formats do not need an additional step, which includes interpretation (Figure 6-3). However, this leads to a format that is not human-readable by itself or with standard editors. We will look at Apache Avro and Protobuf for binary serialized formats. Both formats require a schema; otherwise, the binary data are not decodable [31].

When implementing these formats, you benefit from a clearly typed system with strong types. They are, in general, faster, as they do not need to be interpreted first [30] (Figure 6-3). That generally leads to less necessary processing power to serialize and deserialize. They do not need to read and interpret the data, including the metadata, which is normally simpler when just passing the raw data in binary formats [30]. Binary transmission also leads to less space for storage and transmission, as these formats do not require transmitting text or metadata but just raw binary data [30].

Examples

Now we will briefly look at JSON, YAML, XML, Protobuf, and Apache Avro. The goal is to get a common understanding of the formats and their differences. The first three will represent text-based formats, and the other two will be binary serialized formats.

For each format, we will use the same example of a book entity, with common features used by the online library introduced in Chapter 4. The example is presented in Table 6-1. The goal of that example is to cover the most common cases of an entity. To keep it simple, the binary fields, the cover thumbnail, and the author thumbnail will be reduced to a shorter length, which does not represent a real picture.

Date formats have often limits

Be aware that we are using a date in our example book record: the publication date. Many formats and programming languages use Unix Epoch to store dates. In this format, the seconds are stored as a number since January 1, 1970. This works for file creation dates and other items. However, historical events, like the publication date of a book or a birthday before 1970, are not possible. Therefore, we will not use date types if they use Unix Epoch [34].

Table 6-1 Overview of book example, including data types used in each format [33]

Name	Data type	Required	Example
ISBN 13	13 digits	Yes	9783161484100
Title	String	Yes	The best book
Cover thumbnail in the WebP format	Binary	No	<binary>
Authors	List with Name and optional portrait thumbnail in WebP	Yes	"author 1" with thumbnail in binary, "author 2" (no thumbnail)
Number of text position	0 or positive number	Yes	0
Category	Either nonfictional or fictional	Yes	nonfictional
Tags	Set of string tags	Yes	"DDD," "Modeling"
Rating	Number from 0 to 5 with a precision of 0.1	No	4.5
Has valid license	True or false	Yes	True
Publishing date	Date	Yes	October 1, 2020

In addition to each data sample, we will present a schema format for each text-based format to describe the data. The binary-based formats have a schema format, which will be presented. Serialization and deserialization are done via a compiler.

> ! **Shortened shema samples**
>
> Some schema samples in this section will be abbreviated from the original to improve the readability of the book. The abbreviated places are marked with an ellipsis ("..."). Sometimes, the formatting of the document is updated as well. The full examples are available on GitHub.[2]

JSON

We start with JSON, as it is an intuitive and widespread format. JSON is a data format often used in the web environment as it is a subset of JavaScript [31]. JSON was invented in 2001 and standardized in 2006 and is now standardized by European Computer Manufacturers Association (ECMA), like JavaScript (ECMAScript) [35–38]. It is a pretty simple format and only allows the following types: `object`, `array`, `number`, `string`, `true`, `false`, and `null` [35].

---

[2] https://github.com/Apress/Crafting-Great-APIs-with-Domain-Driven-Design/tree/main/chapter-6-2

```json
{
    "isbn13": "9783161484100",
    "title": "The best book",
    "coverThumbnailWebP": "aW1hZ2UgaW4gYmFzZTY0",
    "authors": [
        {
            "name": "author 1",
            "portraitThumbnailWebP": "aW1hZ2UgaW4gYmFzZTY0"
        },
        {
            "name": "author 2"
        }
    ],
    "numberOfTextPosition": 0,
    "category": "NON_FICTION",
    "tags": ["DDD", "Modeling"],
    "rating": 4.5,
    "hasValidLicense": true,
    "publishingDate": "2020-08-01"
}
```

Listing 6-4 Book example in JSON

By itself it is a schemaless format. However, there are standards for adding a schema definition to JSON. The most widely used one is probably JSON Schema. The standard is officially still in draft form [39, 40].[3] In addition, JSON Schema is probably the most commonly used schema for JSON; it is also important for our next topic that the OAS uses JSON Schema: The working group that specifies the JSON Schema is in contact with the OpenAPI community; they fit together perfectly [41].

However, JSON Schema is not part of a standardization group; they recently decoupled from Internet Engineering Task Force (IETF) and are now a standalone group [42]. The group joined the OpenJS Foundation, which is a subgroup of the Linux Foundation that gives the standard credibility [42–44].

Example
The sample JSON in Listing 6-4 represents the book example from above. The keys are written in `camelCase`. This is not defined by the standard but is a commonly used practice in many guidelines [45–47]. However, others also suggest using `snake_case` [48].

The schema in Listing 6-5 describes the format of the book example. JSON Schema allows you to add descriptions and examples to each property, which adds a lot of value if you need to integrate a new API. It also allows adding of specific formats, for example, for the `publishingDate`, where the format for the

---

[3] For a few years, the drafts have used the publication year instead of a number, for example, like the well-known Draft 7 [41].

type string is specified to `date`. As JSON is used widely on the web, it also has great support for media types: In `coverThubnailWebP`, the `contentMediaType` describes exactly the expected format of the image. `contentEncoding` describes the additional encoding that is used for the image, as JSON does not support native binary formats.

```json
{
  "$id": "https://example.com/book",
  "$schema": "https://json-schema.org/draft/2020-12/schema",
  "title": "Book",
  "description": "A schema representing a book and its metadata for the library for, e.g., search results",
  "type": "object",
  "properties": {
    "isbn13": {
      "type": "string",
      "pattern": "^[0-9]{13}$",
      "description": "The 13-digit ISBN"
    },
    "title": {...},
    "coverThumbnailWebP": {
      "type": "string",
      "contentEncoding": "base64",
      "contentMediaType": "image/webp",
      "description":
        "Base64 encoded WebP image of the book cover thumbnail"
    },
    "authors": {
      "type": "array",
      "items": {
        "type": "object",
        "properties": {
          "name": {
            "type": "string",
            "minLength": 1,
            "maxLength": 100,
            "examples": [ "Eric Evans", "Dr. John Smith"],
            "description": "The full name of the author with titles"
          },
          "portraitThumbnailWebP": {...}
        },
        "required": ["name"]
      },
      "minItems": 1,
      "uniqueItems": true,
      "description": "List of authors of the book"
    },
    "numberOfTextPosition": {...},
    "category": {
      "type": "string",
      "enum": ["NON_FICTION", "FICTION"],
      "description": "The main category of the book"
    },
```

```json
    "tags": {
      "type": "array",
      "items": {...},
      "uniqueItems": true,
      "maxItems": 20,
      "description": "List of tags associated with the book (top 20)"
    },
    "rating": {
      "type": "number",
      "minimum": 0,
      "maximum": 5,
      "multipleOf" : 0.1,
      "description": "The average rating of the book
  (0 to 5 stars, with one decimal place)"
    },
    "hasValidLicense": {
      "type": "boolean",
      "description": "Indicates whether the library has a valid license
  for the book at the moment"
    },
    "publishingDate": {
      "type": "string",
      "format": "date",
      "description": "The date when the book was published",
      "examples": ["2000-01-01"]
    }
  },
  "required": [ "isbn13", "title", ... ],
  "additionalProperties": false
}
```

Listing 6-5 Book schema in JSON Schema

YAML

YAML is not often used for data, mainly configuration files (e.g., Infrastructure as Code (IaC)), as it is a particularly human-friendly data format. YAML was first proposed in 2001 [49]. The authors claim that YAML is a data serialization format, and not a markup language but a human-readable data serialization language. However, they say YAML is "Yet Another Markup Language," which changes, as originally YAML stood for "Yet Another Markup Language" [30, 50, 51].

It is particularly often used to describe OpenAPI and AsyncAPI files. This is also possible in JSON. However, YAML makes it simpler to read, and we will use YAML for these specification files later on.

At the time of writing, YAML is in version 1.2.2 [51]. YAML is simple to read but not amenable to transfer, as it needs indents (spaces or tabs) and newlines, which is comparable to the programming language Python.

To add a schema, JSON Schema can be used for YAML and should also officially be supported by YAML processors [51].

```
# yaml-language-server: $schema=./book.schema.json
isbn13: '9783161484100'
title: The best book
coverThumbnailWebP: "aW1hZ2UgaW4gYmFzZTY0"
authors:
- name: author 1
  portraitThumbnailWebP: "aW1hZ2UgaW4gYmFzZTY0"
- name: author 2
numberOfTextPosition: 0
category: NON_FICTION
tags:
- DDD
- Modeling
rating: 4.5
hasValidLicense: true
publishingDate: '2020-08-01'
```

Listing 6-6 Book example in YAML

Example

Listing 6-6 shows the book example in YAML. It is clearly simple to read for humans. The schema used is the same as in the JSON example: Listing 6-5.

XML

Extended Markup Language (XML) is the oldest of the data formats presented here. It is the successor of Standard Generalized Markup Language (SGML), which was already standardized by ISO in 1986. XML was created in 1996 as a subset of SGML, and it should be a "simple dialect" of SGML [30, 52]. XML itself is standardized by World Wide Web Consortium (W3C) and is still in version 1.0. However, a couple of changes were made that were not properly versioned [53].[4] The standard is famous for being the underlying specification for HTML. However, HTML uses a subset of XML [53, 56].

XML is really powerful compared to all the other standards discussed here. It has many more features. However, this comes at the cost of many security vulnerabilities; here are two examples:

XML eXternal Entity injection (XXE) In XML, it is possible to load subdocuments. This feature allows many attack vectors to be opened:

---

[4] Version 1.0 is at the time of writing in its "fifth edition," which by itself again has errata [54, 55]. This is an example of a version strategy we do not recommend. Each change (semantic or syntactic) should lead to a new version. (We will discuss this in section "Versioning of APIs," in more detail.)

Server Side Request Forgery (SSRF) This can be used to read documents and endpoints from the server that is executing the XML. It can be used, for example, in an application that reads uploaded XML documents. These XML documents can then be used to read internal files of the server and servers in the same network or execute endpoints to which the server has access. The result can then be sent to an attacker with an additional document load containing the first document as a Universal Resource Locator (URL) parameter.

Denial of Service Attack This is an attack in which the attacker tries to overload the server with many requests, in this case, with requests generated from parsing the XML.

The problem is not the specification but the misconfiguration on the server, which allows that. Many programming languages and frameworks, however, allow this in the default configuration. This is why that vulnerability is explicitly mentioned in the OWASP Top Ten list (as part of Security Misconfiguration) [57–59].

XML Entity Expansion Complex schema definitions can cause security problems with recursive schema definitions and huge memory usage [60].

To mitigate these problems, it is important to consider best practices and to use the latest library versions for XML [61].

Multiple schema languages exist. We will focus on XML schema definition (XSD) as a popular schema for XML. It is the one the authors have seen most in projects.

Example

Listing 6-7 shows the example entity represented in XML. It is clear that it is much more verbose than JSON.

XML is a very complex format, as mentioned earlier. We will only look at some basic functionality.

XML has, by default, a clear reference in the data files to the schema. This makes it easy to read them. However, this will be the only format to do this. No other formats have a standardized way of referring to the schema. The referred schema is visible in Listing 6-8

In the example, the XML element tags are defined with pascal case, as is preferred by multiple guidelines [62–64].

```
<?xml version="1.0" encoding="utf-8"?>
<book xmlns:xsi="http://www.w3.org/2001/XMLSchema-instance"
   xsi:noNamespaceSchemaLocation="book.xsd"
   isbn13="9783161484100">
   <title>The best book</title>
   <coverThumbnailWebP>aW1hZ2UgaW4gYmFzZTY0</coverThumbnailWebP>
   <authors>
      <author>
         <name>author 1</name>
         <portraitThumbnailWebP>aW1hZ2UgaW4gYmFzZTY0</portraitThumbnailWebP>
      </author>
      <author>
         <name>author 2</name>
      </author>
   </authors>
   <numberOfTextPosition>4</numberOfTextPosition>
   <category>NON_FICTION</category>
   <tags>
      <tag>DDD</tag>
      <tag>Modeling</tag>
   </tags>
   <rating>4.5</rating>
   <hasValidLicense>true</hasValidLicense>
   <publishingDate>2020-08-01</publishingDate>
</book>
```

Listing 6-7 Book example in XML


```
<?xml version="1.0" encoding="UTF-8"?>
<xs:schema xmlns:xs="http://www.w3.org/2001/XMLSchema">

  <xs:annotation>
    <xs:documentation>
      A schema representing a book and its metadata for the library for,
      e.g., search results
    </xs:documentation>
  </xs:annotation>

  <xs:element name="book" type="BookType"/>

  <xs:complexType name="BookType">
    <xs:sequence>
      <xs:element name="title" type="TitleType"/>
      <xs:element name="coverThumbnailWebP" type="xs:base64Binary"
         minOccurs="0" maxOccurs="1"/>
      <xs:element name="authors" type="AuthorsType"/>
      <xs:element name="numberOfTextPosition"
         type="NonNegativeIntegerType"/>
      <xs:element name="category" type="CategoryType"/>
      <xs:element name="tags" type="TagsType"/>
```

```xml
        <xs:element name="rating" type="RatingType" minOccurs="0"
↪  maxOccurs="1"/>
        <xs:element name="hasValidLicense" type="xs:boolean"/>
        <xs:element name="publishingDate" type="xs:date"/>
    </xs:sequence>
    <xs:attribute name="isbn13" type="ISBN13Type" use="required"/>
</xs:complexType>

<xs:simpleType name="ISBN13Type">
    <xs:annotation>
        <xs:documentation>13-digit International Standard Book
↪  Number</xs:documentation>
    </xs:annotation>
    <xs:restriction base="xs:string">
        <xs:pattern value="[0-9]{13}"/>
    </xs:restriction>
</xs:simpleType>

<xs:simpleType name="TitleType">
    <xs:annotation>
        <xs:documentation>The title of the book</xs:documentation>
    </xs:annotation>
    <xs:restriction base="xs:string">
        <xs:minLength value="1"/>
        <xs:maxLength value="200"/>
    </xs:restriction>
</xs:simpleType>

<xs:complexType name="AuthorsType">
    <xs:sequence>
        <xs:element name="author" type="AuthorType" maxOccurs="unbounded"/>
    </xs:sequence>
</xs:complexType>

<xs:complexType name="AuthorType">
    <xs:sequence>
        <xs:element name="name" type="AuthorNameType"/>
        <xs:element name="portraitThumbnailWebP" type="xs:base64Binary"
↪  minOccurs="0" maxOccurs="1"/>
    </xs:sequence>
</xs:complexType>

<xs:simpleType name="AuthorNameType">
    <xs:annotation>
        <xs:documentation>The full name of the author with titles, e.g.,
↪  "Eric Evans", "Dr. John Smith"</xs:documentation>
    </xs:annotation>
    <xs:restriction base="xs:string">
        <xs:minLength value="1"/>
        <xs:maxLength value="100"/>
    </xs:restriction>
</xs:simpleType>
```

```xml
<xs:simpleType name="NonNegativeIntegerType">
  <xs:restriction base="xs:integer">
    <xs:minInclusive value="0"/>
  </xs:restriction>
</xs:simpleType>

<xs:simpleType name="CategoryType">
  <xs:restriction base="xs:string">
    <xs:enumeration value="NON_FICTION"/>
    <xs:enumeration value="FICTION"/>
  </xs:restriction>
</xs:simpleType>

<xs:complexType name="TagsType">
  <xs:sequence>
    <xs:element name="tag" type="TagType" minOccurs="0" maxOccurs="20"/>
  </xs:sequence>
</xs:complexType>

<xs:simpleType name="TagType">
  <xs:restriction base="xs:string">
    <xs:minLength value="1"/>
    <xs:maxLength value="50"/>
  </xs:restriction>
</xs:simpleType>

<xs:simpleType name="RatingType">
  <xs:restriction base="xs:decimal">
    <xs:minInclusive value="0"/>
    <xs:maxInclusive value="5"/>
    <xs:fractionDigits value="1"/>
  </xs:restriction>
</xs:simpleType>

</xs:schema>
```

Listing 6-8 Book schema in XSD

Protobuf

The Protobuf format was developed by Google in 2001 and was open sourced in 2008 [65, 66]. It is mainly used with gRPC but used in other use cases, for example, Apache Kafka. The Protobuf project is licensed under the 3-Clause BSD License [67–69]. However, the code generated by Protobuf is only licensed by the user generating the code [69].

Protobuf is quite similar to Apache Thrift and was developed at a similar time it was open sourced by Facebook [31]. Thrift, however, is less popular in integration projects that the authors have experienced. Therefore, it will not be covered in this book.

Protobuf uses year-based versioning, for example, 2023, whereas previously numbered versioning was used (`proto2` and `proto3`)[5] [70, 71].

To describe a Protobuf format, the Protobuf IDL is necessary [31].

The Protobuf IDL can be converted to classes in various supported programming languages, which then provide support to serialize and deserialize data [31].

Example

The schema for a book example in Protobuf is shown in Listing 6-9. The example entity itself is not shown, as it is binary encoded and not readable.

```
syntax = "proto3";

package Example.Library;

option csharp_namespace = "Example.Library.Book.Dto.Proto";

// Book represents a book and its metadata for the library, e.g., for
↪    search results
message Book {
  // The 13-digit ISBN
  string isbn13 = 1;

  // The title of the book (1-200 characters)
  string title = 2;

  // WebP image of the book cover thumbnail
  optional bytes cover_thumbnail_webp = 3;

  // List of authors of the book (at least one author)
  repeated Author authors = 4;

  // The number of text positions in the book (non-negative integer)
  uint32 number_of_text_position = 5;

  // The main category of the book
  Category category = 6;

  // List of tags associated with the book (up to 20 unique tags, 1-50
↪    characters each)
  repeated string tags = 7;

  // The average rating of the book (0 to 5 stars, with one decimal place)
  optional float rating = 8;

  // Indicates whether the library has a valid license for the book at the
↪    moment
  bool has_valid_license = 9;
```

---

[5] Some samples still need the old version for compatibility with some tools.

```
  // The date when the book was published, in ISO 8601 format (YYYY-MM-DD)
  // This format allows for representation of dates before the Unix epoch
  // Examples: "2000-01-01", "1850-12-25", "0001-01-01"
  string publishing_date = 10;

  // Author represents an author of a book
  message Author {
    // The full name of the author with titles (1-100 characters)
    // Examples: "Eric Evans", "Dr. John Smith"
    string name = 1;

    // WebP image of the author's portrait thumbnail
    optional bytes portrait_thumbnail_webp = 2;
  }

  // Category represents the main category of a book
  enum Category {
    CATEGORY_NON_FICTION = 0;
    CATEGORY_FICTION = 1;
  }
}
```

Listing 6-9 Book example in Protobuf

Avro

Apache Avro is a binary-serialization format developed in 2009 as a subproject of Hadoop. Protobuf and Apache Thrift were not considered optimal for the needs of the Hadoop project [31, 72]. The Apache Avro project is licensed under the Apache 2.0 license (with some appendices for special libraries included) [73]. At the time of writing, it is in version 1.11.1 [33].

The specification describes not only the data types but also a RPC "Protocol Declaration" and a "Protocol Wire Format," which will not be discussed [33]. Avro schemas can be written in the Avro IDL or as a JSON file.

It allows the following data types:

primitive type   null, boolean, int, long, float, double, bytes, string
complex types   record, enum, array, map, fixed

Along with the binary type, Apache Avro can also be represented in a JSON-like fashion for debugging.

Example

A schema of the example data is shown in Listing 6-10. The example uses pascal case for record types and camel case for field names. This is the same syntax that all examples in the Apache Avro spec uses [33].

```
{
  "type": "record",
  "name": "Book",
  "namespace": "com.example",
  "doc":
↪ "A schema representing a book and its metadata for the library for,
  e.g., search results",
  "fields": [
    {
      "name": "isbn13",
      "type": "string",
      "doc": "The 13-digit ISBN, e.g., 9781234567890"
    },
    {...},
    {
      "name": "coverThumbnailWebP",
      "type": ["null", "bytes"],
      "default": null,
      "doc": "Base64 encoded WebP image of the book cover thumbnail"
    },
    {
      "name": "authors",
      "type": {
        "type": "array",
        "items": {
          "type": "record",
          "name": "Author",
          "fields": [
            {
              "name": "name",
              "type": "string",
              "doc":
↪ "The full name of the author with titles, e.g.: \"Eric Evans\",
              \"Dr. John Smith\""
            },
            {...}
          ]
        }
      },
      "doc": "List of authors of the book"
    },
    {...},
    {
      "name": "category",
      "type": {
        "type": "enum",
        "name": "Category",
        "symbols": ["NON_FICTION", "FICTION"]
```

Data Formats and Their Schemas   145

```
      },
      "doc": "The main category of the book"
    },
    {
      "name": "tags",
      "type": {
        "type": "array",
        "items": "string"
      },
      "doc": "List of tags associated with the book (top 20)"
    },
    {
      "name": "rating",
      "type": ["null", "bytes"],
      "default": null,
      "logicalType": "decimal",
      "precision": 2,
      "scale": 1,
      "doc":
↪ "The average rating of the book (0 to 5 stars, with one decimal place),
    e.g., 4.5"
    },
    {
      "name": "hasValidLicense",
      "type": "boolean",
      "doc":
↪ "Indicates whether the library has a valid license for the book
    at the moment",
      "default": true
    },
    {
      "name": "publishingDate",
      "type": {
        "type": "string",
        "pattern": "^[0-9]{4}-[0-9]{2}-[0-9]{2}$"
      },
      "doc": "The date when the book was published in yyyy-mm-dd format
    ([Date](https://avro.apache.org/docs/1.8.0/spec.html#Date) was not
    used as it does not allow to store values before the unix epoch,
    1 January 1970)"
    }
  ]
}
```

Listing 6-10 Book example in Avro schema (AVSC)

Table 6-2 Data formats and the size of the serialized sample

Format	Size in bytes
JSON	364 bytes
XML	675 bytes
Protobuf	124 bytes
Avro	117 bytes

Comparison of Data Formats

In the code samples is a LibrarySchemaTester sample that shows the usage of the four formats (YAML excluded, as it is not recommended to use to transfer data).[6]

The sample shows how the languages generate code classes and then convert an internal object to each representation (called data transfer object (DTO)) and how to serialize and then deserialize it.

All samples work on Linux, Windows, and Mac, except the creation of XML DTO from the schema. This was never ported from the original .Net Framework to the new .Net. This shows that Microsoft has moved away from XML to other formats.

The sample also briefly sums up how big of a difference it makes depending on which data format is used. The result is presented in Table 6-2.

Conclusion

This section took a brief look at some common data formats used in integration architecture. Developers prefer to start with simple-to-view data types (e.g., YAML or JSON). However, sometimes faster and smaller binary formats need to be used for performance reasons. In addition, schema evolution and the safety of interpretation can be huge junctures for binary formats next to size and speed.

Next to these pure data format considerations, it is also worth looking at the ecosystems and the skills of the developers. Not all data formats represented here have perfect integration in all programming languages, and if a team has experts in one format and already custom libraries, it may be better to keep going with the existing format. In addition, not all code generators are equally powerful. We will look at code generation in Chapter 8, which supports developers dramatically when starting with a new API and using API-First. Such a code generator with good integration and a powerful schema can significantly support the quality goal of self-descriptiveness and user error protection. In addition, there are bigger and smaller tooling ecosystems around the standards, which also makes working with the standards harder or simpler. We will, however, not dive deeply into the tooling and code generator ecosystems as they change quite rapidly and also vary considerably on what a user prefers (UI vs. command line interface (CLI)) and in which ecosystem the user is working.

---

[6] https://github.com/Apress/Crafting-Great-APIs-with-Domain-Driven-Design/tree/main/chapter-6-2/LibrarySchemaTester

Getting started with a new schema can help considerably in converting a sample data entry into a schema with an online or offline converter. It makes it simple to start and generate the first data in the formats or their schema. AI chatbots can also be really helpful in coming up with an initial schema idea if the format is new.

If a new data format needs to be established, it is also necessary to consider the importance of schemas. The authors saw cases where JSON made it very hard to force the teams to stick to schemas as they are an optional component. When using a binary format like Protobuf or Apache Avro, there is no way around a schema. This can bring a lot of benefits, as the format needs to be used properly with a schema. This decision can be compared with the choice of a programming language with enforced types compared to one with dynamic types: When using dynamic types (e.g., with Python or JavaScript), the start is easy. However, problems can crop up later in the runtime that would not happen in a strongly typed language (e.g., C or Java).

Schema evolution

An essential aspect when starting to use a format is to also think about a schema evolution strategy. Schema evolution describes how a schema can change over time and when a breaking change happens. Protobuf and Apache Avro have special mechanisms that allow the modification of property names without breaking consumers' and producers' implementations or runtime behavior [31]. These aspects are particularly important when implementing event sourcing, where you want to be able to read new events and old events that were produced weeks or even years ago in parallel.

However, schema evolution will be discussed in more detail in section "Versioning of APIs".

Idempotency and Guarantees

Before we dive into the definition of APIs, we would like to take a brief look at idempotency and guarantees of messages. This topic is more critical in asynchronous messages than in synchronous ones. However, it is also relevant for synchronous APIs and needs to be discussed.

When we transfer data over a network, we need to assume that every network connection can be unreliable. Data packages can get lost and be reordered because of concurrency, the multipath behavior of network setups, or jitter. In addition, at any time, services can crash during the transmission of messages or in processing.

Therefore, to guarantee a correct state, it is important to understand some fundamental problems and apply some principles to solve the problems.

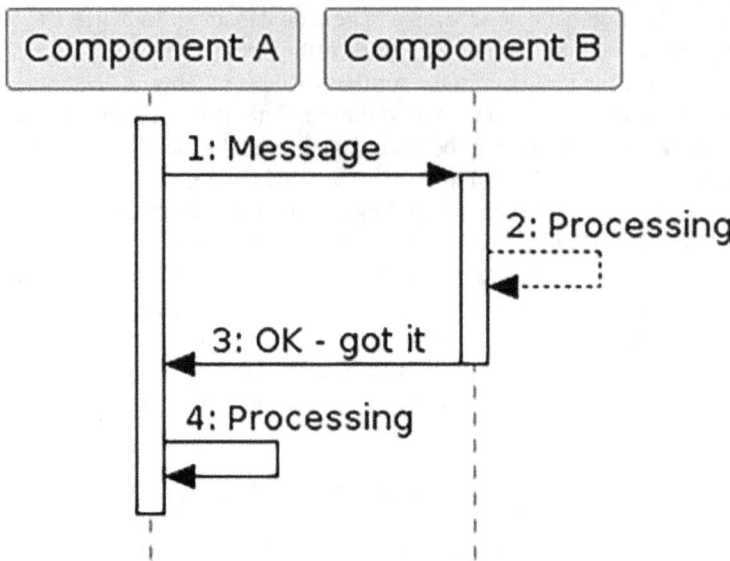

Figure 6-4 Potential problems in synchronous communication between two components

Problems in Transmission

A package can get lost at any time. In synchronous communication, we can look at the details from Figure 2-7 and see what can go wrong. In Figure 6-4, we see the following problems:

1. Component A sends a message to Component B. The message can get lost. Component B never gets the message.
2. Component B processes the message. However, Component B can be unavailable or go down during processing.
3. Component B sends a response to Component A. The message can get lost. Component B sends the response to Component A, but now Component A is unavailable.
4. Component A fails while processing.

In asynchronous communication, the problems might be even more critical because a one-to-one connection between sender and receiver is not given. The situation is shown in Figure 6-5.

- The producer Component A sends a message to a broker. The message can get lost during transmission or the broker may not be available.
- The broker stores the message from Component A.

Idempotency and Guarantees 149

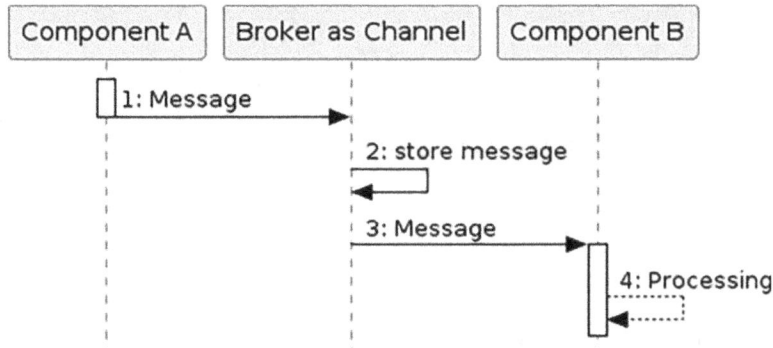

Figure 6-5 Potential problems in synchronous communication between two components

- The broker tries to forward it to Component B. Component B is not available and cannot receive the message from the broker, or the message gets lost during transmission.
- Component B cannot process the message, but usually Component A does not expect a response. The unprocessed message will only be detected as a result of some accident anywhere in the entire business process.

However, these problems can be addressed, as we shall see next.

Mitigating Problems

Data loss or inconsistency is caused by the loss of data packages or the data sequence not being held, for example, by concurrency, the multipath behavior of network setups, and jitter.

Jitter

Jitter refers to a deviation from the ideal timing of an event. The reference event is the differential zero crossing for electrical signals and the nominal receiver threshold power level for optical systems. Jitter is composed of both deterministic and Gaussian (random) content [74]. We mention Jitter here to show that network connections are unreliable. However, protocols, like TCP, have introduced mechanisms to resolve it in a single TCP connection [75].

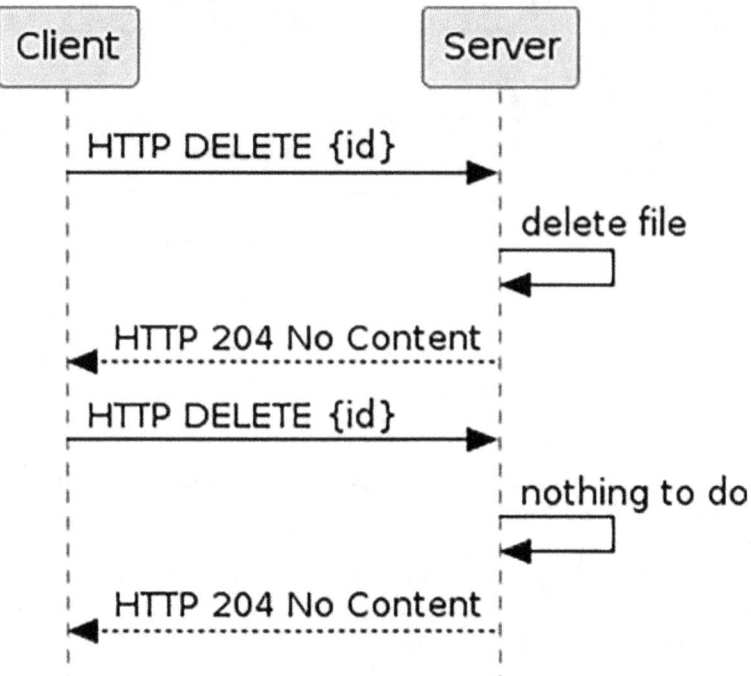

Figure 6-6 Communication for HTTP delete messages

However, separate HTTP requests can still be processed in the wrong order. There are three principal ways to face that problem.

- **Idempotent messages**

    Idempotent message are messages that deliver an equal result independently of how often they are sent.

    A simple example is shown in Figure 6-6, where a delete message in HTTP delivers the same result independently of how often it was sent because the corresponding data are no longer available. Post messages or put messages in HTTP are by their nature not idempotent. However, we should be aware that a put message to change data can be received before the post message to create the corresponding data has been received. Therefore, one should consider changing the data into a sequence of post and delete messages, deleting the old data, and creating a new data set with the changed content. Even when the delete message comes before the creation message, both messages are independent of each other and can be formulated as idempotent [76].

    Let us look at the concrete example shown in Figure 6-7: You send a post message to create a customer with the name *Smith* (1.1). After that, you send a put message to the created customer object to change the name to *Miller* (1.3).

Idempotency and Guarantees

Figure 6-7 Idempotency example

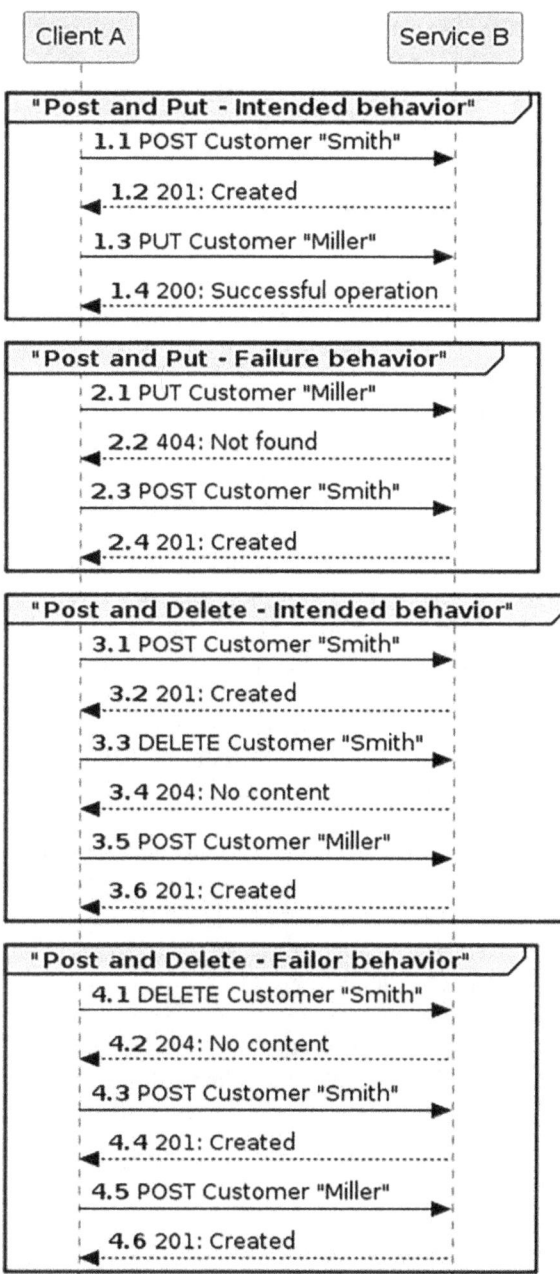

It may happen that the put message is received first (2.1). It is answered with a 404 error message because the customer cannot be found (2.2). Then the post message is received and the customer is created with the name *Smith* (2.3).

To avoid that, the following can be done. First the post message is sent under the name *Smith* (3.1). Then a delete message is sent to delete the customer (3.3). An additional post message is sent to create a customer with the name *Miller* (3.5).

Now, when the delete message is received first, it is answered with a successful message 204, because the customer has not been created (4.2). The customer *Smith* and customer *Miller* are created (4.3 and 4.5). This scenario makes it possible to process messages successfully independently of the sequence in which they are received. However, it requires that certain clean-up procedures be implemented.

- **Concurrency token**

  An alternative implementation for this specific criterion is a concurrency token. A concurrency token allows optimistic locking where every data set handled by the client gets a token (e.g., a timestamp with the last update time on the database or a monotonically increasing version number[7]) from the server. This token is submitted every time the data record is updated. Every update on the databases assures that the client token matches the one in the database record or field, then replaces the concurrency token with a new one. If the concurrency token from the client does not match the database token, the update is rejected, as the client does not have the latest version [77].

- **Clear delivery guarantees for messaging**

  Clear guarantees for messages means that the sender system or broker guarantees that the message will be delivered at least once, at most once, or exactly once.

  - **At least once**

    at least once means that a message is delivered one or multiple times. However, one time is guaranteed. This feature is used when an acknowledgment can be expected. If the acknowledgment is not received after a timeout, the message is delivered again. To support at least once delivery, the message should be idempotent [78, 79].

  - **At most once**

    at most once means that the message can be delivered or not. If the message is lost, it is not redelivered. The producer does not wait for any acknowledgment. That reduces latencies and data loss risk. Messages do not need to be idempotent, as it is guaranteed that they will not be delivered multiple times [78, 79].

---

[7] Timestamps are a simple construct that works great in classic client-server concepts and also helps in error cases to obtain more details; however, in distributed systems, a global time is often not given, and in such cases, a monotonically increasing version number can be preferred. More on that in section "The CAP Theorem and Eventual Consistency."

- **Exactly once**
  exactly once guarantees that a message will be delivered only once. This is desirable, however, as it requires a transactional context if something breaks during the delivery and can be hard to implement. The involved systems must agree to a certain step in the processing, which defines success. Each failure in the delivery before that step needs a redelivery as it was not successful. Each step after that point does not need a re-delivery. In at least once or at most once delivery, this does not need to be defined that hard, as it can be delivered multiple times or not at all.
- The hardest problem is how to mitigate duplicate creations, as there is no key. A solution there is to create the key on the client, e.g., in the form of a universal unique identifier (UUID).

Exactly once is hard or even impossible

As mentioned before, exactly once is very hard. It is also very hard or even impossible to have in a system with distributed states.

Protocols like the Two-Phase Commit protocol can help in distributed systems with a disturbed state. They allow spanning transactions over multiple systems, which is needed for an exactly once guarantee. However, in this scenario, there are edge cases where an exactly once guarantee can fail. Two-Phase Commit enhances the guarantee of getting closer to exactly once; however, in the end, it is still at least once or at most once.

In addition, using exactly once causes considerable overhead, as distributed transactions are expensive and will block multiple systems. The use of at least once should be preferred. We will discuss eventual consistency and what it means to modern software architectures that do not need transactions at all in section "New Requirements and Challenges in the Cloud" [80, 81].

Definition of Synchronous Interfaces

In the previous section "Data Formats and Their Schemas", we saw how a schema for an API can be defined. In this section, we want to discuss how the defined data can be combined with the necessary methods to be applied.

We still follow the API-First approach as presented in section "Collaborative Approaches Improve Processes."

The methods to be applied can be defined directly as in gRPC and partly in GraphQL, or they can be derived from the used protocol as in REST.

In what follows, we want to formulate the interfaces for *Catalog Management* as presented in section "Context Map." Figure 6-8 shows the part of the Visual Glossary as a base for formulating APIs.

In the next section, we formulate the API as gRPC.

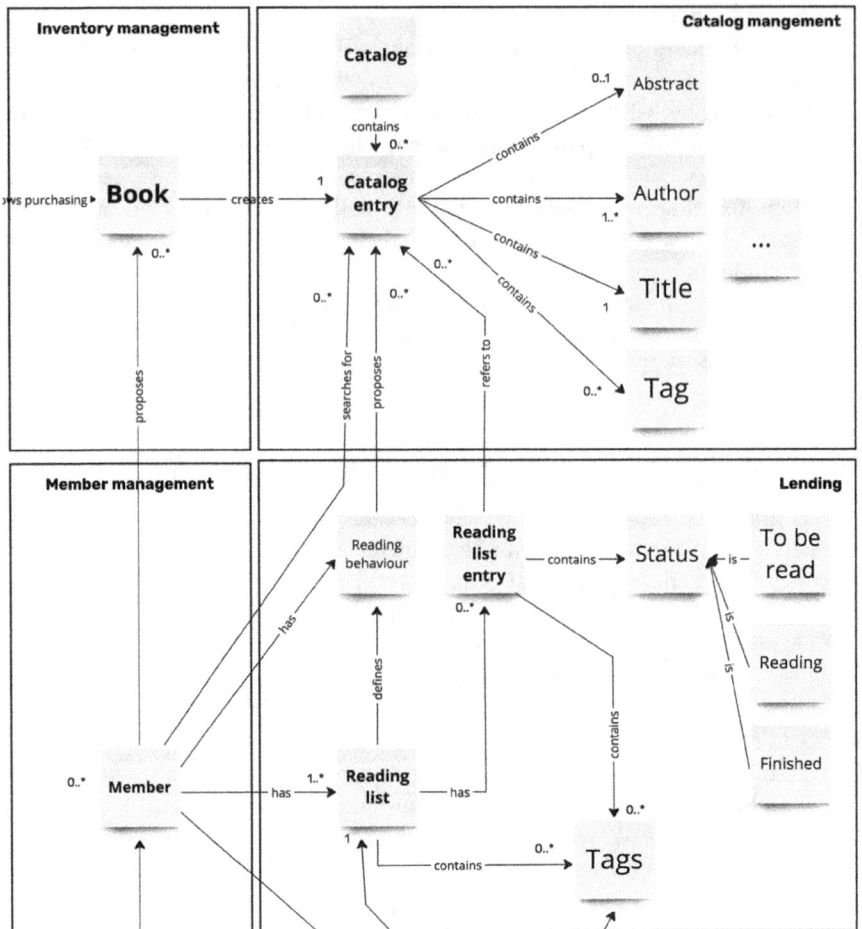

Figure 6-8 Visual Glossary: Catalog Management

Definition as gRPC

gRPC is a modern open-source high-performance remote procedure call (RPC) standard. It can connect services using different programming languages and supports load balancing, tracing, health checking, and authentication [82]. It was first introduced by Google [83]. It uses Protobuf as discussed previously for defining schemas in section "Data Formats and Their Schemas."

It allows a synchronous call to a gRPC server. The defined data structure only needs to be enhanced by the allowed methods, as shown in Listing 6-11.

The definition of the service includes methods to manipulate catalog entries and associated abstracts. Let us look more closely at the methods.

Definition of Synchronous Interfaces   155

```
service CatalogManagement {

    //Creates a new catalog entry
    rpc CreateNewEntry (Book) returns (Identifier) {}

    //Search for catalog entries
    rpc SearchBooksInCatalog(SearchCriteria) returns (CatalogEntries) {}

    //Delivers a catalog entry by its identifier
    rpc GetEntryByIdentifier(Identifier) returns (CatalogEntry) {}

    //Updates a catalog entry by its identifier
    rpc UpdateEntryByIdentifier(CatalogEntry) returns (Identifier) {}

    //Deletes a catalog entry by its identifier
    rpc DeleteEntryByIdentifier(Identifier)
     ↪ returns(google.protobuf.Empty) {}

    //Returns the abstract of a catalog entry by its identifier
    rpc GetAbstractByEntryIdentifier(Identifier) returns (Abstract) {}

    //Add abstract to a catalog entry by a librarian
    rpc AddAbstract(Abstract) returns (Identifier) {}

    //Deletes an abstract of a catalog entry
    rpc DeleteAbstract(Identifier) returns (google.protobuf.Empty) {}

}
```

Listing 6-11 Service definition for Catalog Management using Protobuf

- `CreateNewEntry`
  The method creates a new entry in the catalog and expects a `Book` as input parameter. It returns an identifier that can be used to access the catalog entry directly.
- `SearchBooksInCatalog`
  The method searches for catalog entries that fulfill the given search criteria. The method returns catalog entries, which are the `Book` data enhanced by a catalog identifier as UUID.
- `GetEntryByIdentifier`
  The method makes it possible to access a catalog entry by its identifier.
- `UpdateEntryByIdentifier`
  The method expects a changed catalog entry and returns the associated identifier.
- `DeleteEntryByIdentifier`
  The method will delete a catalog entry identified by its identifier. It returns an empty message to guarantee idempotency.
- `AddAbstract`
  Using this method, an abstract can be added to a catalog entry. The method expects the identifier of the catalog entry and the abstract.

- `DeleteAbstract`
  Using this method, the abstract of the catalog entry is deleted. It returns an empty message.

gRPC is used for fast and reliable server-to-server communication. As we can see, it concentrates on the methods rather to define understandable data structures. Therefore, it is suited for communication between servers.

The gRPC standard alone is unsuitable to implement a web application in the browser because the browser APIs of the HTTP/2 standard does not allow the necessary control over requests. A special proxy would be needed to use HTTP/1 to communicate with the browser [84].

gRPC client implementation exists for web applications, for example, *gRPC-web*, implemented by Google [84]. However, certain APIs for a web application are not implemented, for example, Fetch API. Therefore, *gRPC-web* uses a proxy to map an HTTP/1.1 or HTTP/2 request to a gRPC request [84, 85]. However, a team that needs a high-end performance and type-stable web application should consider a gRPC client, even though it works very differently from a common REST client.

GraphQL can be better used for user interfaces, as we will see in the following section.

Definition as GraphQL

GraphQL is a query language for APIs that makes it possible to retrieve data sets from different servers in a structured way [86]. It is a specification language at its core [87].

The type definitions become the specification, and the client can query the API freely [87]. The catalog entry defined in GraphQL would look like the example in Listing 6-12.

The IDL is programming agnostic. But it is comparable to other IDLs so that the interpretation of the defined object `Book` is not difficult. The texts in quotation marks are descriptions that are added for better understanding in an API-First approach. Required properties are marked by an exclamation mark. Types can be defined freely and reused in complex data structures. Arrays are marked by square brackets.

As in other IDLs as well, a designer can define operations to fetch and manipulate data. GraphQL provides the type `Query` to fetch data. And the type `Mutation` is provided to manipulate data.

Listing 6-13 shows an example of how to query and manipulate date via GraphQL.[8]

Catalog entries can be searched for by given search criteria using `catalogEntries`. A single catalog entry by its identifier is delivered with `catalogEntry`.

---

[8] https://github.com/Apress/Crafting-Great-APIs-with-Domain-Driven-Design/tree/main/Chapter-6-3

```graphql
"A book as entry in the catalog."
type Book {
    # Identifier of the catalog entry
    identifier: ID!
    # The 13-digit ISBN
    isbn13: String!
    # Title
    title: String
    # List of authors of the book - the list cannot be empty
    authors: [String!]!
    # Number of text positions in the book
    numberOfTextPositions: Int
    # Category
    category: Catagory
    # Top twenty individual tags give to the particular book
    tags: [String]
    # Rating of the book between 1 and 5
    rating: Float
    # Date of publication
    publishingDate: String!
    # Abstract written by a librarian
    abstract: String
}

enum Catagory {
    NON_FICTION,
    FICTION
}
```

Listing 6-12 GraphQL definition of a catalog entry

The manipulation of catalog entries follows the principle shown earlier in section, "Definition as gRPC":

- createNewEntry to create a new catalog entry
- updateCatalogEntry to change a catalog entry
- createUpdateAbstract to create or update a particular of a certain entry; an entry can be deleted by updating it with an empty string.

Querying such a defined API is easy. The client programmer only needs to give the query and the data needed. The response only contains the data asked for. In such a way, flexible and powerful APIs can be formulated.

As GraphQL is used in powerful user applications, gRPC is used in server-to-server applications. REST APIs are used widely in user applications as well as in server-to-server communications.

```graphql
type Query {
    "Delivers all books fulfilling the searchCriteria"
    catalogEntries(searchCriteria: String): [Book!]
    "Delivers a book based on its ID"
    catalogEntry(identifier: ID!): Book
}

type Mutation {
    "Creates a new entry of a book in the catalog"
    createNewEntry(book: BookInput!): Book!
    "Updates a catalog entry"
    updateCatalogEntry(identifier: ID!, book: BookInput!): Book!
    "Creates and updates an abstract of a book - deleting
        means an empty abstract"
    createUpdateAbstract(identifier: ID!, abstract: String): ID!
}
```

Listing 6-13 GraphQL definition of a catalog entry

Definition as REST with OpenAPI

Introduction to REST

The presented IDLs GraphQL and gRPC define the methods applied to the schemas directly. REST defines methods applied to data using those methods of the protocol HTTP.

RESTful interfaces use HTTP to transport data to retrieve, create, or change them. REST stands for *Representational State Transfer*, which means that the data are transported directly and not a reference to them, as was usual, for example, in CORBA.

Richardson defines a maturity model of the REST approach [88]. It contains four levels.

- **Level 0**
  Interfaces that simply use HTTP as the transport protocol.
- **Level 1 – Resources**
  A resource is requested via an HTTP request. The resources can be defined in schemas.
- **Level 2 – HTTP Verbs**
  Data retrieval and manipulation are performed using HTTP verbs. The HTTP methods are:[9]
  - GET to retrieve data
  - POST to create data

---

[9] Please be aware that the standard contains more methods with HEAD, CONNECT, OPTIONS, and TRACE as well [29]. However, those are usually not used in REST APIs. Therefore, we do not discuss them.

- PUT to change data
- DELETE to delete data.[10]
- **Level 3 - Hypermedia Controls**
  The last step refers to the architecture approach hypermedia as the engine of application state (HATEOAS).

The last level to use hypermedia controls can sometimes lead to overengineered and badly understood interfaces. Therefore, it should be carefully applied and not seen as a dogmatic rule.

OpenAPI

The parts of the OpenAPI IDL were discussed previously in section "OpenAPI." We concentrate on the `path` part of the IDL and its relation to the interface definition.

A very small version of an OpenAPI is shown in Listing 6-14.

The specification has a couple of parts, which we introduce briefly. The following paragraphs explain them more briefly.

Line 1: openapi   Version of API specification.

Lines 2–5: info   Meta information of API. The version here should be a semantic versioning (to be discussed in section "Versioning of APIs").

Lines 7–11: servers   Shows server information. This API is provided on two servers. The description describes that one is the production environment and one the staging.

Lines 13–26: paths   Describes endpoints and methods allowed in it.

The leading resource in the sample of the `catalog search` is books. To access books, GET can be used, provisioning specific search parameters in a string. The search parameters are provided via the URL as query parameter. To create a new catalog entry, POST is used at books.

An individual book resource can be accessed by an identifier as path parameter. Applying the GET, a user can access a specific catalog entry. Applying the PUT, a catalog entry can be changed. Or it can even be deleted by applying DELETE.

Each of those entries can be bound to the implementing operation by the `operationId`. Access rights to the methods can be controlled by the security scopes such as `book:write`. The entry makes it possible to change a catalog entry.

Success responses contain the search results for the GET operations. Or they contain a link to the just created or changed resource. That is common for security reasons and follows the idea of hypermedia controls.

---

[10] PATCH is often used in non-REST, but HTTP interfaces. PATCH can change data as with PUT, but it does not require the full resource. That is a problem because it is unclear which parts of the defined resource are required. Therefore, it is not recommended to use PATCH in REST interfaces [29].

```
 1  openapi: 3.1.0
 2  info:
 3    title: Sample API
 4    description: Optional multiline or single-line description in
     ↪ [CommonMark](http://commonmark.org/help/) or HTML.
 5    version: 0.1.9
 6
 7  servers:
 8    - url: http://api.example.com/v1
 9      description: Optional server description, e.g. Main (production)
     ↪ server
10    - url: http://staging-api.example.com
11      description: Optional server description, e.g. Internal staging server
     ↪ for testing
12
13  paths:
14    /users:
15      get:
16        summary: Returns a list of users.
17        description: Optional extended description in CommonMark or HTML.
18        responses:
19          "200": # status code
20            description: A JSON array of user names
21            content:
22              application/json:
23                schema:
24                  type: array
25                  items:
26                    type: string
```

Listing 6-14 Example of a simple OpenAPI specification (see [89] adapted to version 3.1.0)

DELETE simply returns a "Successful Operation" string, because the resource no longer exists. A response NOT FOUND should not be returned, to allow an idempotent behavior of delete. When DELETE is sent several times, it always sends back "Successful Operation" as the target of the operation is reached.

An example of what such a definition might look like is given in Listing 6-15.

The access to a single resource is shown in Listing 6-16.

Listing 6-17 shows an additional resource below a specific book defined by its identifier. Such a resource is called a subresource.[11]

---

[11] We use here abstracts as the endpoint. According to the applied guidelines, REST endpoints must be named in the plural [48].

Definition of Synchronous Interfaces

```
1  openapi: 3.1.0
2  ...
3  paths:
4    /books:
5      get:
6        ...
7        parameters:
8          - $ref: '#/components/parameters/SearchCriteriaParameter'
9        responses:
10         200:
11           $ref: '#/components/responses/BookListResponse'
12       ...
13     post:
14       ...
15       requestBody:
16         $ref: '#/components/requestBodies/BookToBeCreatedRequest'
17       responses:
18         201:
19           $ref: '#/components/responses/LinkToBookResponse'
20       ...
21
```

Listing 6-15 Excerpt of a catalog search as OpenAPI for books

```
1  openapi: 3.1.0
2  ...
3  paths:
4    /books/{bookIdentifier}:
5      get:
6        ...
7        parameters:
8          - $ref: '#/components/parameters/BookIdentifierParameter'
9        responses:
10         200:
11           $ref: '#/components/responses/BookResponse'
12       ...
13     put:
14       ...
15       parameters:
16         - $ref: '#/components/parameters/BookIdentifierParameter'
17       requestBody:
18         $ref: '#/components/requestBodies/BookToBeUpdatedRequest'
19       responses:
20         200:
21           $ref: '#/components/responses/LinkToBookResponse'
22       ...
```

Listing 6-16 Excerpt of a catalog search as OpenAPI for individual books

```yaml
openapi: 3.1.0
...
paths:
    /books/{bookIdentifier}/abstracts:
      post:
        ...
        parameters:
          - $ref: '#/components/parameters/BookIdentifierParameter'
        requestBody:
          $ref: '#/components/requestBodies/AbstractRequest'
        responses:
          201:
            $ref: '#/components/responses/LinkToBookResponse'
          ...
      delete:
        ...
        parameters:
          - $ref: '#/components/parameters/BookIdentifierParameter'
        responses:
          200:
            $ref: '#/components/responses/SuccessfulOperation'
        ...
```

Listing 6-17 Excerpt of a catalog search as OpenAPI for abstracts of individual books

abstracts is a subresource of the leading resource books. A subresource can be used when the leading resource is requested at a higher number as certain parts of it, such as book abstracts. The resource enables it to be created via POST or be deleted via DELETE.

Listing 6-15 follows the sound practice of defining the individual parts of requests as components and referring to them. This is valid for request bodies, parameters, and responses, as will be discussed in the following paragraphs.

Request Bodies

Request bodies contain the data that need to be provided as input parameter to the invoked method. Request bodies are necessary for creation POST and changes PUT of resources. They can be defined below the components. An example of the definition of request bodies for the creation and changing of catalog entries is given in Listing 6-18. The request bodies are referenced via the keyword $ref: to defined JSON schemas.

Let us see how the definition of parameters works.

```
 1  components:
 2    requestBodies:
 3      AbstractRequest:
 4        description: Abstract to be updated in a book
 5        content:
 6          application/json:
 7            schema:
 8              $ref: '#/components/schemas/Abstract'
 9
10      BookToBeCreatedRequest:
11        description: Book to be created in catalog
12        content:
13          application/json:
14            schema:
15              $ref: '#/components/schemas/BookToBeCreatedUpdated'
16
17      BookToBeUpdatedRequest:
18        description: Book to be updated in catalog
19        content:
20          application/json:
21            schema:
22              $ref: '#/components/schemas/BookToBeCreatedUpdated'
```

Listing 6-18 Examples for request bodies in catalog search

Parameters

Parameters can be defined in OpenAPI as a query, as header, or as path parameter. How the parameters are provided by the client is defined via the keyword `in:`. An HTTP request can provide multiple parameters, for example, a search parameter as query parameter and an API version in the header. Listing 6-19 shows the parameters of the catalog search as an example.

The schemas of parameters can be defined below `schemas:` as well. However, those parameters are simple types, which makes it more convenient for readers to find the definition directly at the parameter definition.

Responses can be defined as reusable parts, too.

Responses

Responses follow the HTTP response codes defined in the standard [29].

In particular, the error responses for

- `400: Bad Response`,
- `401: Unauthorized`,
- `403: Forbidden`,
- `404: Not found`,
- `500: Service not available`, or
- `default`

```yaml
 1  components:
 2    parameters:
 3      BookIdentifierParameter:
 4        name: bookId
 5        description: Book identifier
 6        required: true
 7        in: path
 8        schema:
 9          type: string
10          format: uuid
11          examples:
12            - 193ad3f6-0c0d-4e57-a4e4-b50766e64f14
13      SearchCriteriaParameter:
14        name: searchCriteria
15        in: query
16        required: true
17        schema:
18          type: string
19          minLength: 2
20          maxLength: 200
21          examples:
22            - Evans
23      ApiVersionParameter:
24        name: apiVersion
25        in: header
26        required: true
27        schema:
28          type: string
29          enum:
30            - 1.0.1
31            - 1.1.1
32          default: 1.1.1
33          examples:
34            - 1.1.1
```

Listing 6-19 Examples of parameters defined for catalog search

can be predefined. Those response codes can be reused over the entire specification.

Messages should generally deliver more information about errors. Therefore, they contain an error object that contains a tracing identifier in addition to the error message, so that the client can map the error to a certain action. It is good practice for the error codes `401: Unauthorized` and `403: Forbidden` not to deliver more detailed information. Thus, a potential invader could retrieve more information when attacking the site. Listing 6-20 shows examples of error responses.

However, successful responses `200: OK` must be explicitly defined because they deliver specific data. Usually, a list of the resources and a single resource is necessary. When creating or changing a resource, it is good practice to send back a link to the newly created or changed resource. Moreover, it is recommended that a

```
 1  responses:
 2  BadRequestResponse:
 3    description: Bad request
 4    content:
 5      application/json:
 6        schema: $ref: '#/components/schemas/Error'
 7  ForbiddenResponse:
 8    description: Forbidden
 9  NotFoundResponse:
10    description: Not found
11    content:
12      application/json:
13        schema: $ref: '#/components/schemas/Error'
14  ServiceNotAvailableResponse:
15    description: Service not available
16    content:
17      application/json:
18        schema: $ref: '#/components/schemas/Error'
19  DefaultResponse:
20    description: Unexpected error
21    content:
22      application/json:
23        schema: $ref: '#/components/schemas/Error'
```

Listing 6-20 Examples of defined error responses

```
 1  responses:
 2  LinkToBookResponse:
 3    description: Successful operation - Catalog entry created
 4    content:
 5      application/json:
 6        schema: $ref: "#/components/schemas/Link"
 7  BookListResponse:
 8    description: List of books in library
 9    content:
10      application/json:
11        schema: $ref: '#/components/schemas/BookList'
12  BookResponse:
13    description: Sends back book
14    content:
15      application/json:
16        schema: $ref: '#/components/schemas/BookDetail'
17  SuccessfulResponse:
18    description: Successful operation
```

Listing 6-21 Examples of successful responses from catalog search

simple response with "Successful operation" be defined to allow certain endpoints to answer without delivering data, for example, a delete operation.

The defined responses can be found in Listing 6-21.

The referenced schemas are defined as JSON schemas in YAML, as already explained in section "Data Formats and Their Schemas."

Mapping JSON definitions to YAML

Here, in the spec section, we use YAML to define the schemas. Earlier we used JSON. When we use references, also to external files, like `$ref: https://example.com/my-components.yaml#/schemas/MySchema`, it does not matter if these are in JSON or YAML. However, if we use it in a file, we need to use a single format.[12]

The format conversion can be done very easily on the CLI, for example, using the tool "yq."[13] The tool makes it possible to convert JSON to YAML with the command `yq -Poy input.json`.

YAML to JSON works with `yq -Poj input.yaml`.

Security

Even security schemes can be defined [90]. Security schemes can be used below the `path` or `path item` like a `get:` or `put:` definition as the security requirement to the associated request.

Today, different security types are supported:

- API Keys
- HTTP Authentication
- Mutual TLS
- OAuth 2.0
- OpenID Connect

API Keys

An API Key is an alphanumeric string that can be used to control access to an API [91]. It works in the end as a password for API access.

The server validates the authenticity of the requester by the API Key.

The API Key scheme definition contains a description, the type `apiKey`, the name to be referenced to, and where it can be found in the request. An example is given in Listing 6-22.

---

[12] Technically, some formats allow embedding other formats in it, for example, JSON can be embedded in YAML files as YAML is a strict superset of JSON [51].

[13] https://github.com/mikefarah/yq

Definition of Synchronous Interfaces 167

```
1  components:
2    securitySchemes:
3      DefaultAPIKey:
4        description: API key provided via header
5        type: apiKey
6        name: api-key
7        in: header
```

Listing 6-22 Example of API Key security scheme

```
1  components:
2    securitySchemes:
3      BasicHttpAuthentication:
4        description: Basic HTTP Authentication
5        type: http
6        scheme: Basic
7      BearerHttpAuthentication:
8        description: Bearer token using a JWT
9        type: http
10       scheme: Bearer
11       bearerFormat: JWT
```

Listing 6-23 Example of HTTP Authentication security scheme

HTTP Authentication
OpenAPI supports HTTP Authentication [29,90] implementing the `Authorization` header. It differentiates between the basic scheme and the bearer format, as shown in Listing 6-23 [90].

Mutual TLS
Mutual authentication via Transport Layer Security (TLS) (mutual TLS (mTLS)) is a common approach. To define it, only the type needs to be defined as `mutualTLS`.

OAuth 2.0
OAuth 2.0 is quite popular for securing API access [92]. It is developed within the IETF Working Group. It describes the security scheme defined by the `OAuth flow object`. That flow can be an implicit flow, a password flow of the resource owner, a client credential flow, or authorization code flow.
  The flow object contains the following fixed fields [24]:

- `authorizationUrl`
- `tokenUrl`
- `refreshUrl` optional
- `scopes|`

```
 1  components:
 2    securitySchemes:
 3      OAuth2AuthorizationCode:
 4        type: oauth2
 5        flows:
 6          AuthorizationCode:
 7            authorizationUrl:
    ↪  https://idp.online-library.org/oauth/2.0/auth
 8            tokenUrl: https://idp.online-library/oauth/2.0/auth
 9            scopes:
10              book:write
11              book:read
```

Listing 6-24 Example of OAuth 2.0 authorization code flow

```
1  components:
2    securitySchemes:
3      OpenIdConnect:
4        type: openIdConnect
5        openIdConnectUrl:
   ↪  https://idp.online-library.org/openid-configuration
```

Listing 6-25 Example of OpenID Connect security scheme

The scopes define the roles that can be emmployed by users to access the API endpoints. They are usually defined by the scope and the resource to be accessed separated by a colon.

An example of such a security definition is given in Listing 6-24 [92].

OpenID Connect
OpenID Connect is a specific profile of OAuth 2.0. It is more complex than OAuth2.0.

It can be externalized completely based on a discovery mechanism [90]. It is extremely easy to define because only this configuration endpoint needs to be defined, as shown in Listing 6-25.

In the path configuration, specific security scopes can be required, which are defined in the referenced configuration. In our examples, we will usually make reference to OpenID Connect as the security scheme for OpenAPI specifications.

With the presented parts of the OpenAPI specification, an interface can be defined well. The entire OpenAPI specification can be found on GitHub.[14]

---

[14] https://github.com/Apress/Crafting-Great-APIs-with-Domain-Driven-Design/tree/main/Chapter-6-3

Comparison of gRPC, GraphQL, and HTTP REST

In the following paragraphs, we compare the different styles of gRPC, GraphQL, and REST [93, 94].

- **Architectural principles**
  - *gRPC*
    Procedure calls, which can be stateful or stateless. Using Protobuf as schema definition, it uses a typed system.
  - *GraphQL*
    Fetching only those data that are needed over a single HTTP[15] endpoint.[16] It is a query language, and it uses a strongly typed system. Output is given in JSON. All requests are made via a POST request. The client defines what to fetch and change. Batching of requests is possible.
  - *HTTP REST*
    Resources are accessed from a decoupled client to a server. The server is stateless and does not hold any client state (if implemented correctly). Server responses might be cached. Intermediaries as gateways might exist between client and server and are transparent to client and server. Usually, JSON is used as the data schema. However, XML and others can also be used. It uses standard HTTP verbs: GET, POST, PUT, and DELETE.
- **Communication**
  - *gRPC*
    The client invokes a procedure on the server and vice versa [96]. A stub makes the function call, whereas on the server side, a stub invokes the actual function.
  - *GraphQL*
    The response is shaped as it is defined in the request. It can combine different schemas.
  - *HTTP REST*
    Client calls resource on server via HTTP.
- **Parameters**
  - *gRPC*
    Client sends parameters when invoking the procedure. The structure is defined by the server and must be enforced by the client and server.
  - *GraphQL*
    Parameters are defined in the query sent.[17]

---

[15] Technically. GraphQL is transport-independent. We use the most common approach here, over HTTP [95].

[16] Even though GraphQL allows subscriptions, we concentrate on the synchronous capabilities of the specification language.

[17] Technically, the parameters do not need to be embedded in the query. They can also be defined as placeholders in the query and then send the parameters and query separately in the same request.

- *HTTP REST*
  Client sends parameter via standardized interfaces. The data structure is determined by the server. Several formats are possible in one interface.
- **Versioning**
  - *gRPC*
  Versioning of breaking changes is done over package versioning; nonbreaking changes are made via Protobuf versions [97].
  - *GraphQL*
  Generally, no API versioning is required as only the needed fields are requested. Objects can get versioned, however (though this is not common).
  - *HTTP REST*
  Supports multiple API versions
- **When to use**
  - *gRPC*
  This is to be used when an action result is necessary for further processing, for example, complex calculations, or when higher performance for actions is necessary. It allows for the generation of servers and clients.
  - *GraphQL*
  This is suitable for rapid development because the fetched data can be easily adapted. It can be used nicely for applications with constantly changing data requirements. In environments where data from quite different services are needed, a GraphQL system might be beneficial. Implementing security requirements is difficult because the server does not control the fetched data. For the same reason, it is not easy to cache. One must consider performance requirements carefully because the combined different data sources could react with quite different response times. It allows for the generation of servers and clients.
  - *HTTP REST*
  When data access or change is necessary. It is most suited for security and monitoring, as we will discuss in Chapter 8, "Developer Experience and API Implementation." REST allows for the generation of servers and clients using the API specification.

In our example, we will concentrate on HTTP REST. Intensive calculations are not necessary in the online library. The data are well defined via Visual Glossary, and floating data requirements are not expected.

As synchronous communications is discussed in this section, the next section will define events for asynchronous communication.

Definition of Asynchronous Interfaces

In the previous section, we defined synchronous interfaces for HTTP in OpenAPI, gRPC, and GraphQL. In this section, we will focus on how we can define asynchronous interfaces. As discussed in section "Communication and APIs", asynchronous APIs often use a broker as a middleman. We need to take the broker into account in the definition of the interfaces, as it is a central part of where and how the interfaces will be used.

AsyncAPI

To define the interfaces, we will stick to AsyncAPI [28]. AsyncAPI is heavily indebted to OpenAPI[18] [99, 100]. However, it is not protocol-dependent, as OpenAPI is on HTTP. AsyncAPI is an independently designed protocol and supports AMQP, MQTT, WebSockets, Apache Kafka, HTTP, and others [99]. The protocol-specific add-ons are described with so-called protocol bindings.

AsyncAPI was created in 2017 and is licensed under Apache License 2.0. The standard is developed as the "AsyncAPI Initiative" under the Linux Foundation umbrella [100].

AsyncAPI provides a specification standard for asynchronous APIs focusing on EDA; next to the standard, the community builds web viewers, a studio/IDE, and code generators.[19]

Specification in Detail

Figure 6-9 visualizes a comparison between the naming of AsyncAPI vs. Apache Kafka. This should help in getting used to the naming. With this knowledge, we can dive into Listing 6-26, showing a small sample of an AsyncAPI containing only some essential parts. Let us go over the different parts and their meanings:

Line 1: asyncapi Version of API specification.

Lines 2–5: info Meta information of API. The version here should be a semversion (to be discussed in section "Versioning of APIs").

Lines 6–15: servers Shows server information. Here, we see that the API is provided on a Kafka Cluster. However, the API can also be provided on multiple Kafka Clusters or on different brokers with different protocols.

Starting at line 11, Kafka-specific bindings are visible. These are extensions of the AsyncAPI specification to allow the Kafka-specific information to be passed. The Schema Registry provides the up-to-date schema version of the events or messages.

Lines 16–26: channels These describe the channels over which the data flow (e.g., HTTP, message queue, or topic) [101].

Lines 27–37: operations The operations describe the behavior of how applications interact with channels. An operation describes an action or interaction that can be performed on a channel by an application (sending, receiving, requesting, or replying) [102].

Lines 38–61: components Here, the components referred to earlier are listed:

---

[18] The similarity was greater initially; with the current version, version 3, AsyncAPI is starting to drift away from OpenAPI [98, 99].

[19] https://www.asyncapi.com

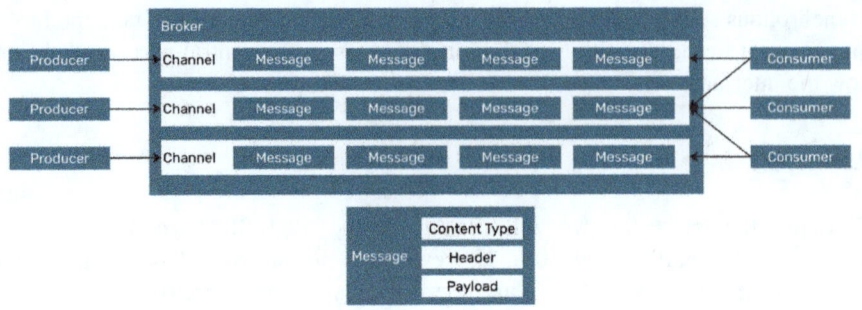

(a) AsyncAPI naming visualized

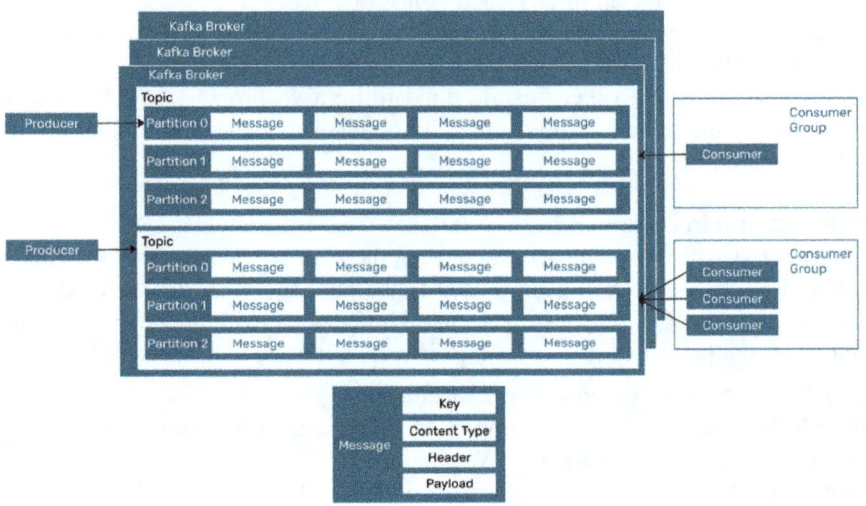

(b) Kafka naming visualized

Figure 6-9 Naming in AsyncAPI compared to Kafka

> messages These are the messages or events that are transmitted.
> schemas Here, the schemas are listed (only JSON Schema is allowed in this section).

```
1  asyncapi: 3.0.0
2  info:
3    title: Book Purchased API
4    version: 1.0.0
5    description: The API notifies when a book get purchased.
6  servers:
7    event-broker-library:
8      host: broker.online-library.org:8092
9      protocol: kafka
```

Definition of Asynchronous Interfaces

```yaml
      description: Event broker of the online library
      bindings:
        kafka:
          schemaRegistryUrl: 'https://online-library-schema-registry.com'
          schemaRegistryVendor: 'confluent'
          bindingVersion: '0.5.0'
channels:
  BookPurchasedEventChannel:
    description: Channel where messages are stored when a book was
      purchased
    address: book-purchased
    messages:
      BookPurchasedEvent:
        $ref: '#/components/messages/BookPurchasedEvent'
    bindings:
      kafka:
        topic: 'book-purchased'
        bindingVersion: '0.5.0'
operations:
  BookPurchasedEventReceive:
    action: "receive"
    channel:
      $ref: "#/channels/BookPurchasedEventChannel"
    bindings:
      kafka:
        bindingVersion: '0.5.0'
        groupId:
          type: string
          pattern: ^cg-book-purchased-.*$
components:
  messages:
    BookPurchasedEvent:
      headers:
        $ref: '#/components/schemas/MessageHeader'
      payload:
        $ref: '#/components/schemas/BookPurchased'
      bindings:
        kafka:
          bindingVersion: '0.5.0'
          key:
            type: string
            format: uuid
  schemas:
    MessageHeader: {}
    BookPurchased:
      description: Book was purchased by a librarian
      type: object
      required: [author, title]
      properties:
        authors:
          description: Authors of the book with ...
```

```
62        type: array
63        items: {}
```

Listing 6-26 Book Purchased API

```
 1  components:
 2    messages:
 3      BookPurchasedEvent:
 4        contentType: application/json
 5        headers:
 6          $ref: '#/components/schemas/MessageHeader'
 7        payload:
 8          $ref: '#/components/schemas/BookPurchased'
 9        bindings:
10          kafka:
11            bindingVersion: '0.5.0'
12            key:
13              type: string
14              format: uuid
15    schemas:
16      MessageHeader:
17        type: object
18        properties:
19          orderDate:
20            type: string
21            format: date-time
22        required:
23          - orderDate
24      BookPurchased:
25        $id: https://example.com/book
26        $schema: https://json-schema.org/draft/2020-12/schema
27        title: Book
28        description: A schema representing a book and its metadata for the
      library for, e.g., search results
29        type: object
30        properties:
31          isbn13:
32            type: string
33            pattern: ^[0-9]{13}$
34            description: The 13-digit ISBN
```

Listing 6-27 AsyncAPI scaffold with the most important keys and description

The most important parts are the operations, channels, and components. The part `operations` contains those messages or events than can be sent or received by the service applied to the specification, the part `channels` shows where this happens, and the part `components` contains the referable parts, which generally contain message definitions and other definitions. Listing 6-27 shows an AsyncAPI scaffold with the important compontent types in minimal form.

Next to the default fields in an AsyncAPI, it is also essential to use the protocol-specific bindings, as they are needed to represent some implementation-specific and

```
channels:
  BookPurchasedEventChannel:
    description: Channel where messages are stored when a book was
    ↪ purchased
    address: book-purchased
    messages:
      BookPurchasedEvent:
        $ref: '#/components/messages/BookPurchasedEvent'
    bindings:
      kafka:
        bindingVersion: '0.5.0'
        partitions: 10
        replicas: 2
        topicConfiguration:
          cleanup.policy: ["delete", "compact"]
          retention.ms: 604800000
          retention.bytes: 1000000000
          delete.retention.ms: 86400000
          max.message.bytes: 1048588
```

Listing 6-28 AsyncAPI channel binding with Kafka-specific fields

sometimes business-critical information. An example is the retention period of a topic in Apache Kafka. This shows how long a published message will be available. Suppose this number is smaller than the potential downtime of a consumer. The consumer may lose messages.

Next, we will look at the different parts of the specification in more detail.

Channels

The channels describe where the data flow over (e.g., HTTP, message queue, or topic). This can be an HTTP connection, but in a transmission with a broker it is mostly a durable form of storage, like a message queue or a topic [101].

The channel part gets quite complex in the implementation phase when all protocol-specific details are added. When a single AsyncAPI specification is used for multiple protocols, the complexity of the channel part gets multiplied. Listing 6-28 shows such an example. It adds, for example, important information on how many partitions are in the topic. This can be relevant for scaling consumers.

Operations

The operations describe the behavior of applications in their interactions with channels. They describe an action or interaction that can be performed on a channel by an application. These can normally be grouped into sending, receiving, requesting, or replying [102].

Requesting and replying are, however, not part of EDA. This is a messaging, not an eventing, pattern.

```
 1  components:
 2    messages:
 3      BookPurchasedEvent:
 4        contentType: application/json
 5        headers:
 6          $ref: '#/components/schemas/MessageHeader'
 7        payload:
 8          $ref: '#/components/schemas/BookPurchased'
 9        bindings:
10          kafka:
11            bindingVersion: '0.5.0'
12            key:
13              type: string
14              format: uuid
15    schemas:
16      MessageHeader:
17        type: object
18        properties:
19          orderDate:
20            type: string
21            format: date-time
22        required:
23          - orderDate
24      BookPurchased:
25        $id: https://example.com/book
26        $schema: https://json-schema.org/draft/2020-12/schema
27        title: Book
28        description: A schema representing a book and its metadata for the
  ↪  library for, e.g., search results
29        type: object
30        properties:
31          isbn13:
32            type: string
33            pattern: ^[0-9]{13}$
34            description: The 13-digit ISBN
```

Listing 6-29 AsyncAPI snipped with a message referring to a JSON Schema and headers

Message and Schema Definition

An important part of the API is the definition of the messages (or events). We will look at how to define a message payload schema for JSON and Apache Avro. Along with the payload, we can also define headers and binding-specific fields, like the `partitionkey` in Apache Kafka.

JSON Message and Schema

In Listing 6-4 in section "JSON," a JSON Schema was introduced. This schema can be used directly in the AsyncAPI; see Listing 6-29.

Avro Message and Schema

In Listing 6-10 in section "Avro," an Apache Avro schema was introduced. Listing 6-30 shows the Apache Avro message in the AsyncAPI. The Apache Avro schema can be used in a message; however, in version 3.0.0 it cannot be used in the schema part (without problems) [103].

Apache Avro is as of this writing in version 1.11. However, in the example, we use version 1.9 as the AsyncAPI tooling still only works with this version, and there are no important changes between the versions for this use case [104].

```
components:
  messages:
    BookPurchasedEventAvro:
      headers:
        $ref: '#/components/schemas/MessageHeader'
      payload:
        schemaFormat: 'application/vnd.apache.avro+yaml;version=1.9.0'
        schema:
          type: record
          name: Book
          namespace: com.example
          doc: A schema representing a book and its metadata for the
            library for, e.g., search results
          fields:
            - name: isbn13
              type: string
              doc: The 13-digit ISBN, e.g., 9781234567890
            - name: title
              type: string
              doc: The title of the book
            - name: coverThumbnailWebP
              type:
                - "null"
                - bytes
              default: null
              doc: Base64 encoded WebP image of the book cover thumbnail
            - name: authors
              type:
                type: array
                items:
                  type: record
                  name: Author
                  fields:
                    - name: name
                      type: string
                      doc: 'The full name of the author with titles, e.g.:
                        "Eric Evans", "Dr. John Smith"'
                    - name: portraitThumbnailWebP
                      type:
                        - "null"
                        - bytes
                      default: null
                      doc: Base64 encoded WebP image of the author's
                        portrait thumbnail
```

```
42            doc: List of authors of the book
43          - name: numberOfTextPosition
44            type: long
45            doc: The number of text position in the book
46          - name: category
47            type:
48              type: enum
49              name: Category
50              symbols:
51                - NON_FICTION
52                - FICTION
53            doc: The main category of the book
54          - name: tags
55            type:
56              type: array
57              items: string
58            doc: List of tags associated with the book (top 20)
```

Listing 6-30 AsyncAPI snipped with Avro schema

Protbuf Message and Schema

Listing 6-31 represents a snipped AsyncAPI that contains the Protobuf example from section "Protobuf" (Listing 6-9). Here it is notable that the Protobuf code is still in the IDL as this cannot be converted to YAML or JSON.

Headers

In Listing 6-27 we saw a message `BookPurchasedEvent` with headers and a payload; in Listing 6-29, we see a more realistic example.

Headers are defined in JSON Schema; they are, however, normally mapped to native headers with key values[20] [28].

Security Schemes

Like OpenAPI, AsyncAPI also supports security schemas. An example is shown in Listing 6-32. AsyncAPI supports User/Password, API key, certificates, end-to-end encryption, HTTP Basic authentication, API Keys over HTTP headers, JWT Bearer over HTTP, OAuth 2.0, and SASL [105].

Different Scenarios for AsyncAPI

In section "Communication and APIs," we discussed the different interfaces and where they appear. All AsyncAPI examples until now were based on a producer exposing events on a broker to third parties. We only described the view of a potential consumer: The AsyncAPI in Listing 6-26 only has an operation `BookPurchasedEventReceive`. The operation `BookPurchasedEventPublisch`

---

[20] According to the specification, "The headers MAY be subdivided into protocol-defined headers and header properties defined by the application which can act as supporting metadata" [28].

Definition of Asynchronous Interfaces

```yaml
channels:
  BookPurchasedEventProtobufChannel:
    description: Channel where messages are stored when a book was
      purchased in Protobuf
    address: book-purchased-protobuf
    messages:
      BookPurchasedEventProtobuf:
        $ref: '#/components/messages/BookPurchasedEventProtobuf'
    bindings:
      kafka:
        topic: 'book-purchased-protobuf'
        bindingVersion: '0.5.0'
components:
  messages:
    BookPurchasedEventProtobuf:
      headers:
        $ref: '#/components/schemas/MessageHeader'
      payload:
        schemaFormat: application/vnd.google.protobuf;version=3
        schema: |
          syntax = "proto3";

          package Example.Library;

          // Book represents a book and its metadata for the library,
            e.g., for search results
          message BookPurchasedEventProtobuf {
            // The 13-digit ISBN
            string isbn13 = 1;

            ...
          }
          ...
```

Listing 6-31 AsyncAPI snipped with a Protobuf schema

obviously needs to exist as well. This is, however, what the other side needs to do, and it was intentionally not included. Like this, the AsyncAPI describes what consumers can do. This is visualized in Figure 6-10. It is the beginning of a so-called service mesh architecture. Each team exposes its data for which it is responsible.

This is a mindset shift where each team is held responsible for clean and well-versioned data. This is the countermovement from centralized data teams and data lakes, where a single team pulls all the data together and cleans them to finally provide them again.

The benefit is that these Data Products are exposed by the teams that know the data best and, if they see their Data Products as first-class citizens, they will try to create clean and usable data. This should eliminate the part of data cleaning and interpretation that is a time-consuming part of a central data lake approach [66, 106].

```yaml
asyncapi: 3.0.0
info:
    title: Streetlights Kafka API
    version: 1.0.0
servers:
    scram-connections:
        host: 'test.mykafkacluster.org:18092'
        protocol: kafka-secure
        description: Test broker secured with scramSha256
        security:
            - $ref: '#/components/securitySchemes/saslScram'
    mtls-connections:
        host: 'test.mykafkacluster.org:28092'
        protocol: kafka-secure
        description: Test broker secured with X509
        security:
            - $ref: '#/components/securitySchemes/certs'
components:
    securitySchemes:
        saslScram:
            type: scramSha256
            description: Provide your username and password for SASL/SCRAM
↪ authentication
        certs:
            type: X509
            description: Download the certificate files from service
↪ provider
```

Listing 6-32 AsyncAPI with security schema from documentation [105]

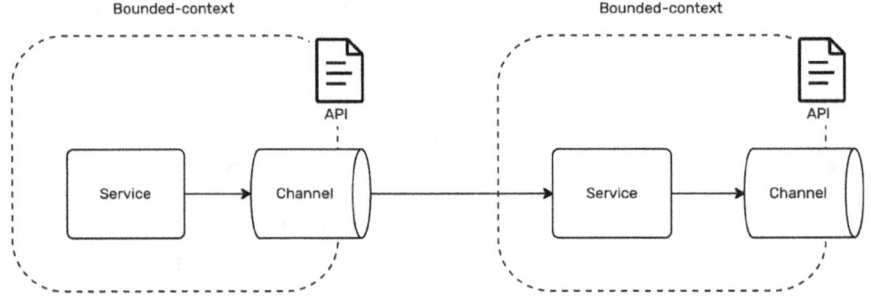

Figure 6-10 Data products: Each context exposes its own data as a contract

> **More on that later**
>
> We will discuss the data mesh concept in section "Data Mesh Concept."

Along with the explained approach, others can be taken:

Broker perspective It is also possible to describe all possible operations on a message broker. This is, for example, what would happen if an automated export were used to export all channels on a broker (e.g., all topics on a Apache Kafka broker)?
This view is very technical, as it focuses on the infrastructure that provides something.

Consumer perspective An AsyncAPI can also be described for a consumer and only show what a consumer is using, with multiple brokers and multiple bounded contexts included.
This approach would not be reusable. It would only be usable for a single consumer who only has a documentation perspective to know what is consumed. This can be useful, if an automated report is possible, to know the state of a system for security reviews, for example.

Producer perspective An AsyncAPI can also be described from the producer's point of view. It would then only show how the producer publishes to a broker. This would be the same as the consumer perspective, and it probably only makes sense if it can be automated for, for example, security reviews.

Brokerless Async Communication

Besides broker-based asynchronous communication, sometimes asynchronous communication is also used without a broker, as discussed in section "Communication and APIs."

An example of this is a Webhook in HTTP or an AMQP implementation without a broker (however, these brokerless/peer-to-peer uses that are possible in AMQP do not seem to be widely adopted).

Schema Registry in AsyncAPI

In section "Schema Registry," we will extend the AsyncAPI with Schema Registry information. However, this is done not in the design phase but in the implementation phase and will therefore be performed later.

What AsyncAPI Misses

Even though AsyncAPI is currently the best standard to represent asynchronous APIs, it is not perfect.

Before we look at different viewpoints, note that an AsyncAPI can have (section "Different Scenarios for AsyncAPI"). Having different possible viewpoints of a specification file without explicit guidance is not trivial, and the lack of an explicit recommendation or labeling makes understanding the files more complex.

Delivery guarantees are very important when handling asynchronous messages. Some protocols define delivery guarantees explicitly as a quality requirement (e.g., MQTT has the Quality of Service (QoS) field [79, 81, 107]). But these are at most an afterthought when using AsyncAPI [81, 108]. In section "Schema Registry," we will look at how to handle schema versioning. We will also see that this is not perfectly handled with AsyncAPI.

This results in some problems not being perfectly fixed with the AsyncAPI. Therefore, it is important to document these important parts as descriptions.

CloudEvents

CloudEvents is a standard introduced by the Cloud Native Computing Foundation (CNCF), part of the Linux Foundation. The standard proposes a standard field in an event and has mappings to protocols to map these fields. The goal is to have simplified event routing and transformation between different protocols. Besides the standard, the Serverless Working Group of CNCF works on various SDKs to make it easier to work with the standard [80, 109]. The standard has adopted by prominent vendors with products like Amazon EventBridge, Azure Event Grid,[21] Google Cloud Eventarc, Alibaba Cloud EventBridge, and SAP S/4HANA [109–114].

An example of a CloudEvent serialized in JSON is visualized in Listing 6-33. By default, the specification establishes an envelope around an event with metadata. Bindings can be used with a protocol. Binary bindings will remove this JSON-style envelope and pass the fields in protocol-native ways. However, there is also a "structured mode," which allows for the passing of events, as in Listing 6-33 in the JSON format. This is, however, an overhead as we have two schemas in one (in the example, an XML inside a JSON).

One useful example is the mapping of an event from Apache Kafka to HTTP to send an event from the internal infrastructure to a partner. The standard has bindings for both protocols and allows the mapping of all specific fields to other fields (in binary mode). Listing 6-34 shows the HTTP binding of an event, and Listing 6-35 is the Apache Kafka counterpart. Besides these standard fields, there are also clear definitions of what to do if some fields are missing. For example, if an event has no explicit `partitionkey` (key in Apache Kafka), the CloudEvents `id` should be used.

CloudEvents can be integrated in multiple ways into an AsyncAPI depending on the mode (binary or structured) and with specific properties depending on the

---

[21] They use their own envelope format https://aws.amazon.com/blogs/compute/sending-and-receiving-cloudevents-with-amazon-eventbridge/.

```
1   {
2       "specversion" : "1.0",
3       "type" : "com.github.pull_request.opened",
4       "source" : "https://github.com/cloudevents/spec/pull",
5       "subject" : "123",
6       "id" : "A234-1234-1234",
7       "time" : "2018-04-05T17:31:00Z",
8       "comexampleextension1" : "value",
9       "comexampleothervalue" : 5,
10      "datacontenttype" : "text/xml",
11      "data" : "<much wow=\"xml\"/>"
12  }
```

Listing 6-33 Example from specification of a CloudEvent serialized as JSON [115]

```
1   POST /someresource HTTP/1.1
2   Host: webhook.example.com
3   ce-specversion: 1.0
4   ce-type: com.example.someevent
5   ce-time: 2018-04-05T03:56:24Z
6   ce-id: 1234-1234-1234
7   ce-source: /mycontext/subcontext
8   .... further attributes ...
9   Content-Type: application/json; charset=utf-8
10  Content-Length: nnnn
11
12  {
13      ... application data ...
14  }
```

Listing 6-34 Example from specification of binary mode mapping of an event with an HTTP POST request [116]

protocol binding. An unofficial guide can be found on GitHub[22] as the official way of integrating it remains a matter of discussion at the time of writing.[23]

Advantages and Disadvantages of Different Schemas and Protocols

Earlier, in sections "Data Formats and Their Schemas, Definition of Synchronous Interfaces, and Definition of Asynchronous Interfaces", we discussed the data formats and the protocols on how data exchanges can be defined.

Now let us see how the data exchanges can be used in combination with data formats and protocols.

---

[22] https://github.com/Lazzaretti/asyncapi-with-cloudevents-traits
[23] https://github.com/cloudevents/spec/issues/1276

```
1   ------------------ Message -------------------
2
3   Topic Name: mytopic
4
5   ------------------ key ----------------------
6
7   Key: mykey
8
9   ------------------ headers ------------------
10
11  ce_specversion: "1.0"
12  ce_type: "com.example.someevent"
13  ce_source: "/mycontext/subcontext"
14  ce_id: "1234-1234-1234"
15  ce_time: "2018-04-05T03:56:24Z"
16  content-type: application/avro
17  .... further attributes ...
18
19  ------------------ value --------------------
20
21  ... application data encoded in Avro ...
22
23  ---------------------------------------------
```

Listing 6-35 Example from specification of binary mode mapping of an event to Kafka [117]

Table 6-3 Comparison of data formats across communication protocols

	Synchron			Asynchron	
	REST	GraphQL	gRPC	AMPQ	Kafka
JSON	Native	Native		Native	Common
XML	Common			Common	Common
Protobuf	Common		Native	Common	Common
AVRO				Common	Native

Table 6-3 shows the possible combinations of protocols and schemas. Let us discuss the use cases in which the respective combinations can be used.

The advantages and disadvantages of specific approaches can only be used under certain circumstances. Therefore, we will use some typical situations where the data formats and the protocols can be used effectively.

Let us assume the following development situations:

- **Development of prototypes in a small team**
  The team members know REST, GraphQL, and RabbitMQ.
  The team develops small prototypes in the financial consultancy area. The data are volatile and the interfaces unreliable.
- **Development of an enterprise product in multiple teams**
  The members of the teams know Apache Kafka, Apache Avro, REST, and gRPC.

The teams are highly independent in their design decisions. What's more, they work in a globally distributed company. This means that the teams have a hard time communicating with each other.

- **Development of an anticorruption layer in a platform team**
  The team members know Apache Kafka, Apache Avro, REST, and GraphQL. The platform team supports the steady-stream teams, but it is hard to be business-agnostic. But only the steady-stream teams are permitted to implement business logic.
- **Development of a data-driven application in an expert team**
  The application is data-driven, and the data are not well defined. The data definitions are constantly changing. The team knows GraphQL and REST. The application collects the data assets of a large enterprise. What they mean and in what combination the data can be used are unclear.

Schemaless Versus Schemafull with Respect to the Single Data Formats

As discussed in section "Data Formats and Their Schemas,", schemafull definitions should be preferred over schemaless definitions for maintainable and reliable software.

Schemaless interfaces might be preferable when a schema cannot be predicted, such as in data-driven applications with volatile data definitions. Thus, using GraphQL in such situations might be helpful. In a data-driven application, all data can be provided. The client's request alone defines the data used by the client. Those approaches must be used carefully because data-access rights can only be applied with considerable effort and contradict the freedom the access protocol offers.

Schemafull approaches like REST or gRPC are more suitable for well-defined data with high requirements for data access rights.

Schemaless approaches are better suited for the following scenarios:

- Development of prototypes in a small team
- Development of a data-driven application in an expert team

Schemafull approaches must be used in the following scenarios:

- Development of an enterprise product in multiple teams
- Development of an anticorruption layer in a platform team

Read and Write Functions and Their Advantages and Disadvantages

Usually, web applications need to read data more than write them. Imagine the online library where a user selects books from a catalog with thousands of books. The books in the online library must be fetched to show them to the user. The user chooses a couple of them for reading. Read functions are usually needed in user interactions, which await an intermediate response. Those behaviors should be implemented with synchronous calls.

The reading of books according to the search criteria must be fast, whereas writing books for the user's reading list can take longer. The user needs time to navigate to the book for reading.

Such a behavior is supported primarily by REST and GraphQL. It can be applied in all four scenarios.

We have already discussed the advantages of asynchronous behavior. Information can be written fast when using a broker as an intermediary for events or messages. Fast writing is one of the most valuable advantages of brokers.

> **Name of Kafka broker**
>
> Kafka as an event broker was developed by *Jay Kreps* in 2011. He named the broker Kafka after the German author *Franz Kafka*, because "it is a system for writing" and he liked Kafka's work [118].

Asynchronous approaches should be used when highly decoupled service-to-service communication is necessary or when the business process is asynchronous. Writing new catalog entries to the catalog when a new book arrives is an example of both. The purchasing system needs to be highly decoupled from the catalog management system, and the purchase business process should behave asynchronously with the catalog system.

Tooling for Linting

The next point in selecting the correct protocol and schema should be the availability of linters. A linter is a tool that tests the syntactical correctness of specifications based on a rule set. It tests the specification statically, so no additional runtime or implementation is necessary. Syntactic correctness covers not only the requirements of the standards but custom rules as well. Those rules are usually formulated in API guidelines. A linter can enforce guidelines.

OpenAPI
Several linter tools exist for OpenAPI [119]. Spectral provides the most widely used one. It provides predefined rule sets covering most cases needed [120].

GraphQL
The definitions in GraphQL need to follow specific rules enforced by a linter. GraphOS and Inigo offer linters with a primary set of rules that can be enhanced [121, 122].

gRPC

Linting gRPC can be performed either directly or via Protobuf linting [123, 124]. The tool support is not as broad as for OpenAPI, but it is sufficient.

AsyncAPI

The linters for AsyncAPI are available and reliable. However, it depends on the message format, for example, not all versions of Apache Avro are supported. The tools are ready for use, and the rule sets can be adapted.

The AsyncAPI initiative offers a rule set by themselves, which makes the start especially easy [125]. Redocly offers another easy-to-use linter via CLI [126].

The common linter for OpenAPI Spectral can also be used for AsyncAPI [127]. That makes the introduction of such a toolset easy.

> **More on that later**
>
> We will dive a bit deeper into what linters are, where they should be used, and how to apply them in section "Static Code Tests and Linting." This section only focused on the tooling available for the protocols mentioned in this section.

Compression During Transit

During transit, data can be or, better, should be compressed. Binary data are an exception. Because they are usually highly efficiently packed, compression cannot be efficiently applied.

For HTTPs, the most popular compression standard is Gnu ZIP (gZIP). It was originally written by *Jean-Loup Gailly*. *Marc Adler* wrote the decompression part [128].

Usually developers do not need to perform compression themselves because it is done by servers and clients automatically [129].

For Kafka, Apache Avro can be used uncompressed; since it is in binary format, it is already highly efficient. The message size can be reduced by 50% in comparison to JSON [130].

Compression in AMQP is not delivered out of the box, though certain sample implementations exist to use compression standards such as gZIP [131].

Messages should be compressed in service communications to optimize bandwidth usage, enlarge transportation speed, and reduce latency. Compressing can be left out in prototype environments, when it is important to follow message paths.

North–South and East–West Communication

Discussions on which protocol would be appropriate must include examining whether the communication contains a user interaction. The discussion can be

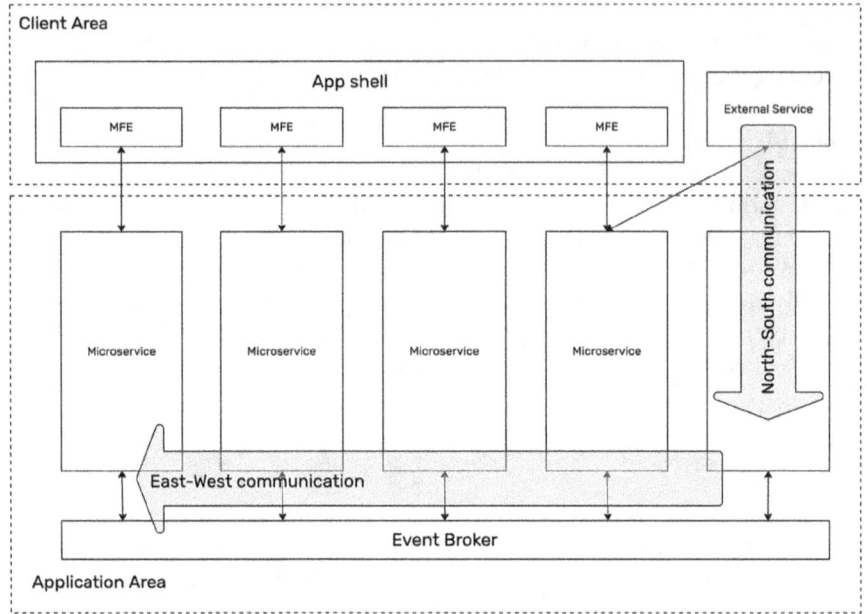

Figure 6-11 North–south and east–west communication types

supported by imagining a north–south and east–west map of a standard architectural model (Figure 6-11).

Usually, clients (frontends, typically web pages) are drawn at the upper border of a model, business logic in the middle layer, and the persistence layer at the lower end, as a map shows, from north to south. Services are located from left to right or, as shown on a map, from west to east. Therefore, we can speak about north–south communication, including user interactions, and east–west communication for server-to-server communication [132].

Synchronous communication is suited for north–south communication because it includes user interactions: Requests receive synchronous feedback. Asynchronous communication is especially suited for east–west communication because a direct response is usually not necessary, and the requester's process flow of the requester should not be disturbed by the invoked server's unavailability. Such exceptions to the rules need to be considered, too.

Let us see how all the perspectives discussed in selecting the proper data format and protocol can be combined for a decision.

Decision Tree Use of Schemas and Protocols

Knowing what protocol and schema can be used in a particular scenario is essential. Figure 6-12 shows a decision tree depending on the nature of the application to be implemented.

Advantages and Disadvantages of Different Schemas and Protocols

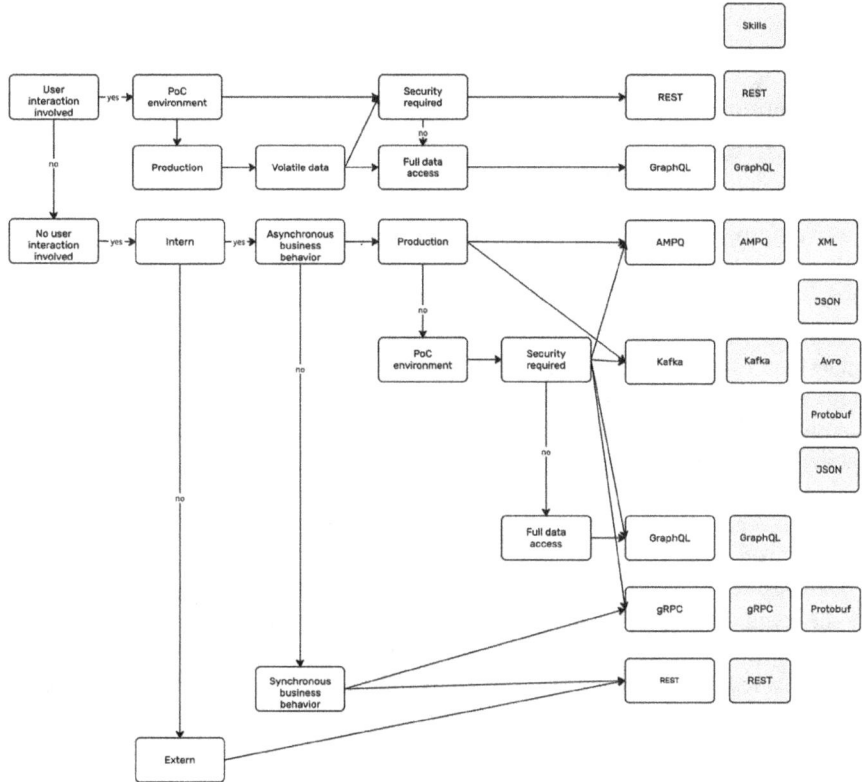

Figure 6-12 Decision tree for format and protocol

The first question converns the involvement of users.

If user interaction is involved, the next question is whether the current implementation is a Proof of Concept (PoC) or intended to be deployed in a production environment.

If volatile data need to be handled in a production environment and no security is required, GraphQL can be used. If data access security is necessary in either environment, REST must be used.

If no user interaction is required, the question is whether the communication is internal or external to other applications or companies. For external communication, REST must be used.

The business process can be synchronous or asynchronous. For synchronous business processes, gRPC or REST can be used.

For asynchronous business processes in a production environment, Apache Kafka or AMQP is recommended. Even gRPC can be used for PoC implementations. Without data-access security requirements, a PoC implementation can use GraphQL even for asynchronous business processes.

If we apply the decision tree to the four scenarios, we get the following decisions:

- **Development of prototypes in a small team**
  The team settled on REST and RabbitMQ with JSON. They work in a financial consultancy area with high security requirements. Even though it is a proof of concept, GraphQL seems unsuitable.
- **Development of an enterprise product in multiple teams**
  The team members know Apache Kafka, Apache Avro, REST, and gRPC.
  The teams are highly independent in their design decisions. Even more, they work in a globally distributed company. This means that the teams have difficulties communicating with each other. The teams are open to using gRPC or REST for service-to-service communication. When they want to use gRPC, the teams responsible for the services' communicating need to synchronize. However, for long running connections and heavy calculations, it might be the better choice compared to events. For asynchronous communication, they must use Apache Kafka. They can opt for JSON or Apache Avro as the message format for asynchronous communication, although Apache Avro is highly recommended.
- **Development of an anticorruption layer in a platform team**
  The team members know Apache Kafka, Apache Avro, REST, and GraphQL. The platform team supports the steady-stream teams, but it is challenging to be business-agnostic. Only steady-stream teams are permitted to implement business logic. The platform team decides to use Apache Kafka with Apache Avro. For teams that need to use the platform but do have no knowledge of Apache Kafka, they provide libraries and how-tos.
- **Development of a data-driven application in an expert team**
  The application is data-driven, and the data are not well defined. The data definitions are constantly changing. The team knows GraphQL and REST. The application collects the data assets of a large enterprise. What they mean and in what combination the data can be used are unclear. The team settles on GraphQL. The provided data are not highly sensible. Sensible data are filtered out on the server side and inaccessible via the GraphQL interface.

Typical Antipatterns

After demonstrating how to design and specify an API, we want to look at some antipatterns that should generally be avoided.

Synchronous Events

Often, events or other asynchronous communications are submitted synchronously. This happens mainly because if business people are asked, "Can you accept a small delay until the data are written in the target system?" the answer is: "No! We need to be sure that it is there immediately!"

Typical Antipatterns

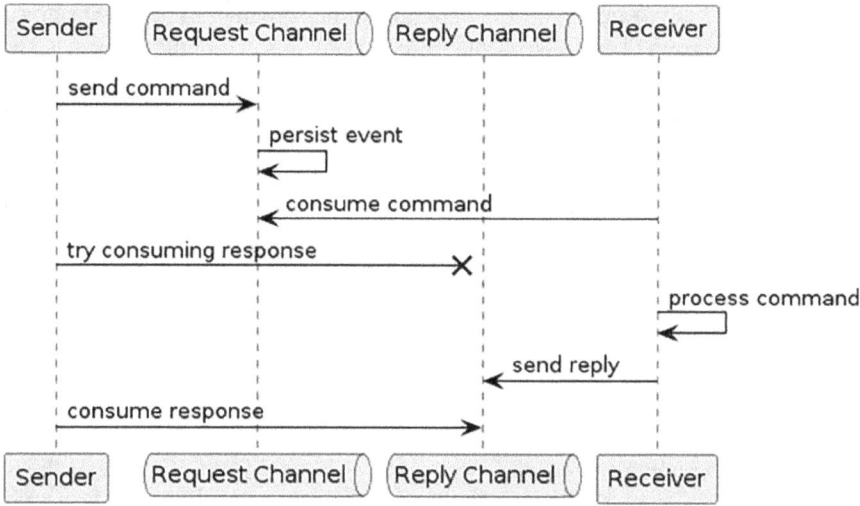

Figure 6-13 Sequence diagram of antipattern synchronous events

Such a conversation often leads to the wrong assumption that synchronous communication is required. However, if you asked, "Should the process stop when the other system is not working?," then the answer would be: "No, it must always work!" This would be an indicator that asynchronous communication is preferred.

Asynchronous Commands

An antipattern can, however, also be seen in the other direction. Synchronous commands that need an answer are submitted over an asynchronous channel. These are then systems that, in code, are not blocked by the sending but actively wait for an answer by polling for a response and are blocked during that period. This result is visualized in Figure 6-13.

This pattern is called "Request-Reply" [133]. There are marginal cases where this is useful; however, a synchronous call would be the more straightforward and elegant solution in many cases.

Messages as Events

Another common mistake is sending messages (but not events) that are called events, for example, a command. These are often people new to the EDA that call clear point-to-point messages events. These messages are designed only for a single consumer and can probably never be used by another system but are called events.

An example here can be a message/command that is called `createThumbnail`. The `createThumbnail` message comes after a book is listed in the library catalog, and the thumbnail needs to be processed. It is a command in an orchestration-style integration, as the sender clearly says what the consumer needs to do. The message is intended for an image-processing function. This can be made asynchronous as it may take some time, and the caller does not need the result.

However, this is not an event; it is a point-to-point message. If this should be represented as an event, it may be reformulated as a `catalogEntryCreated` event. This event happens when a book gets listed in the catalog. The image processor can now take this event and know that the book must get a thumbnail. This will now be a choreography.

Huge Events

A frequent problem of teams is that they want to put everything into an event. The event then gets huge, containing, for example, images and Portable Document Format (PDF) attachments.

The problem is that an event is no longer usable for many consumers, as much data are passed that are not needed. On all common message brokers, a message must be read as a whole, and parts cannot be skipped.

A solution here is to use the Claim-Check pattern [133]. The pattern addresses the problem by storing only an external reference to the big data content. In the event itself, a reference is passed to a large amount of data. The most needed fields can still be passed in the event itself; however, they are not big attachments. The big attachments are then either pulled from the original service by the reference (e.g., a key or a URL). Even better is to use an external storage (e.g., an object storage like an Amazon S3) to store the external file. This still helps to reduce the coupling between the services. Figure 6-14 visualizes the Claim-Check pattern.

What is too big and what is not too big is really difficult to say in general. Many providers have limits on their infrastructure to guarantee performance, for example, Amazon Kinesis Streams with 1 MB payload [134], or Confluent Cloud Kafka has a default limit of 2 MB [135]. In general, the authors recommend that the content of an event be used by 80% of the default processing of each consumer. If this is not the case, there should be a discussion about whether the data are needed in the event or if they can be loaded later on with a Claim-Check pattern or over another API. The discussion can be done using the Event Storming workshop format, as discussed in section "Event Storming".

AI Generation of API Definitions: Chances and Limitations

Generative AI is widely used in programming. It is used as an assistant in programming and makes suggestions for better formulations of common problems.

AI Generation of API Definitions: Chances and Limitations

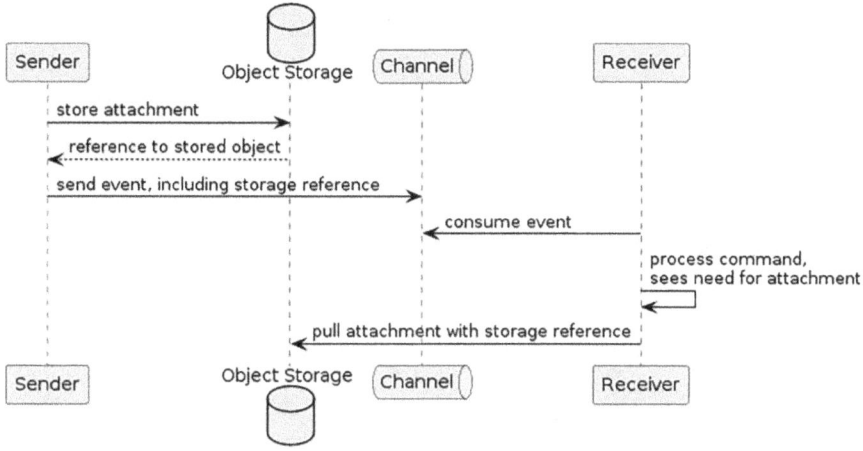

Figure 6-14 Sequence diagram of Claim-Check pattern

Using AI generating APIs specifications is not common. However, it is quite useful, as we already showed in section "Generation of API Definitions with AI."

API definitions like OpenAPI or AsyncAPI can be created from the information of the discussed DDD process (Chapter 5, "API Design Supported by Domain–Driven Design").

The definitions can be generated using a general chatbot of an large language model (LLM). To set the input first, the stage has to be set by a prompt (Listing 6-36).

After prompting with the stage, the API can be generated. The API can be prompted with pictures of the DDD process (Figures 5-17 and 5-39). To do so, the LLM must be able to interpret pictures as input.

The entire result is a rather good starting point and can be used for further work. The created example misses good descriptions and needs to be tweaked to be used as a useful API definition. The generated file is accessible on Github.[24]

AsyncAPIs can be generated in a comparable manner.

AI is useful for creating starting points of an API specification. It needs to be limited to the knowledge of the corresponding specification language, for example, OpenAPI. The content can be delivered by the images created during the DDD process.

To produce meaningful results, the AI needs good templates. Those templates can be delivered additionally by language-specific skeletons.

---

[24] https://github.com/Apress/Crafting-Great-APIs-with-Domain-Driven-Design/blob/main/Chapter-6-8/CatalogManagement.yaml

```
Prompt: You are an OpenAPI and DDD expert

AI:
I'll help create a detailed OpenAPI specification for the Catalog
    Management domain based on the provided artifacts. Let's analyze
    the domain model and create a specification that aligns with DDD
    principles.

From the visual glossary and context map, we can identify key
    aggregates and entities:
- Catalog (root aggregate)
- CatalogEntry (entity with relationships)
- Author (entity)
- Title (value object)
- Tag (value object)
- Abstract (value object)
- SearchCriteria (value object)

Let me create an enhanced OpenAPI specification based on the skeleton:



I've created a comprehensive OpenAPI specification that reflects the
    domain model shown in the diagrams. Key aspects of the design
    include:

1. Bounded Context:
- Focuses on Catalog Management context
- Clear aggregate boundaries around Catalog and CatalogEntry

2. Domain Model Alignment:
- Preserves relationships from visual glossary
- Maintains cardinality constraints (e.g., 0..*, 1..*)
- Implements value objects as nested schemas

3. Strategic Design:
- Separates search functionality as its own tag/resource
- Implements proper CRUD operations for aggregates
- Adds proper validation constraints

4. Technical Considerations:
- OpenID Connect security
- Proper versioning
- Comprehensive error responses
- UUID-based identification

Would you like me to explain any specific part of the specification in
    more detail or make any adjustments to better align with your
    requirements?

```

Listing 6-36 Example of AI chat creating an OpenAPI [136]

An OpenAPI and AsyncAPI skeleton is delivered in the code samples of this book, see code samples on GitHub.[25]

Conclusion

In the previous chapter, we saw that modularization was necessary for software development. Interface definitions have existed almost since the beginning. The term "interface" has existed since the 1960s.

Definitions of interfaces that communicate over a network have existed since the 1980s. Today we use, among others, OpenAPI and AsyncAPI as IDLs.

Moreover, to use an interface, we need to define not only the method applied to the data but also the data structures. Data structures used for interfaces can be text-based formats or binary serialized formats. Text-based formats are XML, JSON, and YAML. Apache Avro and Protobuf were introduced as binary serialized formats.

Besides data structures, protocols must be defined by the IDL. For synchronous interfaces, gRPC, GraphQL, or REST can be used. Services-to-service communication can be nicely implemented with gRPC. Web applications, including user interaction, are best implemented with REST.

Asynchronous communications are best defined using AsyncAPI. AsyncAPI supports several protocols, including Apache Kafka and AMQP.

As we have seen, several points must be considered when deciding on a data format and protocol. In the next chapter, we will show how to apply the decision tree and the specifications to the online library example.

Points to Remember

We discussed the following points in this chapter:

- Interfaces hide implementations to service clients.
- Interfaces can be defined by IDLs.
- OpenAPI is a common IDL used to define HTTP interfaces.
- AsyncAPI is a common IDL used to define asynchronous interfaces.
- The used data formats transferred via APIs are JSON, XML, Protobuf, and Apache Avro.
- Synchronous interfaces can be defined with gRPC, GraphQL, or REST.
- The specification defines, besides the data format, the applicable methods, whereas the methods can be protocol specific like HTTP verbs in REST, or they can be defined as procedure names as in gRPC or GraphQL.
- Asynchronous interfaces can be defined using a broker or without one (broker-less).
- A typical event broker is Apache Kafka.

---

[25] https://github.com/Apress/Crafting-Great-APIs-with-Domain-Driven-Design/tree/main/Chapter-6-8

- Because AsyncAPI is independent of the protocol used, specific protocol bindings must be added if the specification is to be used with a certain protocol.
- The specification for asynchronous communication defines, besides the data format, the channels over which the messages are transported and the operations used to process the corresponding messages.
- For synchronous and asynchronous communication, standards must be applied. For example, HTTP response codes must be defined for REST. And for Apache Kafka, corresponding topics and retention period must be defined. The CloudEvents standard was discussed to define the meta information of an event.
- The decision for a data format and a protocol depends on the application, the skills of the teams, and the architectural structures. For user interactions, synchronous communication is to be preferred. For server-to-server interactions, asynchronous communication is to be preferred.

Review Questions

6.1 What kind of specification language is used to define network-based interfaces?

(a) Interfaces can only be defined inside of the program code.
(b) Interfaces are always defined in Python.
(c) Interfaces are always defined in Java.
(d) Interfaces can be defined by an Interface Description Language.

6.2 What specification language is currently used to define synchronous network-based interfaces?

(a) OpenAI
(b) XML
(c) Avro
(d) OpenAPI

6.3 What specification language is currently used to define asynchronous networked-based interfaces?

(a) OpenAI
(b) AsyncAPI
(c) Avro
(d) OpenAPI

6.4 Does AsyncAPI require a specific protocol?

(a) Yes, Kafka
(b) Yes, AMQP
(c) No
(d) Only if the communication is brokerless

6.5 Does north–south communication typically include user interactions?

(a) Yes
(b) No
(c) No, only east–west communication includes user interactions
(d) Sometimes yes, sometimes no, it depends on the architectural approach.

6.6 What influences the decision for a data format and a protocol?

(a) Communication with or without user interaction, team skills
(b) Development environment and data characteristics
(c) Business behavior and external or internal communication
(d) All of the above

References

1. Samokhin V (2017) Liskov substition principle. [Online] Available: https://medium.com/@wrong.about/liskov-substitution-principle-a982551d584a. Visited on 03 Nov 2024
2. R. 1050 (1988) Rpc: Remote procedure call protocal specification. [Online] Available: https://datatracker.ietf.org/doc/html/rfc1050. Visited on 03 Nov 2024
3. Introduction to ONC RPC (2018). [Online] Available: https://www.ibm.com/docs/en/cics-ts/5.5?topic=interfaces-introduction-onc-rpc. Visited on 09 Nov 2024
4. Boughton W (1997) Som and object REXX. [Online] Available: http://www.edm2.com/index.php/SOM_and_Object_REXX. Visited on 03 Nov 2024
5. Rouse M (2011) System object model. [Online] Available: https://www.techopedia.com/definition/1315/system-object-model-som-ibm. Visited on 09 Sep 2024
6. Mansfield J, Clothier J (1995) Distributed computing environment: An architecture for supporting change? Department of Defence, Salisbury, Technical Report
7. Shirley J, Rosenberry W (1995) Microsoft RPC Programming Guide. Nutshell Handbook. O'Reilly and Associates, Sebastopol. ISBN: 978-15-65920-70-5
8. Remote procedure call (RPC) (2022). [Online] Available: https://learn.microsoft.com/en-us/windows/win32/rpc/rpc-start-page. Visited on 09 Nov 2024
9. Sheldon R (2022) Distributed component object model (DCOM). [Online] Available: https://www.techtarget.com/whatis/definition/DCOM-Distributed-Component-Object-Model. Visited on 09 Nov 2024
10. Morrissey W (1996) The Object Technology Casebook: Lessons from Award-Winning Business Applications. Wiley, Hoboken. ISBN: 978-04-71147-17-6
11. Corba history (2024). [Online] Available: https://www.corba.org/history_of_corba.htm. Visited on 09 Nov 2024

12. Christensen E, Curbera F, Meredith G, Weerawarana S (2000) Web services description language (WSDL) 1.0. [Online] Available: https://xml.coverpages.org/wsdl20000929.html. Visited on 03 Nov 2024
13. XML WSDL (2024). [Online] Available: https://www.w3schools.com/xml/xml_wsdl.asp. Visited on 09 Nov 2024
14. Turner D, Oeschger I (2003) Creating XPCOM components, Appendix B: XPCOM API reference. [Online]. Available: https://www-archive.mozilla.org/projects/xpcom/book/cxc/html/appb. Visited on 03 Nov 2024
15. Prunicki A (2011) Apache thrift. [Online]. Available: https://web.archive.org/web/20110723051326/ http://jnb.ociweb.com/jnb/jnbJun2009.html. Visited on 03 Nov 2024
16. Apache thrift (2024). [Online] Available: https://thrift.apache.org/. Visited on 09 Nov 2024
17. Seely S (2022) SOAP Cross Web Service Development Using XML. Pretence Hall PTR, Upper Saddle Rive. ISBN: 01-3090-763-4
18. Arman (2024) Introduction to raml - essential things you need to know! . [Online] Available: https://testfully.io/blog/raml/. Visited on 10 Nov 2024
19. Unlock Raml's benefits with teh best api design and mangement platform (2024). [Online]. Available: https://www.mulesoft.com/api/design/what-is-raml. Visited on 10 Nov 2024
20. Raml version 1.0: Restful API modeling language (2021). [Online] Available: https://github.com/raml-org/raml-spec/commits/master/versions/raml-10/raml-10.md. Visited on 10 Nov 2024
21. Ralphson M (2018) A brief history of the openapi specification (2018). [Online] Available: https://dev.to/mikeralphson/a-brief-history-of-the-openapi-specification-3g27. Visited on 10 Nov 2024
22. Team P (2023) Openapi vs. swagger. [Online] Available: https://blog.postman.com/openapi-vs-swagger/. Visited on 10 Nov 2024
23. Openapi specification v3.0.0 (2017). [Online] Available: https://spec.openapis.org/oas/v3.0.0.html. Visited on 10 Nov 2024
24. OpenAPI Initiative (2021) Openapi specification v.3.1.0. [Online] Available: https://spec.openapis.org/oas/latest.html. Visited on 14 July 2024
25. Gardiner M (2023) Openapi moonwalk 2024. [Online] Available: https://www.openapis.org/blog/2023/12/06/openapi-moonwalk-2024. Visited on 10 Nov 2024
26. Davis C (2019) Cloud Naitive Patterns. Manning, Shelter Island. ISBN: 978-16-17294-29-7
27. Cormier J (2022) What is asyncapi? . [Online] Available: https://bump.sh/blog/what-is-asyncapi. Visited on 10 Nov 2024
28. Asyncapi 3.0.0 (2023). [Online]. Available: https://www.asyncapi.com/docs/reference/specification/v3.0.0. Visited on 10 Nov 2024
29. Fielding RT, Nottingham M, Reschke J (2022) HTTP Semantics, RFC 9110. https://doi.org/10.17487/rfc9110. [Online] Available: https://www.rfc-editor.org/info/rfc9110. Visited on 29 July 2024
30. Oswalt M (2023) Network Programmability and Automation: Skills for the Next-Generation Network Engineer, Adell C, Lowe SS, Edelman J (eds.), 2nd edn. O'Reilly, Beijing, 799 pp. Includes index. ISBN: 978-10-98110-83-3
31. Kleppmann M (2017) Designing Data-Intensive Applications: The Big Ideas Behind reliable, Scalable, and Maintainable Systems, 1st edn. O'Reilley, Beijing, 1590 pp. Hier auch später erschienene, unveränderte Nachdrucke. ISBN: 14-9190-311-2
32. Josefsson S (2006) The Base16, Base32, and Base64 Data Encodings, RFC 4648. https://doi.org/10.17487/rfc4648. [Online] Available: https://www.rfc-editor.org/info/rfc4648. Visited on 09 Sep 2024
33. Specification lapache avro (2022). [Online] Available: https://avro.apache.org/docs/1.11.1/specification/. Visited on 24 Aug 2024
34. Posix.1-2024 - the open group base specifications issue 8 (2024). [Online] Available: https://pubs.opengroup.org/onlinepubs/9799919799/. Visited on 13 Sep 2024
35. Ecma-404, 2nd edn (2017). [Online]. Available: https://ecma-international.org/wp-content/uploads/ECMA-404_2nd_edition_december_2017.pdf. Visited on 20 Aug 2024

References

36. Ecma-262, 15th edn (2024). [Online] Available: https://ecma-international.org/wp-content/uploads/ECMA-262_15th_edition_june_2024.pdf. Visited on 20 Aug 2024
37. Javascript technologies overview - javascript (2024). [Online] Available: https://developer.mozilla.org/en-US/docs/Web/JavaScript/JavaScript_technologies_overview Visited on 20 Aug 2024
38. Amundsen M (2017) RESTful Web Clients: Enabling Reuse Through Hypermedia, 1st edn. O'Reilly Media, Sebastopol, 1348 pp. Includes bibliographical references and index. ISBN: 978-14-91921-85-2
39. Wright A, Andrews H, Hutton B, Dennis G (2022) Json schema: A media type for describing json documents. [Online] Available: https://json-schema.org/draft/2020-12/json-schema-core. Visited on 17 Aug 2024
40. Wright A, Andrews H, Hutton B (2022) Json schema validation: A vocabulary for structural validation of JSON. [Online]. Available: https://json-schema.org/draft/2020-12/json-schema-validation. Visited on 17 Aug 2024
41. JSON schema - specification links (2024). Source on https://github.com/json-schema-org/website/blob/6a544fa173a6691c616fe358eb35c7076286f316/pages/specification-links.md?plain=1. [Online] Available: https://json-schema.org/specification-links. Visited on 17 Aug 2024
42. Decoupling from IETF (2022). [Online] Available: https://github.com/json-schema-org/json-schema-spec/blob/0c9a8a05fa931defb5b4399bf145233efe09f35b/adr/2022-09-decouple-from-ietf.md. Visited on 17 Aug 2024
43. Node.js foundation and JS foundation merge to form openjs foundation (2019). [Online] Available: https://www.linuxfoundation.org/press/press-release/node-js-foundation-and-js-foundation-merge-to-form-openjs-foundation. Visited on 17 Aug 2024
44. JSON-schema-joins-openjs-foundation (2022). [Online] Available: https://openjsf.org/blog/json-schema-joins-openjs-foundation. Visited on 17 Aug 2024
45. Google json style guide (2018). See source: https://github.com/google/styleguide. [Online] Available: https://google.github.io/styleguide/jsoncstyleguide.xml. Visited on 24 Aug 2024
46. Microsoft azure rest API guidelines (2024). [Online] Available: https://github.com/microsoft/api-guidelines/blob/0674b4c3b39b9b400510aace0c06b795ce6138cc/azure/Guidelines.md. Visited on 24 Aug 2024
47. JSON schema naming and design rules technical specification (2022). Found over https://unece.org/trade/documents/2023/07/session-documents/item-5-c-standards-noting-api-technical-specification. [Online] Available: https://unece.org/sites/default/files/2023-11/API-TECH-SPEC_JSON_Schema_NDR_version1p0.pdf. Visited on 24 Aug 2024
48. Zalando restful API and event guidelines (2024). Source at https://github.com/zalando/restful-api-guidelines. [Online] Available: https://opensource.zalando.com/restful-api-guidelines/. Visited on 24 Aug 2024
49. Evans CC (2001) Yaml draft 0.1. [Online] Available: https://web.archive.org/web/20010603012942/. http://groups.yahoo.com/group/sml-dev/message/4710. Visited on 30 Nov 2024
50. Ingerson B, Evans CC, Ben-Kiki O (2001) Yet another markup language (YAMl) 1.0. [Online] Available: https://yaml.org/spec/history/2001-08-01.html. Visited on 20 Aug 2024
51. YAML ain't markup language (YAML™) version 1.2 (2021). [Online] Available: https://yaml.org/spec/1.2.2/. Visited on 19 Aug 2024
52. Bray T, Sperberg-McQueen CM (1996) Extensible markup language (XML). [Online] Available: https://www.w3.org/TR/WD-xml-961114.html. Visited on 30 Nov 2024
53. Cowan J et al (2008) Extensible Markup Language (XML) 1.0, 5th edn. [Online] Available: https://www.w3.org/TR/xml/. Visited on 23 Nov 2024
54. Bray T, Paoli J, Sperberg-McQueen CM, Maler E, Yergeau F (2013) Extensible markup language (XML) 1.0, 5th edn. [Online] Available: https://www.w3.org/TR/2008/REC-xml-20081126/. Visited on 17 Aug 2024
55. Walsh N (2009) Errata in rec-xml-20081126. [Online] Available: https://www.w3.org/XML/xml-V10-5e-errata. Visited on 17 Aug 2024

56. Harold ER, Means WS (2007) XML in a Nutshell, 3rd edn. [Repr.] O'Reilly, Beijing, 689 pp. ISBN: 978-05-96007-64-5
57. XML External Entity Prevention Cheat Sheet, OWASP Foundation (2024). [Online] Available: https://cheatsheetseries.owasp.org/cheatsheets/XML_External_Entity_Prevention_Cheat_Sheet.html. Visited on 24 Nov 2024
58. Morgan TD, Ibrahim OA (2014) XML schema, DTD, and entity attacks. [Online] Available: https://www.nccgroup.com/media/bndihh0n/xmldtdentityattacks.pdf. Visited on 24 Nov 2024
59. Owasp top ten, OWASP Foundation (2021). [Online] Available: https://owasp.org/www-project-top-ten/. Visited on 14 Aug 2024
60. XML Security Cheat Sheet, OWASP Foundation (2024). [Online] Available: https://cheatsheetseries.owasp.org/cheatsheets/XML_Security_Cheat_Sheet.html. Visited on 24 Nov 2024
61. Xml security - owasp cheat sheet series (2024). [Online] Available: https://cheatsheetseries.owasp.org/cheatsheets/XML_Security_Cheat_Sheet.html. Visited on 17 Aug 2024
62. Xml naming and design rules for CCTS 2.01 (2021). Found in: https://unece.org/trade/uncefact/xml-ndr. [Online] Available: https://unece.org/sites/default/files/2023-10/XMLNamingAndDesignRulesV2.1.1.pdf. Visited on 24 Aug 2024
63. Xml naming and design rules technical specification (2009) Found in: https://unece.org/trade/uncefact/xml-ndr. [Online] Available: https://unece.org/DAM/cefact/xml/UNCEFACT+XML+NDR+V3p0.pdf. Visited on 24 Aug 2024
64. Google XML Document Format Style Guide (2008). [Online] Available: https://google.github.io/styleguide/xmlstyle.html. Visited on 26 Nov 2024
65. History |protocol buffers documentation. [Online] Available: https://protobuf.dev/history/. Visited on 30 Nov 2024
66. Bellemare A (2023) Building an Event-Driven Data Mesh, Patterns for Designing and Building Event-Driven Architectures, 1st edn. O'Reilly, Beijing, 1 p. ISBN: 10-9812-757-9
67. License for protobuf. [Online] Available: https://ptolemy.berkeley.edu/ptolemyII/ptII11.0/ptII/lib/protobuf-license.htm. Visited on 20 Aug 2024
68. The 3-clause bsd license. [Online] Available: https://opensource.org/license/BSD-3-Clause. Visited on 20 Aug 2024
69. Protobuf/license at protocolbuffers/protobuf (2018). [Online] Available: https://github.com/protocolbuffers/protobuf/blob/9e080f7ac007b75dacbd233b214e5c0cb2e48e0f/LICENSE. Visited on 19 Aug 2024
70. Protocol buffers edition 2023 language specification (2024). Source https://github.com/protocolbuffers/protocolbuffers.github.io/blob/main/content/reference/protobuf/edition-2023-spec.md. [Online] Available: https://protobuf.dev/reference/protobuf/edition-2023-spec/. Visited on 17 Aug 2024
71. Protobuf editions overview (2023). [Online] Available: https://protobuf.dev/editions/overview/. Visited on 17 Aug 2024
72. Cutting D (2009) [proposal] new subproject: Avro, Apache Mail Archives. [Online] Available: https://lists.apache.org/thread/z571w0r5jmfsjvnl0fq4fgg0vh28d3bk. Visited on 11 Sep 2024
73. Avro/license.txt at apache/avro (2024). [Online] Available: https://github.com/apache/avro/blob/97c8ba829b4d264844bbd92588b9c562ed6efeee/LICENSE.txt. Visited on 19 Aug 2024
74. Fibre channel - methodologies for jitter specification (1998). [Online] Available: https://web.archive.org/web/20160731025808/. http://www.t11.org/ftp/t11/member/fc/jitter_meth/98-055v6.pdf. Visited on 22 Feb 2025
75. Thombre S (2018) Network jitter analysis with varying TCP for internet communications. In: 2018 3rd International Conference for Convergence in Technology. [Online] Available: https://ieeexplore.ieee.org/document/8529816/authors#authors. Visited on 21 Feb 2025
76. Delete - http |MDN (2024). Source https://github.com/mdn/content/blob/main/files/en-us/web/http/methods/delete/index.md?plain=1. Mozilla Foundation. [Online] Available: https://developer.mozilla.org/en-US/docs/Web/HTTP/Methods/DELETE. Visited on 24 Feb 2025

References

77. Handling concurrency conflicts, Microsoft. [Online] Available: https://learn.microsoft.com/en-us/ef/core/saving/concurrency?tabs=data-annotations. Visited on 24 Feb 2025
78. Message delivery guarantees (2025). [Online] Available: https://docs.confluent.io/kafka/design/delivery-semantics.html. Visited on 21 Feb 2025
79. What is MQTT quality of service (QOS) 0,1, & 2? - MQTT essentials: Part 6, HiveMQ (2024). [Online] Available: https://www.hivemq.com/blog/mqtt-essentials-part-6-mqtt-quality-of-service-levels/. Visited on 15 Feb 2025
80. Basig L, Lazzaretti F, Aebersold R, Zimmermann O (2021) Reliable event routing in the cloud and on the edge: an internet-of-things solution in the agetech domain. In: Software Architecture. Springer International Publishing, Berlin, pp 243–259. ISBN: 978-30-30860-44-8. https://doi.org/10.1007/978-3-030-86044-8_17
81. Basig L, Lazzaretti F (2011) Reliable Messaging Using the Cloudevents Router. OST - University of Applied Sciences of Eastern Switzerland. Visited on 15 Feb 2025
82. gRPC Authors, Grpc. [Online] Available: https://grpc.io/. Visited on 17 Nov 2024
83. Sazanavets F (2022) Microservices Communication in .NET Using gRPC. Packt, Birmingham. ISBN: 978-18-03236-43-8
84. Branadhorst J (2019) The state of GRPC in the browser. [Online] Available: https://jbrandhorst.com/post/state-of-grpcweb/. Visited on 21 Dec 2019
85. GRPC-web: Typed frontend development (2021). [Online] Available: https://github.com/improbable-eng/grpc-web. Visited on 22 Dec 2024
86. Graphql - a query language for your API (2024). [Online] Available: https://graphql.org/. Visited on 17 Nov 2024
87. Lyon W (2022) Full Stack GraphQL Applications. Manning, Shelter Island. ISBN: 978-16-17297-03-8
88. Fowler M (2010) Richardson maturity model. [Online]. Available: https://martinfowler.com/articles/richardsonMaturityModel.html. Visited on 25 Jun 2024
89. Openapi 3.0 tutorial (2023). [Online] Available: https://support.smartbear.com/swaggerhub/docs/en/get-started/openapi-3-0-tutorial.html. Visited on 22 Dec 2024
90. Describing API security (2023). [Online] Available: https://learn.openapis.org/specification/security.html. Visited on 21 Dec 2024
91. What is an API key? (2024). [Online] Available: https://aws.amazon.com/what-is/api-key/. Visited on 21 Dec 2024
92. Oauth2.0 (2024). [Online] Available: https://oauth.net/2/. Visited on 21 Dec 2024
93. What the difference between RPC and rest? (2024). [Online] Available: https://aws.amazon.com/compare/the-difference-between-rpc-and-rest/. Visited on 22 Dec 2024
94. Vadecha A (2024) Graphql vs. rest APIS: A comprehensive comparison for developers (2024). [Online] Available: https://hygraph.com/blog/graphql-vs-rest-apis. Visited on 22 Dec 2024
95. Serving over http, GraphQL Foundation (2025). [Online] Available: https://graphql.org/learn/serving-over-http/. Visited on 11 April 2025
96. Core concepts, architecture and lifecycle (2024). [Online] Available: https://grpc.io/docs/what-is-grpc/core-concepts/. Visited on 22 Dec 2024
97. Newton-King J, Picket W, Latham L (2024) Versioning GRPC services. [Online] Available: https://learn.microsoft.com/en-us/aspnet/core/grpc/versioning?view=aspnetcore-9.0. Visited on 21 Dec 2024
98. Coming from openapi lasyncapi initiative for event-driven APIS (2024). [Online] Available: https://www.asyncapi.com/docs/tutorials/getting-started/coming-from-openapi. Visited on 05 Dec 2024
99. About lasyncapi initiative for event-driven APIS (2024). Source: https://github.com/asyncapi/website/blob/800b17c055423c96931eaf3821213da208c5cc4b/markdown/about/indAug. [Online] Available: https://www.asyncapi.com/about. Visited on 05 Dec 2024
100. Gornicki L (2020) Happy birthday asyncapi (week 47, 2020). [Online] Available: https://www.asyncapi.com/blog/status-update-47-20. Visited on 05 Dec 2024

101. Channel |asyncapi initiative for event-driven APIS (2024). [Online] Available: https://www.asyncapi.com/docs/concepts/channel. Visited on 19 Dec 2024
102. Adding operations |asyncapi initiative for event-driven APIS (2024). [Online] Available: https://www.asyncapi.com/docs/concepts/asyncapi-document/adding-operations. Visited on 19 Dec 2024
103. Lazzaretti F, Moya S (2024) Avro specification inside asycnapi file #1015 (2024). [Online] Available: https://github.com/asyncapi/spec/issues/1015. Visited on 18 Dec 2024
104. Bodiachevskii P, Lane D, Gornicki L, Méndez F, Lagoni J, Wichmann A (2024) Extend avro and openapi schema versions #1051. [Online] Available: https://github.com/asyncapi/spec/issues/1051. Visited on 20 Dec 2024
105. Server security |asyncapi initiative for event-driven APIS, A model for optimizing real-time data services (2024). [Online] Available: https://www.asyncapi.com/docs/concepts/asyncapi-document/server-security. Visited on 20 Dec 2024
106. Dulay H, Mooney S (2023) Streaming Data Mesh: A Model for Optimizing Real-Time Data Services, Mooney S (ed), 1st edn. O'Reilly, Beijing, 1 p. ISBN: 10-9813-069-3
107. Higginbotham J (2021) Principles of Web API Design: Delivering Value with APIs and Microservices. Pearson Education, London. ISBN: 978-01-37355-63-1
108. Hagen N (2022) Asyncapi and apicurio for asynchronous APIS. [Online] Available: https://www.asyncapi.com/blog/asyncapi-and-apicurio-for-asynchronous-apis. Visited on 08 Feb 2025
109. Cloudevents (2024). [Online] Available: https://cloudevents.io/. Visited on 17 Dec 2024
110. Boyne D (2024) Sending and receiving cloudevents with amazon eventbridge. [Online] Available: https://aws.amazon.com/blogs/compute/sending-and-receiving-cloudevents-with-amazon-eventbridge/. Visited on 17 Dec 2024
111. Cloudevents v1.0 schema with azure event grid - azure event grid |microsoft learn (2024). [Online] Available: https://learn.microsoft.com/en-us/azure/event-grid/cloud-event-schema. Visited on 17 Dec 2024
112. Eventarc overview |google cloud (2024). [Online] Available: https://cloud.google.com/eventarc/docs/. Visited on 17 Dec 2024
113. Eventbridge: Serverless event bus service - alibaba cloud (2024). [Online] Available: https://www.alibabacloud.com/en/product/eventbridge. Visited on 17 Dec 2024
114. Maradiaga A (2024) Cloudevents at sap. [Online] Available: https://community.sap.com/t5/application-development-blog-posts/cloudevents-at-sap/ba-p/13620137. Visited on 17 Dec 2024
115. Cloudevents - version 1.0.2 (2023). [Online] Available: https://github.com/cloudevents/spec/blob/v1.0.2/cloudevents/spec.md. Visited on 17 Dec 2024
116. Http protocol binding for cloudevents - version 1.0.2 (2022). [Online] Available: https://github.com/cloudevents/spec/blob/v1.0.2/cloudevents/bindings/http-protocol-binding.md. Visited on 17 Dec 2024
117. Kafka protocol binding for cloudevents - version 1.0.2 (2022). [Online] Available: https://github.com/cloudevents/spec/blob/v1.0.2/cloudevents/bindings/kafka-protocol-binding.md. Visited on 17 Dec 2024
118. Palino T, Narkhede N, Shapira G (2017) Kafka: The Definitive Guide. O'Reilly, Boston. ISBN: 978-14-91936-11-5
119. Doerrfeld B (2019) 8+ openapi linters. [Online] Available: https://nordicapis.com/8-openapi-linters/. Visited on 22 Dec 2024
120. Spectral - an open-source API style guide enforcer and linter (2024). [Online] Available: https://stoplight.io/open-source/spectral. Visited on 22 Dec 2024
121. Schema linting (2024). [Online] Available: https://www.apollographql.com/docs/graphos/platform/schema-management/linting. Visited on 22 Dec 2024
122. Murphy E (2024) Graphql schema linting: Ensuring consistency across subgraphs. [Online] Available: https://inigo.io/blog/linting-inigo. Visited on 22 Dec 2024
123. Schulz L (2022) Grpc_linting. [Online] Available: https://github.com/lotharschulz/grpc_linting. Visited on 22 Dec 2024.

References

124. Linting - tutorial (2024). [Online] Available: https://buf.build/docs/lint/tutorial/. Visited on 22 Dec 2024
125. Asyncapi studio (2024). [Online]. Available: https://studio.asyncapi.com/. Visited on 22 Dec 2024
126. Lint asyncapi with redocly cli (2024). [Online] Available: https://redocly.com/docs/cli/guides/lint-asyncapi. Visited on 22 Dec 2024
127. Spectral (2024). [Online] Available: https://github.com/stoplightio/spectral. Visited on 22 Dec 2024
128. Gnu gzip (2020). [Online] Available: https://www.gnu.org/software/gzip/. Visited on 22 Dec 2024
129. Compression in http (2024). [Online] Available: https://developer.mozilla.org/en-US/docs/Web/HTTP/Compression. Visited on 22 Dec 2024
130. Garbes G (2024) Maximizing message efficiency in kafka: Comparing JSON and avro (2024). [Online] Available: https://medium.com/@galvao.gabrielg/maximizing-message-efficiency-in-kafka-comparing-json-and-avro-6e0ab3d4f069. Visited on 22 Dec 2024
131. Papadopoulos T (2024) Compress rabbitmq messages. [Online] Available: https://dzone.com/articles/compress-rabbitmq-messages. Visited on 22 Dec 2024
132. Junker A (2023) Event-getriebene intergrationsarchitekturen. [Online] Available: https://www.informatik-aktuell.de/entwicklung/methoden/eventgetriebene-integrationsarchitekturen.html. Visited on 27 Dec 2024
133. Hohpe G, Woolf B (2013) Enterprise Integration Patterns: Designing, Building, and Deploying Messaging Solutions. The Addison-Wesley Signature Series, 17. print. Addison-Wesley, Boston, 683 pp. ISBN: 032-1200-68-3
134. Quotas and limits - amazon kinesis data streams (2024). [Online] Available: https://docs.aws.amazon.com/streams/latest/dev/service-sizes-and-limits.html. Visited on 23 Dec 2024
135. Apache kafka message size limit: Best practices & config guide (2024). [Online] Available: https://www.confluent.io/learn/kafka-message-size-limit/. Visited on 23 Dec 2024
136. Chat with claude 2024-12-23 (2024). [Online] Available: https://claude.site/artifacts/abefee1c-e8c4-4ab0-b780-49e226042288. Visited on 23 Dec 2024

Defining the Online Library Interfaces   7

In previous chapters, we learned how to select appropriate API architecture approaches. We learned how to select synchronous data exchanges and how to specify them. We learned how to specify events for particular services. We want to apply the experience from Chapter 5, "API Design Supported by Domain-Driven Design," to the online library.

Introduction

We will review the defined services to apply our knowledge about the design of APIs to the designed online library.

The API design should be done in a workshop format as well. Usually, it is done as a tactical design by the team implementing the corresponding bounded context. The collaborative format of workshops guarantees the common understanding of business experts and IT specialists. For this book, the authors take over the function of implementation teams. The basis for the workshops is the context map, which was created in the Event Storming workshops defining the bounded contexts and the team boundaries. In Chapter 9, "Collaborative Design and Agility", we will discuss in greater detail how to setup those workshops. The goal is to define the APIs so that later on the teams can start implementing the defined endpoints.

We show the design using the example of the online library and its context map, shown again in Figure 7-1. The teams developing the online library decided to use Kafka as an event broker. The synchronous APIs are mostly APIs to communicate with a user interface. Therefore, the teams decided to use REST to implement those synchronous interfaces. For both technologies, the teams possess the necessary skills.

In the examples, we will apply certain API guidelines. We use here the Zalando guidelines [1]. However, the rules are adapted to the necessities of this book. We will indicate when we those rules are being applied. We might even apply additional rules; we will discuss them at the appropriate places.

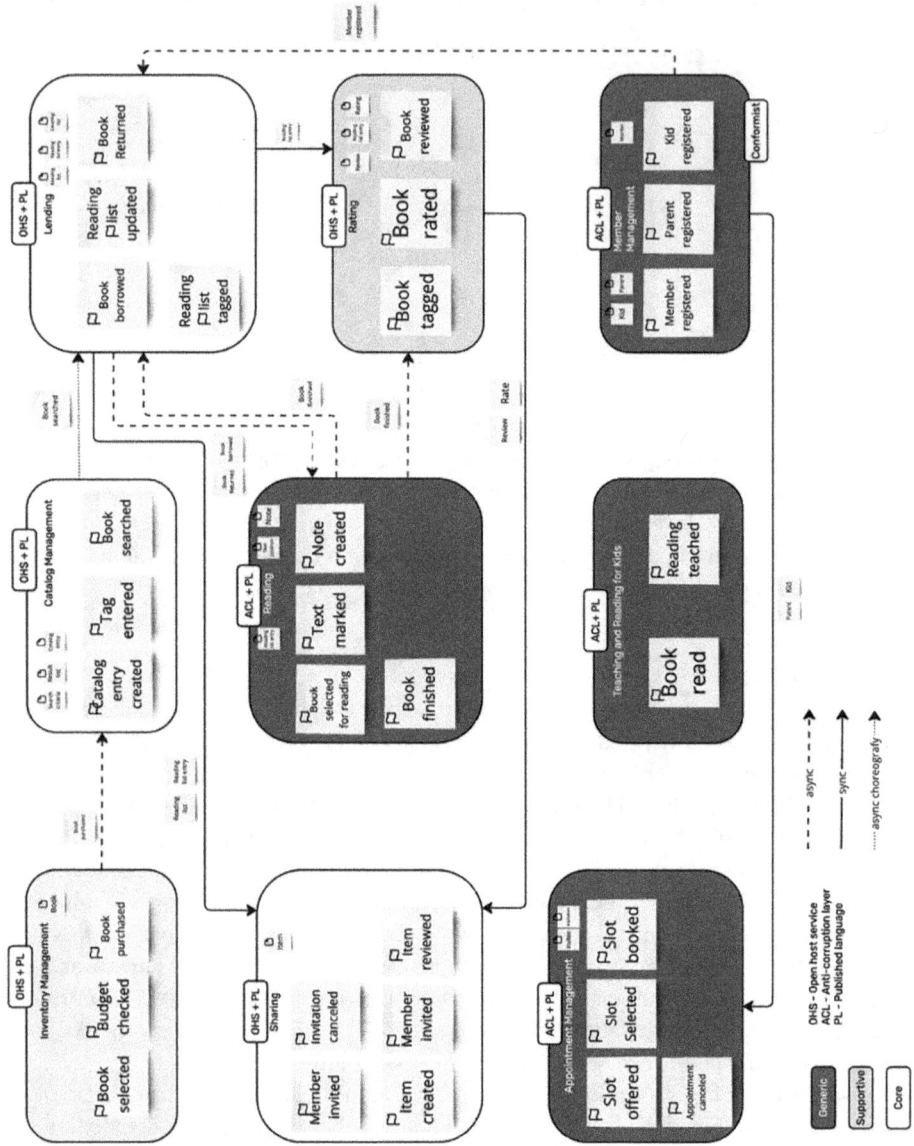

Figure 7-1 Context map of online library

The designed events use JSON as the data format because it is easier to read compared to Apache Avro. Therefore, we use JSON here in our examples.

Moreover, certain architectural decisions need to be made in the tactical design of the services. We will discuss them in detail for each service. As a result of each service, an Architecture Communication Canvas and a Bounded Context Canvas

Introduction

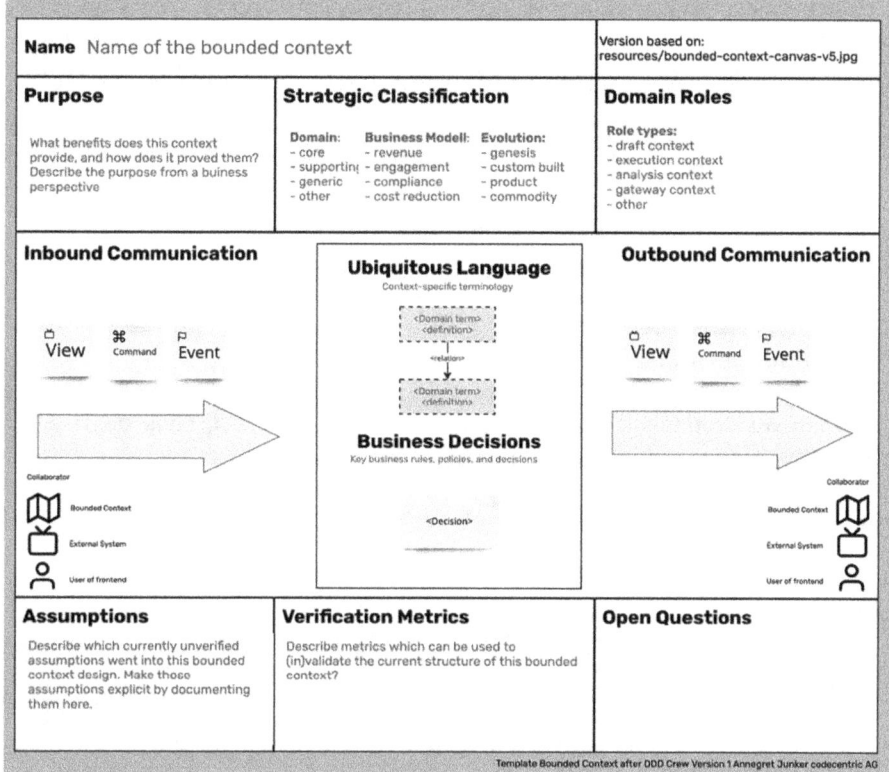

Figure 7-2 Bounded Context Canvas template

for each service will be created. Both canvases are explained in the following subsections. Moreover, the discussion result of the API and event design are documented in corresponding canvases. The canvases can be used for discussion in the workshops and for documentation purposes.

Bounded Context Canvas

A Bounded Context Canvas can describe a bounded context very closely [2]. It is used to show and discuss the essence of a bounded context. The perspective is closer to the business purpose and functionality. To create such a canvas, a template in Miro can be used [3, 4] (Figure 7-2).

It contains the following areas:

- **Name**
  The area contains the name of the bounded context.
- **Strategic Classification**
  The strategic classification contains the type of subdomain to which the bounded context belongs. The subdomain can be core, supportive, generic, or another.

The business model can be revenue, used for customer engagement, compliance reasons, or cost reduction. The evolution stage complies with the Wardley map: genesis, custom-built, product, or commodity (section "Qualifying Business Ideas").
- **Domain Roles**
  The following role types can use the bounded context:
  – Draft context – for sketches
  – Execution context – for the execution of capabilities
  – Analysis context – for analyzing business or technology
  – Gateway context – for compliance or security reasons
  – other
- **Inbound Communication**
  This area describes the collaborators of the bounded context described, triggering an inbound communication. Collaborators can be bounded contexts, external systems, or users via a front end.
- **Ubiquitous Language**
  The area contains a description of the ubiquitous language using a visual glossary.
- **Business Decisions**
  This area contains the essential business decisions belonging to the corresponding bounded context. Invariants of the bounded context in particular need to be listed.
- **Outbound Communication**
  This area contains an outbound communication description comparable to that of the inbound communication area.
- **Assumptions**
  This area contains the assumptions belonging to the bounded context.
- **Verification Metrics**
  This area describes the most critical metrics that can be used to validate the structure of the bounded context. If those metrics show the structure is no longer valid, the architecture must be reengineered.
- **Open Questions**
  This area lists the open questions belonging to the bounded context. Open questions can be questions regarding business or architecture.

Besides the Bounded Context Canvas, the Architecture Communication Canvas helps to understand the functionality and architecture of a bounded context. The canvas can give a very succinct overview of the bounded context and can help outsiders to quickly understand the core functionality of the bounded context.

Architecture Communication Canvas

An Architecture Communication Canvas is a reduced form to describe the architecture of an application (Figure 7-3). It focuses on architecture-related decisions and the quality requirements of the bounded context. It is highly suited to introducing the

Introduction

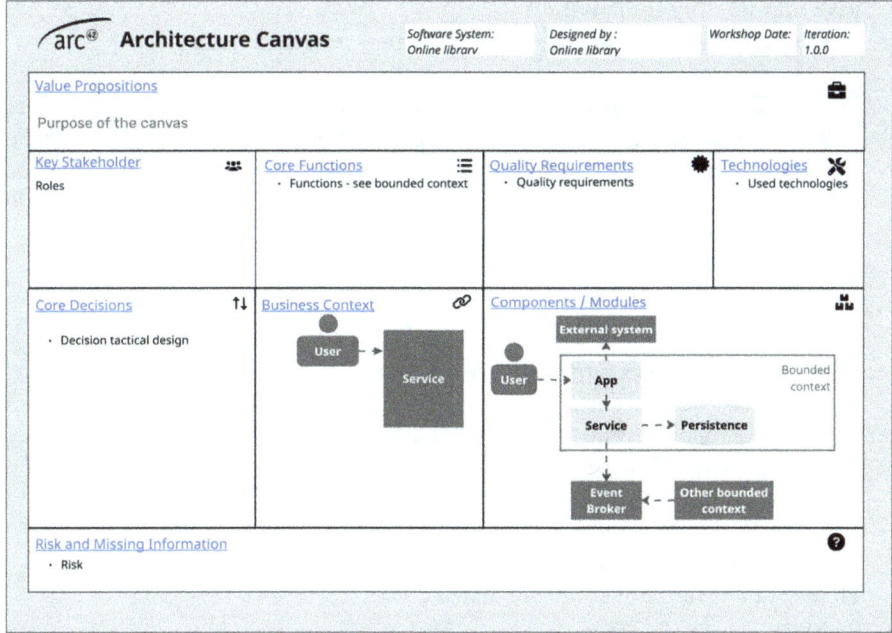

Figure 7-3 Architecture Communication Canvas

architecture of a bounded context and documenting the most important architectural decisions [5]. It was developed by the arc42 group, which created an architectural documentation template standard in Europe [6].

The following information is provided:

- **Value propositions**
- **Key stakeholders**
  Key stakeholders are listed by their roles.
- **Core functions**
  Core functions are oriented toward the business capabilities collected in the business capability map (section "Qualifying Business Ideas").
- **Quality requirements**
  The quality requirements that must be applied specifically to the corresponding bounded context are listed in this area.
- **Technologies**
  This area lists the technologies used in the bounded context.
- **Core decisions**
  This area lists the core decisions when designing the bounded context.
- **Business context**
  This area shows the context level of the C4 model [7].
- **Components/Modules**
  This area shows the containers using the second level of the C4 model [7].

- **Risk and missing information**
  The risks of implementing the bounded context are listed here.

Both the Bounded Context Canvas and the Architecture Communication Canvas describe the bounded context from different perspectives. A bounded context should be described with both. We will discuss how to use them in a real development journey in Chapter 9, "Collaborative Design and Agility." Additionally, an API product canvas can be helpful.

API Product Canvas

However, this book concentrates on APIs. Therefore, the authors invented an API product canvas. Such a canvas can help to design and document the synchronous API and asynchronous events of a bounded context.

The canvas is shown in Figure 7-4.

The canvas contains the following sections:

- **Name of bounded context**
  This section is the name of the bounded context.
- **Version**
  This section contains the version of the canvas. It should also apply to the API version when the canvas is used for documentation.
- **Value propositions**
  This section contains a statement about the value propositions of the bounded context. It can be copied from the Architecture Communication Canvas when the canvas is available.
- **Core functions**
  This section contains a list of the core functions provided by the bounded context. It can be copied from the Architecture Communication Canvas when the canvas is available.
- **Contact**
  The name of the person who should be contacted when questions to the bounded context APIs arise.
- **Sync protocol**
  This contains the name of the protocol used for synchronous communication, for example, HTTP.
- **Architectural pattern**
  This section contains the architectural approach to synchronous communication, for example, RESTful oriented.
- **Server**
  This contains the server URL or base URL of the synchronous interface.
- **Aggregates and entities**
  This section lists aggregates and entities exposed via the synchronous interface. Additionally, each aggregate or entity is enhanced by a list of operations applied

Introduction

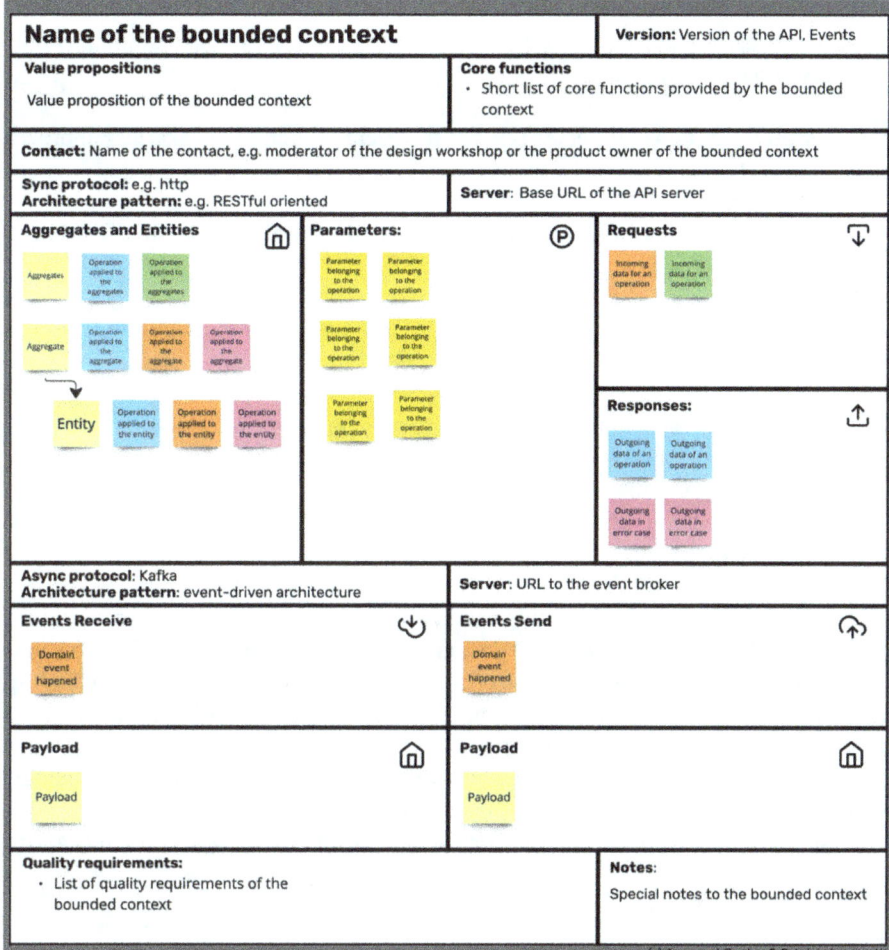

Figure 7-4 API Product Canvas

to it. For example, those operations could be HTTP verbs. Or it can be the defined operations of a gRPC interface.
- **Parameters**
  The necessary parameters must be given for each aggregate or entity in the list.
- **Requests**
  This section contains the data that must be provided for certain operations.
- **Responses**
  This section contains the date that will be provided in the synchronous API responses.
- **Async protocol**
  This section contains the asynchronous protocol that is used, for example, Apache Kafka.

- **Architectural pattern**
  This section contains the architectural pattern of asynchronous communication, for example, EDA.
- **Server**
  The section contains the server URL of the event broker. If a brokerless approach is applied, the section can be empty.
- **Events receive**
  The section contains the domain events (from Event Storming session) consumed by the bounded context.
- **Payload** belonging to receive
  This section contains the entities and aggregates assigned to the aforementioned events as payload.
- **Events send**
  This section contains the domain events (from Event Storming session) produced by the bounded context.
- **Payload** belonging to send
  This section contains the entities and aggregates assigned to the aforementioned events as payload.

The canvas can be used for design workshops as well as for documentation. We will discuss how to set up an API design workshop in Chapter 9, "Collaborative Design and Agility." This discussion includes why we discuss the exposure of synchronous APIs. However, we do not include consuming APIs in the canvas as we do with events in the event sections.

We will use the canvases introduced here in the following sections to document the design decisions made for the bounded contexts of the online library.

First, we want to apply our experience to the bounded context `Inventory Management`.

Inventory Management

The DDD process showed that `Inventory Management` was a supportive subdomain. A standard product customized to the necessities of an online library can be used. The product is encapsulated for use as an open host service.

Besides the apparent properties of a book, the object contains information about the authors and the editor. Additional details about digital licenses are not necessary. This information is collected in the corresponding bounded context canvas (Figure 7-5).

Definition of AsyncAPI

The service must only publish the domain event `BookPurchased` as shown in the communication canvas. All other events belong to the service itself. The payload of

Inventory Management

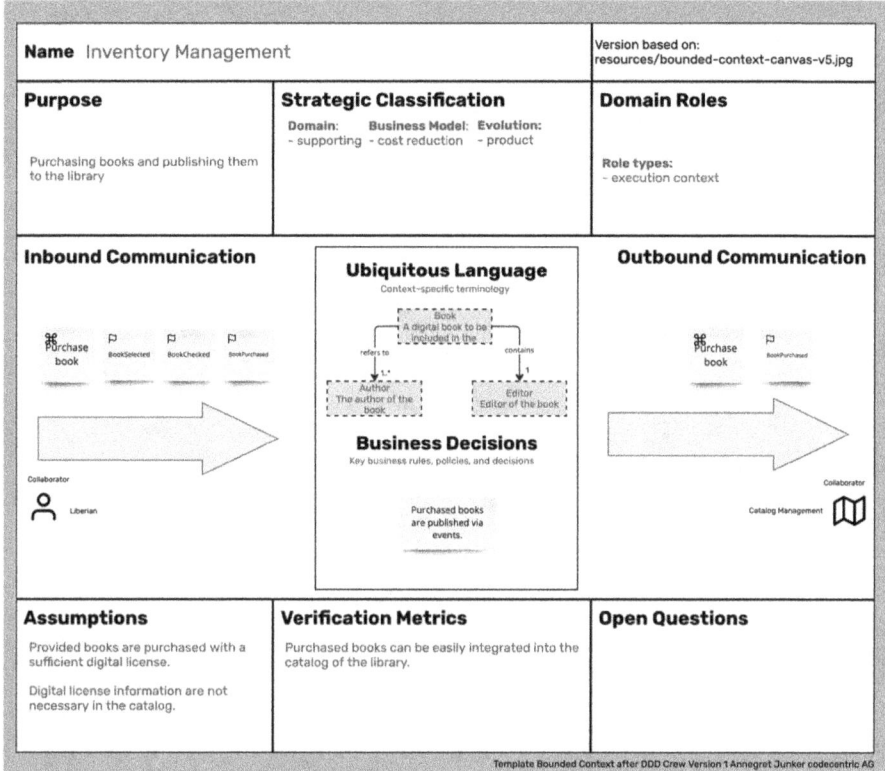

Figure 7-5 Bounded context inventory management

the event contains the purchased book as published language. However, the book contains more information than is needed in the domain event.

The event contains brief information about the authors. Here, brief information means names and identifiers. Additional information, like author biographies and portraits, can be read later via the synchronous interface.

The definition of the `AuthorShort` is shown in Listing 7-1.

The book is a short version as well, because the book does not contain a cover image. In addition to typical metadata such as authors and titles, it contains a link and format to the book's digital media. However, the long version of the author and book can be read synchronously when necessary. That is the 80% solution discussed in the previous chapter.

Even though the long versions of `Book` and `Author` do not appear in the asynchronous definition, they are defined as `BookShort` and `AuthorShort` to ensure consistency with the synchronous API definition.

To avoid difficulties with special characters, the image must be encoded. To formulate such requirements, the string type can be enhanced by the format `binary` for binaries like the image.

```
 1  AuthorShort:
 2    description: Author of a book
 3    type: object
 4    required:
 5      - authorId
 6      - givenName
 7      - familyName
 8    properties:
 9      authorId:
10        description: Identifier of author
11        type: string
12        format: uuid
13        examples:
14          - 08a13654-05f3-4664-8c8d-fdd9114ec3be
15      givenName:
16        description: Given name of author
17        type: string
18        minLength: 2
19        maxLength: 55
20        examples:
21          - Kent
22      familyName:
23        description: Family name of author
24        type: string
25        minLength: 2
26        maxLength: 55
27        examples:
28          - Beck
29
```

Listing 7-1 Brief information about author for an event

Inheritance using allOf

The extended version of `Book` inherits information from the short version. The inheritance can be defined by `allOf`. Below the keyword, all types are listed from which the corresponding type inherits information. The different generators react differently to `allOf`. The Typescript generator does not generate an inheritance. It simply creates different objects [8].

Comparable issues were also reported for Java [9]. Using the Maven plugin `openapi-generator-maven-plugin` in version 7.8.0 creates classes with all properties but without an inheritance. An example of those generated classes using a Baeldung package [10] is given using the example of `Catalog Management`.[1]

---

[1] https://github.com/DDAPID/CatalogManagement/tree/main/target/generated-sources/openapi/src/main/java/com/baeldung/openapi/model

```
1   AuthorFull:
2     description: Full description of an author
3     type: object
4     allOf:
5       - $ref: '#/components/schemas/AuthorShort'
6     properties:
7       authorPortraitWebP:
8         description: Portrait of author as WebP
9         type: string
10        format: binary
11      authorBiography:
12        description: Short bio of author
13        type: string
14        minLength: 10
15        maxLength: 2048
```

Listing 7-2 Enhanced information about author – used for synchronous APIs of purchased product

However, the use of an inheritance is great and should be done when defining APIs. It helps to understand the API better. An example of this enhanced information about an author is given in Listing 7-2.

The inheritance with multiple objects does not work at all, because neither Typescript nor Java supports multiple inheritances.

To allow synchronous access to enhanced author information, authors are modeled as entities with an identifier. The aggregate of the bounded context is Book. An aggregate is an entity by definition so that it can be identified.

Utilities

We discussed in the preceding chapter that external references to specification documents create a coupling between schema definitions. In our example, we see that Book is used in slightly changed form in several bounded contexts. To avoid mistakes, providing the schema information for those widely used objects as a utility is recommended. Utility definitions can be reused by copying. An example of utilities is given on Github.[2] Utilities do not need to be a valid specification. That specification only provides parts to be reused.

The events defined in the messages section acquire the name of the corresponding domain event BookPurchased. The associated channel enhances the event's name by Channel. The associated operation acquires the channel name enhanced by the associated action. In the case of BookPurchased, the action is "send" because

---

[2] https://github.com/Apress/Crafting-Great-APIs-with-Domain-Driven-Design/blob/main/Chapter-7/Utilities/Utilities.yaml

the event is produced by the `InventoryManagement` service. Recall the structure of the AsyncAPI as discussed in section "Components of an Interface Definition" and shown in Figure 6-1.

The operation and the event broker obtain the associated Kafka bindings, as it was stated earlier that the teams possessed the necessary skills.

The entire definition of the asynchronous communication of the `Inventory Management` service can be found on GitHub.[3]

Definition of OpenAPI

The REST API only needs endpoints to expose the long versions of `Book` and `Author`. In section "Lending," we will go into more detail on the entity's long and short versions. However, in this bounded context, we will only use the long version. Both can be accessed via an identifier.

Further endpoints are not necessary because the user interface is provided by the purchasing product customized for the online library. The endpoints send back the full variation of `author` or `book`. The types are named with a `full` postfix to allow consistency with the asynchronous specification.

The paths and the successful answer can be seen in Listing 7-3, and the entire definition of the synchronous interface can be found on Github.[4]

The designed events and API can be documented using the API Product Canvas. The result is shown in Figure 7-6.

The presented APIs contain a couple of architectural decisions that can be documented in the Architecture Communication Canvas. The associated canvas is shown in Figure 7-7.

The Inventory Management covers the purchase and storage of books, but books need to be searchable and need to be found by library members. The bounded context `Catalog Management` is responsible for those functionalities. We will discuss this in the following section.

Catalog Management

`Catalog Management` is one of the core components of the online library. Librarians can manage the catalog, automatic entries can be made. Members can search for books provided by the library.

Figure 7-8 shows the Bounded Context Canvas.

---

[3] https://github.com/Apress/Crafting-Great-APIs-with-Domain-Driven-Design/blob/main/Chapter-7/Purchase/InventoryManagement.aas.yaml

[4] https://github.com/Apress/Crafting-Great-APIs-with-Domain-Driven-Design/blob/main/Chapter-7/Purchase/InventoryManagement.oas.yaml

Catalog Management

```yaml
1   paths:
2     /authors/{authorId}:
3       get:
4         description: Delivers the full information of an author
          identified by its ID
5         operationId: getAuthorById
6         parameters:
7           - $ref: '#/components/parameters/BookIdentifierParameter'
8           - $ref: '#/components/parameters/VersionParameter'
9         responses:
10          '200':
11            $ref: '#/components/responses/AuthorResponse'
12
13    /books/{bookId}:
14      get:
15        description: Delivers the full information of a book identified
          by its ID
16        operationId: getBookById
17        parameters:
18          - $ref: '#/components/parameters/AuthorIdentifierParameter'
19          - $ref: '#/components/parameters/VersionParameter'
20        responses:
21          '200':
22            $ref: '#/components/responses/BookResponse'
```

Listing 7-3 REST API – the stripped down paths

The catalog information is directly created based on purchased books. Otherwise, librarians handle catalog entries. In addition to the editorial information, librarians can add tags and subtracts to the books. Using tags, librarians create lists for interested library members, for example, a list of classic mysteries. Librarians create subtracts to help members find the right book to read.

For the moment, we assume that the library contains only one catalog. The catalog contains several entries, enhanced by an abstract and up to 25 tags.

When members search for a book, they get a list of results and can select a book to read. The search is drafted as a full-text search over all catalog entries. Therefore, the domain event Book searched is renamed to Book selected as a better description of the event.

Catalog Management contains a couple of writing challenges and challenges to the search. The search must be fast, especially so that members can use the software in a relaxed way. Therefore, specific architectural policies must be introduced to meet the requirements. Those decisions are documented in the architectural communication canvas shown in Figure 7-9.

The Catalog Management service is implemented following the Command and Query Responsibility Segregation (CQRS) pattern [11]. CQRS means that the operations for writing and reading data are separated. Management operations performed by librarians are separate from searches conducted by members. The management applications for librarians use a standard relational SQL database,

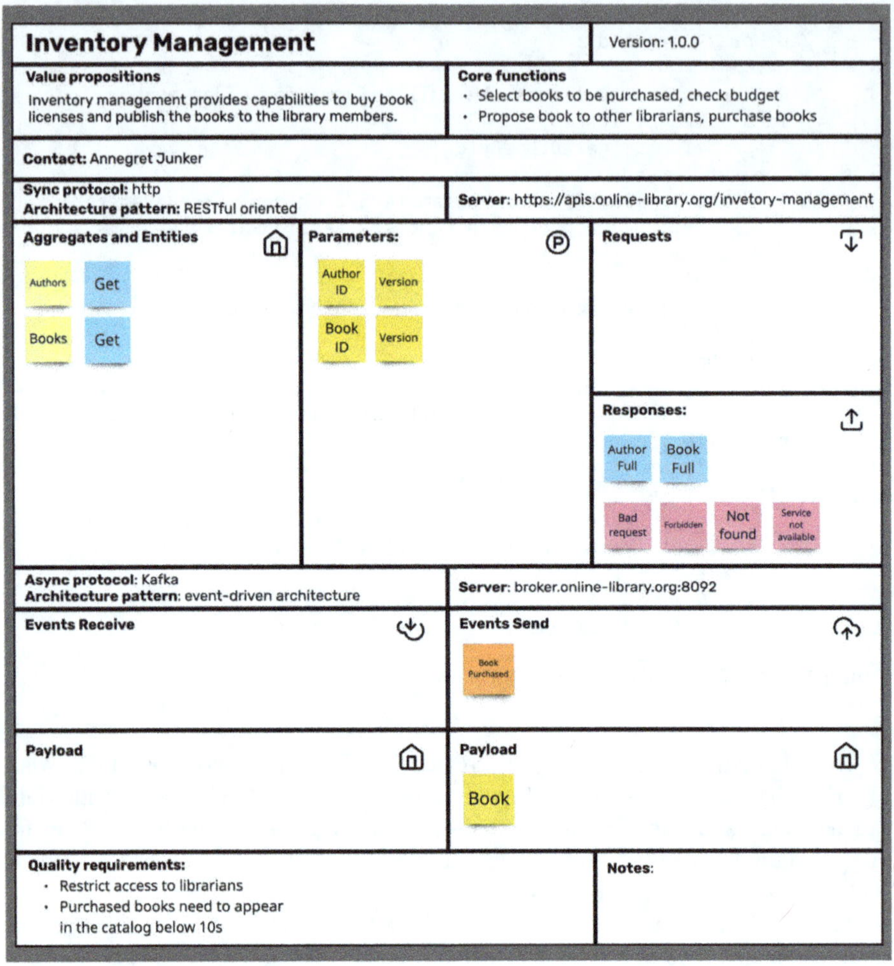

Figure 7-6   API Product Canvas: Inventory Management

whereas the search application is supported by a database tuned for full-text searches.

Be careful with the use of CQRS

Be careful when applying CQRS. The use case described here for CQRS is to have different databases in the same bounded context with different purposes: storage, update (command), and search (query). The authors have seen widespread misuse of CQRS. Be careful to use the CQRS pattern to replicate too much data. When data updates or deletions are not always replicated to the second data storage, it is harder to make changes, and more errors can crop up. In addition, this increases the solution's infrastructure costs.

Catalog Management

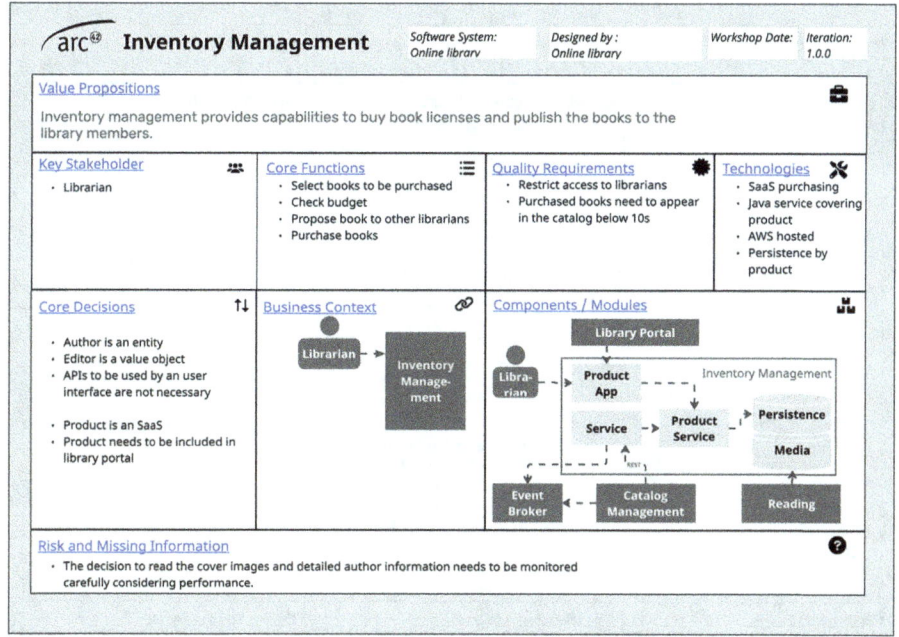

Figure 7-7 Architecture Communication Canvas: Inventory Management

CQRS can also be used between bounded contexts to simplify querying over multiple services. This can lead to a distributed monolith when data ownership is unclear and makes changes harder. A composition of several APIs instead of one can sometimes be a better solution [12].

Therefore, `Catalog Management` is implemented as two services. Both services provide a REST interface and produce and consume events.

Management Application

The management application provides a typical REST API, which contains elements to create and change catalog entries. A search entry via `get` is not provided because it is provided via the search application.

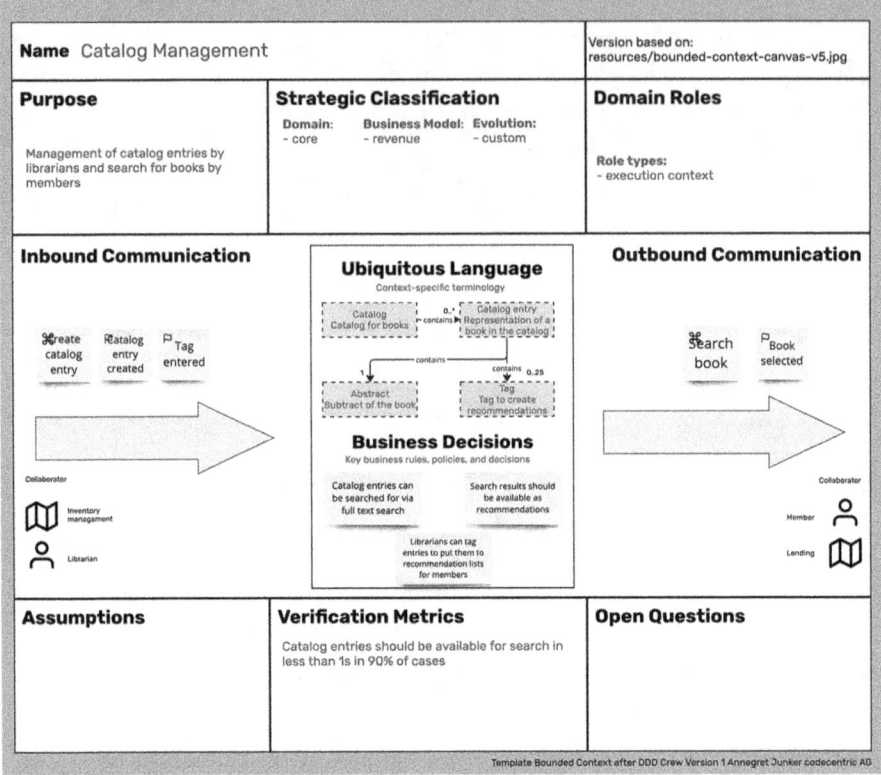

Figure 7-8 Bounded Context Canvas: Catalog Management

The elements `abstract` and `tag` are modeled as subresources below the aggregate `catalog-entries`. Therefore, they get endpoints below the `/catalog-entries/{catalogEntryId}`, which can be used to access a single catalog entry in its complete form. Both subresources get the HTTP verbs `put` and `delete`. A `post` is unnecessary because they are modeled as value objects that do not have their own identity and lifecycle.

The catalog entries resemble the `Book` aggregate of `Inventory Management`. However, they are slightly different, for example, the extended version contains, besides the cover, the abstract and tags.

The resulting API Product Canvas is shown in Figure 7-10.

A shortened version of the API is presented in Listing 7-4. The full OpenAPI specification can be found on Github.[5]

---

[5] https://github.com/Apress/Crafting-Great-APIs-with-Domain-Driven-Design/blob/main/Chapter-7/CatalogManagement/CatalogManagement.oas.yaml

Catalog Management

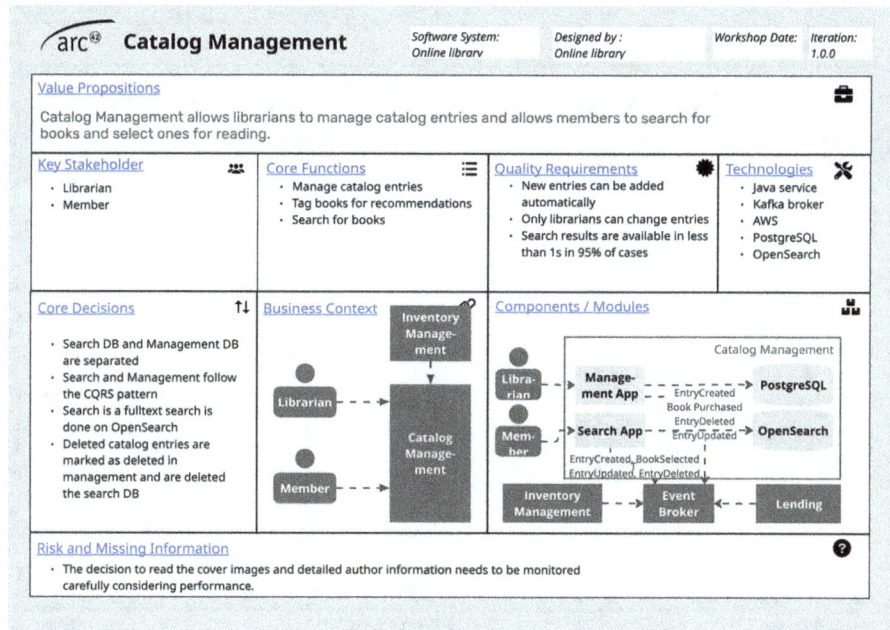

Figure 7-9 Architectural communication canvas: Catalog Management

The `Catalog Management` consumes the event `Book purchased`. Based on this event, new entries are created in the catalog. This produces the events `CatalogEntryCreated`, `CatalogEntryChanged`, and `CatalogEntryDeleted`.

Those events are necessary for the `Search` application of `Catalog Management`. The events are not consumed by other components of the online library. They are not domain events defined in the DDD process. They are technical events necessary to implement the CQRS pattern.

The full AsyncAPI specification for `Catalog Management` can be found on Github.[6]

Search Application

The `Search App` provides a REST interface and events.

The REST interface provides an endpoint for a full-text search and direct access to a specific catalog entry in its complete form with abstract and tags. Additionally, it provides an endpoint to access the complete information of authors, including their biography and portrait photo.

---

[6] https://github.com/Apress/Crafting-Great-APIs-with-Domain-Driven-Design/blob/main/Chapter-7/CatalogManagement/CatalogManagementLibrarian.aas.yaml

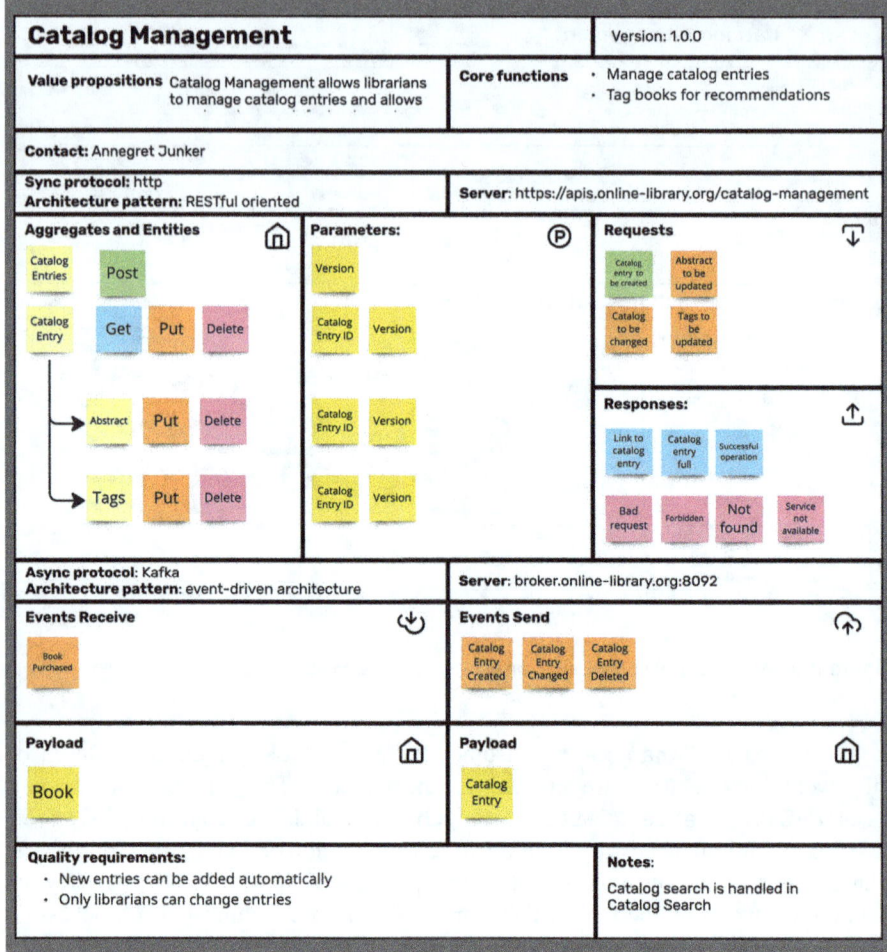

Figure 7-10 API Product Canvas: Catalog Management

Because the search can provide a high number of results, it is recommended to apply a paging mechanism. Using the OpenAPI specification, even paging behaviors can be specified. A rather simple method for paging is the provisioning of an offset and limit of a page [13].

The resulting API Product Canvas is shown in Figure 7-11.

Those parameters can be provided via a header as defined in the sample (Listing 7-5).

To avoid mistakes in sorted lists, the header information is sent back in the search result (Listing 7-6).

The result is given as an object because it contains, in addition to the results, the search meta data. Also, it is recommended to send back result sets as objects and not

Catalog Management

```yaml
paths:
  /catalog-entries:
    post:
      description: Creates a new catalog entry
      requestBody:
        $ref: '#/components/requestBodies/CatalogEntryToBeCreatedRequest'
      responses: {'200': { $ref: '...LinkToCatalogEntryResponse' }}
  /catalog-entries/{catalogEntryId}:
    get:
      description: Sends back a single catalog entry based on its identifier
      parameters: ['...CatalogEntryIdentifierParameter', '...VersionParameter']
      responses: {'200': { $ref: '...CatalogEntryResponse' }}
    put:
      description: Changes a catalog entry with all possible entries
      parameters:['...CatalogEntryIdentifierParameter', '...VersionParameter']
      requestBody:
        $ref: '#/components/requestBodies/CatalogEntryToBeChangedRequest'
      responses: {'200': { $ref: '...LinkToCatalogEntryResponse' }}
    delete:
      description: Deletes a catalog entry
      parameters:
        - $ref: ['...CatalogEntryIdentifierParameter', '...VersionParameter']
      responses: {'200': { $ref: '...SuccessfulOperationResponse' }}
  /catalog-entries/{catalogEntryId}/abstract:
    put:
      description: Adds or changes the abstract of a catalog entry
      parameters:
        - $ref: ['...CatalogEntryIdentifierParameter', '...VersionParameter']
      requestBody:
        $ref: '#/components/requestBodies/AbstractUpdateRequest'
      responses: {'200': { $ref: '...LinkToCatalogEntryResponse' }}
    delete:
      description: Deletes an abstract from a catalog entry
      parameters:
        - $ref: ['...CatalogEntryIdentifierParameter', '...VersionParameter']
      responses: {'200': { $ref: '...SuccessfulOperationResponse' }}
  /catalog-entries/{catalogEntryId}/tags:
    put:
      description: Adds or updates a set of tags to a catalog entry
      parameters:
        - $ref: ['...CatalogEntryIdentifierParameter', '...VersionParameter']
      requestBody:
        $ref: '#/components/requestBodies/TagsUpdateRequest'
      responses: {'200': { $ref: '...LinkToCatalogEntryResponse' }}
    delete:
      description: Deletes the entire tag list from a catalog entry
      parameters:
        - $ref: ['...CatalogEntryIdentifierParameter', '...VersionParameter']
      responses: {'200': { $ref: '...SuccessfulOperationResponse' }}
```

Listing 7-4 Reduced Catalog Management API

as an array. This will make it easy to enhance search results later on by additional information without causing breaking changes.

The `Search` application consumes the technical events of the CQRS pattern and produces the domain event `Book selected`. This event is consumed by the `Lending` component, which we discuss in the following section.

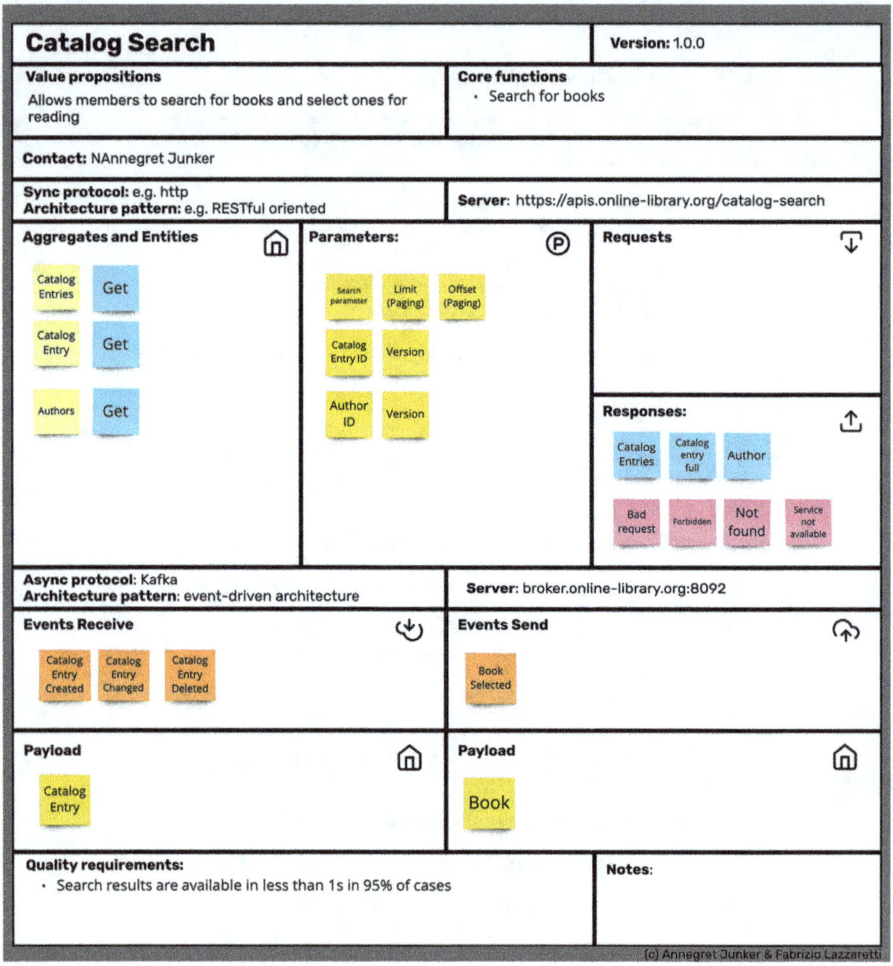

Figure 7-11 API Product Canvas: catalog search

Lending

As `Catalog Management` is the application for librarians, `Lending` is the application for members. The functionalities are shown in Figure 7-12.

Members can borrow books they selected previously. They can create individual reading lists by tagging them and assigning books to them or by tagging books directly. They can share those reading lists with other members.

Moreover, members can rate or review books. When a member publishes a review, all members of the library can read the review. Other members can read

```
 1  OffsetParameter:
 2      name: offset
 3      description: Offset of paging in a search result
 4      in: header
 5      required: false
 6      schema:
 7          type: integer
 8          format: int32
 9          minimum: 0
10          maximum: 10000
11          default: 0
12  LimitParameter:
13      name: limit
14      description: Limit of paging of search results
15      in: header
16      required: false
17      schema:
18          type: integer
19          format: int32
20          minimum: 0
21          maximum: 250
22          default: 30
```

Listing 7-5 Header information for paging large search results

```
 1  CatalogEntries:
 2      description: List of short catalog entries as a result list of a
        ↪ search
 3      type: object
 4      required:
 5          - searchMetaData
 6          - catalogEntries
 7      properties:
 8          searchMetaData:
 9              $ref: '#/components/schemas/SearchMetaData'
10          catalogEntries:
11              description: Search result
12              type: array
13              minItems: 0
14              maxItems: 10000
15              items:
16                  $ref: '#/components/schemas/CatalogEntryShort'
```

Listing 7-6 Paging information in result set

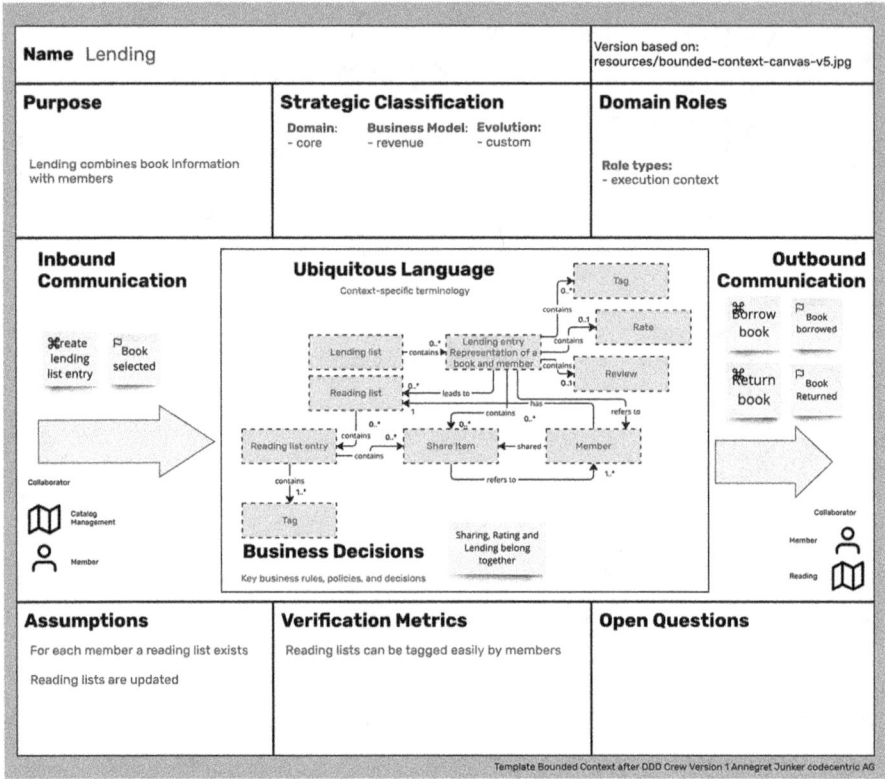

Figure 7-12 Bounded Context Canvas: Lending

unpublished reviews, whether the book is part of a shared reading list or is shared directly.

The sharing, rating, and review functionalities were identified as different bounded contexts during the DDD process. When discussing the Lending bounded context in detail, the functions appear to be thoroughly coupled. To implement them in different bounded contexts and, therefore, in different teams would entail deeply dependent teams that cannot work independently.

That is not efficient. Therefore, it was decided that the functions should be put together in one service. Those decisions cannot be made only by the team responsible for Lending. Changes in service architecture must be made by an overarching architecture team. The architecture team can be built by team architects synchronizing together. We will discuss organizational aspects of the process later in Chapter 9, "Collaborative Design and Agility," in more detail.

The resulting API Product Canvas is shown in Figure 7-13.

The corresponding Architecture Communication Canvas is shown in Figure 7-14.

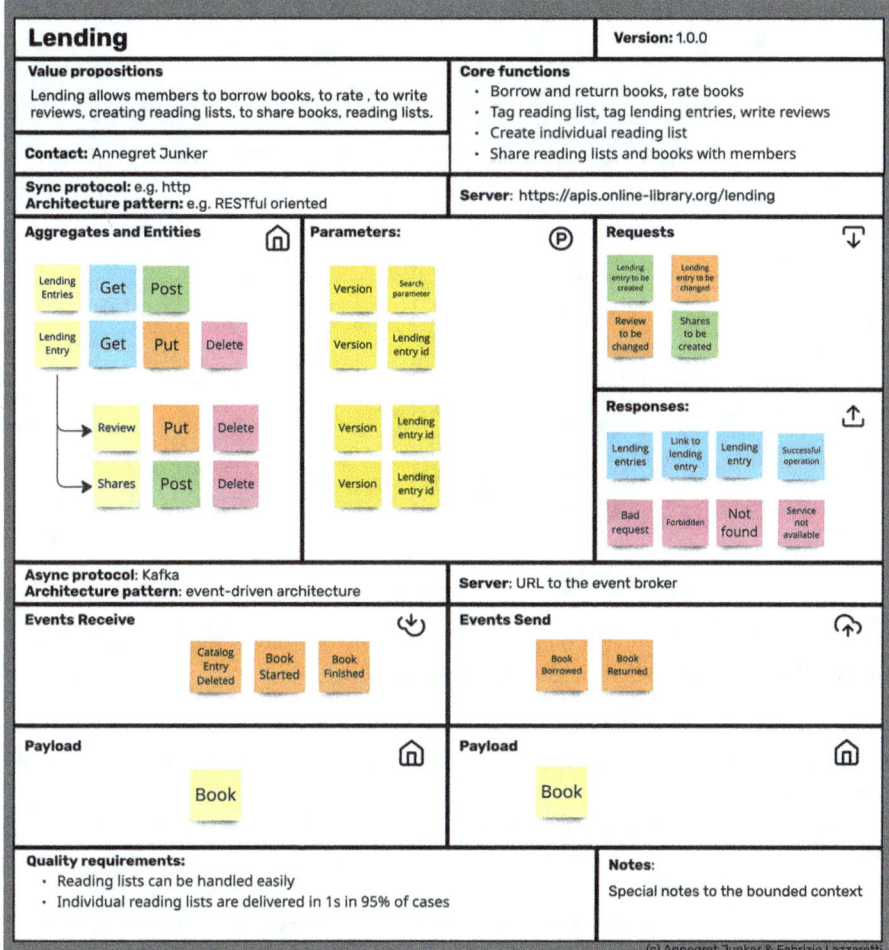

Figure 7-13 API Product Canvas: Lending

As the canvas shows, the services `Lending`, `Sharing`, and `Rating` are still separate deployment units. However, they work together on one database in one bounded context service.

To share reading lists or books with other members, `Sharing` accesses `Member Management` synchronously. When discussing the bounded context, a team member decides against replicating user data to avoid the misuse of member data. Those decisions can be made by the implementing team. An overarching architecture synchronization is not necessary. When accessing data synchronously within a defined user context, the allowed data can be better filtered directly at `Member Management`. Alternatively, the access rules would be implemented in

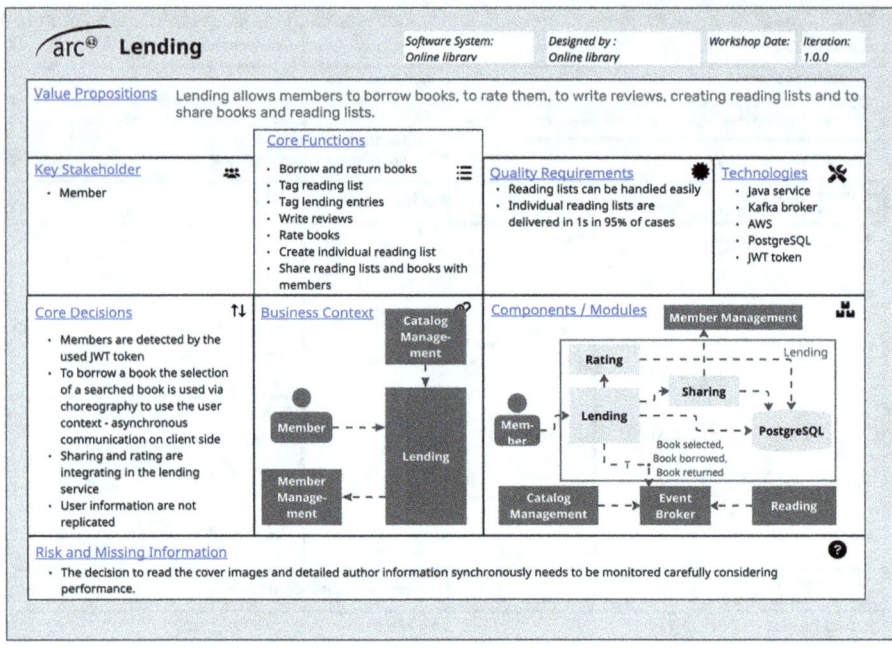

Figure 7-14 Architecture Communication Canvas: Lending

`Sharing`, and maintaining consistency with `Member Management` would be highly complex.

Consequently, the REST API is longer. It contains the following endpoints:[7]

- `/lending-entries`:
  Contains a `post` to create a new lending entry for a member who wants to read a book and a `get` to search for particular lending entries.
- `/lending-entries/{lendingEntryId}`:
  The `get` allows accessing a lending entry defined by its identifier. Using `put` can change a lending entry. Furthermore, a lending entry can be deleted with `delete`. Deletion here means to set the book as returned. The book no longer appears in the member's current reading list. However, individual reading lists of the members still contain the book.
- `/lending-entries/{lendingEntryId}/review`:
  Review is a subresource of the lending entry. It can be created by a `put` and deleted by `delete`. Only the member owning the corresponding lending entry can create, change, and delete the review. The review can be published so that other members can access it.

---

[7] The entire definition of the REST API can be found in Github: https://github.com/DDAPID/code-samples/blob/main/Chapter-7/Lending/LendingRestApi.yaml.

```
 1  {
 2      "iss": "online-library",
 3      "iat": "1737223122",
 4      "exp": "1737224022",
 5      "sub": "lending-entries",
 6      "aud": "member",
 7      "identifier": "7a23d9ab-8973-406c-b1fa-0dadccfd046c",
 8      "name": "Junker",
 9      "firstName": "Annegret",
10      "role": "member"
11  }
```

Listing 7-7 Example payload of JWT token

- /lending-entries/{lendingEntryId}/shares:
  A member owning the lending entry can add other members to the lending entry using post. Members can be deleted from the sharing list using delete.
- /reading-lists:
  Members can create a reading list by selecting their lending entries and assigning them to a tagged reading list. Alternatively, they can create a reading list with tags, and its tags will automatically filter the lending entries. Using get, members can access the reading lists they created or that are shared with them.
- /reading-lists/{readingListId}:
  The endpoint allows the access and change to a reading list accessed by its identifier. It even allows deleting a reading list. However, shared lists cannot be deleted.
- /reading-lists/{readingListId}/tags:
  The endpoint makes it possible to change the tag lists of a reading list. A tag list needs at least one entry; otherwise, it is the current reading list of a member handled in the lending entries.
- /reading-lists/{readingListId}/shares:
  In the same manner as lending entries, reading lists of members can be shared with other members. Accordingly, shares can be deleted.
- /shares/{memberId}:
  This endpoint allows access to all a member's shares. Shares are items shared by other members or by direct access by the member to reading lists or books.

The endpoint does not contain the current member using the application as a parameter. Members can be identified by their JSON Web Token (JWT) [14]. The token contains, besides a member identifier, additional information like role and name (Listing 7-7) The token content has the function of an API and requires the same level of carefulness. The information is transmitted via header in standard processes and does not need additional parameters.

JWT Content

JWTs usually contain standard information. At minimum, they need some information about validity and claim. The following information should be provided:

- About the issuer "iss" as a string or URI
- Time when it was issued "iat"
- Time when it is expired "exp"
- Subject "sub"
- Audience "aud"

Event though those values are optional in the specification [15], it is highly recommended to use them [14].

When accessing the shares of a member, those shares can be books from the lending entries, or they can be reading lists. Polymorphism can be used to express such behavior.

Polymorphism with oneOf or anyOf

Polymorphism in object-oriented languages means an object can have different behaviors depending on its type. Usually, all objects inherit from a virtual object, so all classes have a standard root class from which all others inherit.

"Polymorphism" means that the sender of a stimulus does not need to know the receiving instance's class. The receiving instance can belong to an arbitrary class [16].

In our case, we need to respond to the /shares: request with either a book or a reading list. Using the keyword `oneOf` allows for creating a collection of entries containing books and reading lists (Listing 7-8).

Again, a disadvantage of this procedure is the different behaviors of generators. The Maven plugin openapi-generator-maven-plugin generates an interface that is not implemented so that the behavior of the different items cannot be used directly; see example on Github (Figure 7-15).[8]

A better version might be two arrays in the shares – one with `Book` items and one with `Reading list` items. Both arrays need to be checked regarding emptiness separately to avoid null pointer exceptions. When more different objects need to be handled, it can cause difficult bugs, because certain checks are not implemented correctly.

---

[8] https://github.com/DDAPID/Lending/blob/main/target/generated-sources/openapi/src/main/java/com/baeldung/openapi/model/ShareOfAMemberItem.java

Lending

```yaml
ShareOfAMember:
  description: Item that is shared with a member
  type: object
  required:
    - isOwner
    - item
  properties:
    isOwner:
      description: Flag is set to true if member is the owner of the
        corresponding item
      type: boolean
      default: true
      examples:
        - true
    item:
      description: Item that is shared with member
      oneOf:
        - $ref: '#/components/schemas/LendingEntry'
        - $ref: '#/components/schemas/ReadingList'
```

Listing 7-8 Share of a member

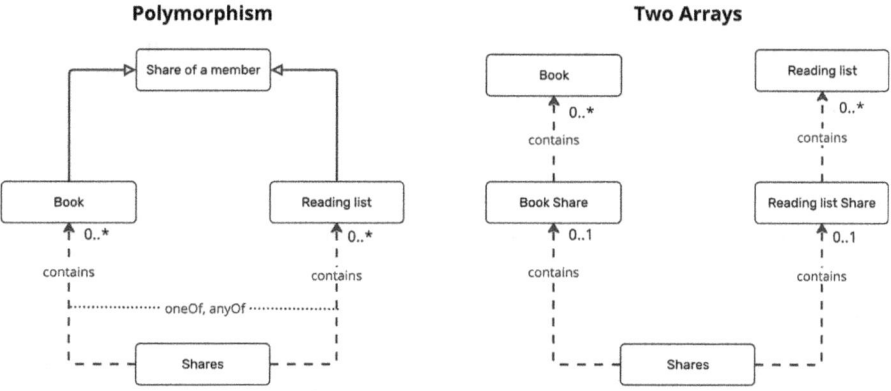

Figure 7-15 Solution using polymorphism or arrays

The lending entry contains information about a member and a catalog entry. Both are identified by their unique identifier. At any rate, on a client, one want to display member names, book titles, and authors. Therefore, the lending entry contains not only the identifiers but also members' names and very concise catalog entry information. Thus, it is easier to show the necessary information without rereading the information via synchronous request. The drawbacks of this solution are two more events: lending needs to consume a changed catalog entry and a deleted event. The information is updated in the lending and reading lists. Member information is updated when the member is logged the next time.

The lending service consumes and produces the following events:[9]

- Consume `CatalogEntryChanged`
  The event is produced by `Catalog Management` and is already consumed by `Catalog Search`. It can be consumed by `Lending` too. It is an event triggered by the architectural solution and was not detected through the DDD process.
- Consume `CatalogEntryDeleted`
  The event is produced by `Catalog Management` and is already consumed by `Catalog Search`. It is an event triggered by the architectural solution and was not detected through the DDD process. It can be consumed by `Lending` too.
- Consume `BookFinished`
  The reading application raises the event when a user has finished her book.
- Produce `BookBorrowed`
  The event is raised when a user borrows a book selected previously in the catalog search.
- Produce `BookReturned`
  When members return a book, it disappears from their lending list. It is not directly accessible and must be borrowed again.

The book information contains concise member and catalog entry information for the convenience of client developers so that this information will be available for display without further requests to the associated bounded contexts. This information is replicated.

Be aware that the ubiquitous language of `Lending` contains the term `Book`, which is used differently than in the bounded context `Inventory Management`. In `Inventory Management`, `Book` means the description of an inventory item, and in `Lending` it means the combination of member and catalog entry. When lists are involved, the term `Lending list entry` is more convenient, because it differentiates the general lending list from individual reading lists.

Reading is a bounded context that should contain a standard reading application. It requires an anticorruption layer to hide the external language from the ubiquitous language of the online library reading. We will discuss this in the following subsection.

Reading

Even though reading is a commodity, it defines customer satisfaction similarly to search or lending. When the product was defined, the product for the reading component was not selected. However, whether the selected product contains a text

---

[9] The entire defintion of the asynchronous API can be found in Github: https://github.com/Apress/Crafting-Great-APIs-with-Domain-Driven-Design/blob/main/Chapter-7/Lending/Lending.aas.yaml

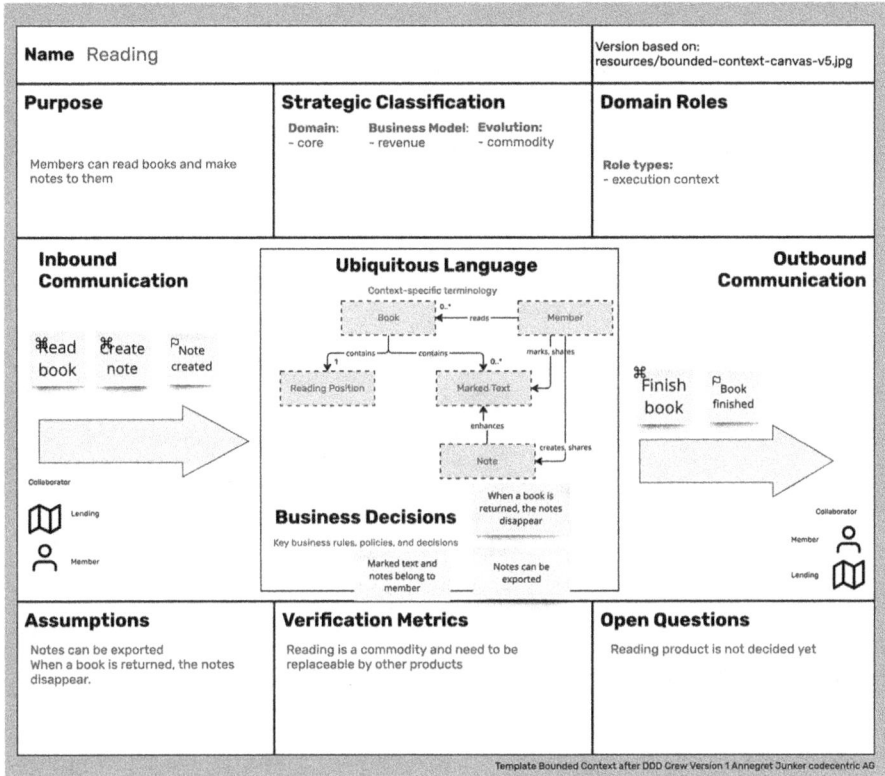

Figure 7-16 Bounded Context Canvas: Reading

mark and note management is unclear. The functionalities of the bounded context are shown in the Bounded Context Canvas (Figure 7-16).

Although an API can be defined that covers the functionality, the API must be implemented by the integration team that integrates the selected product into the online library and maps the provided API to the defined API in an anti-corruption layer (ACL) (Figure 7-17).

The REST API contains the following endpoints;[10] see Figure 7-18 too:

- `/books`:
  The endpoint contains a `get` to get all books on the member's shelf.
- `/books/{bookId}`:
  The endpoint contains a `get` endpoint to get the full book, including the cover image and fulltext. The information is fetched from the inventory management,

---

[10] The entire definition can be found on GitHub: https://github.com/Apress/Crafting-Great-APIs-with-Domain-Driven-Design/blob/main/Chapter-7/Reading/Reading.oas.yaml

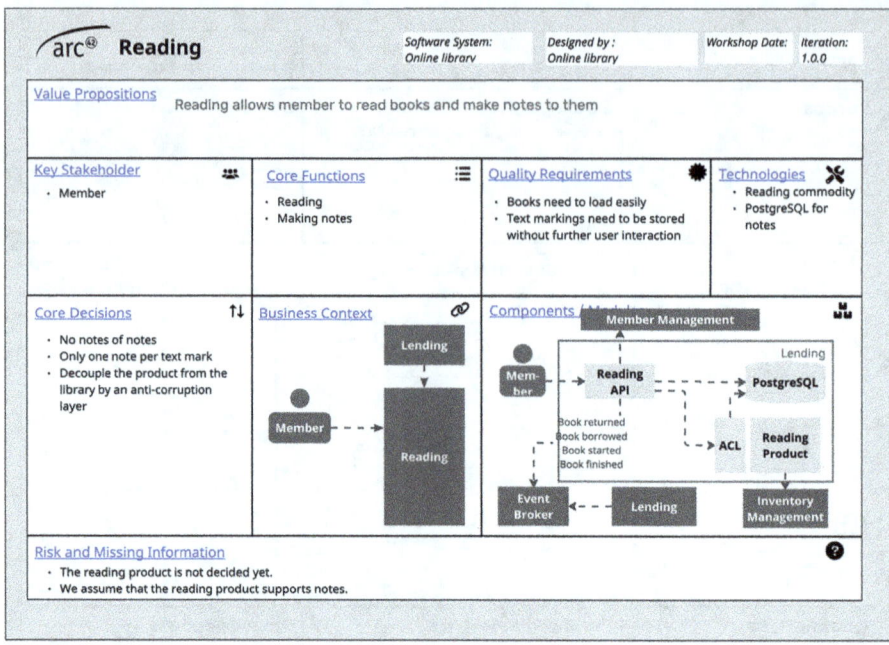

Figure 7-17 Architecture Communication Canvas: Reading

which provides access to the inventory. However, following the DDD process, the inventory management excludes librarians. Members cannot access inventory management. To control the access management easily, members can only access books in full text via `Reading` management.

- `/books/{bookId}/marked-texts`:
  Marked text is a subresource of `books`, including defined shares. With `get`, all marked text passages of the book can be provided. With `post`, a new marked text passage is created.
- `/marked-text/{markedTextId}/notes`:
  A note can be created for each marked text. It enhances the marked text. The text can be changed with `put` or deleted with `delete`.
- `/books/{bookId}/notes`:
  Using this convenient solution, a client can access all book notes. The endpoint only delivers text passages where the notes are not empty.
- `/books/marked-text/{markedTextId}/shares`:
  Using `post`, the owner of a text passage can define new shares or, using `delete`, delete them.

Because of the status of a lending entry in `Lending`, an additional event, `Book started`, is necessary. Thus, the status can be changed from `TO_BE_READ`

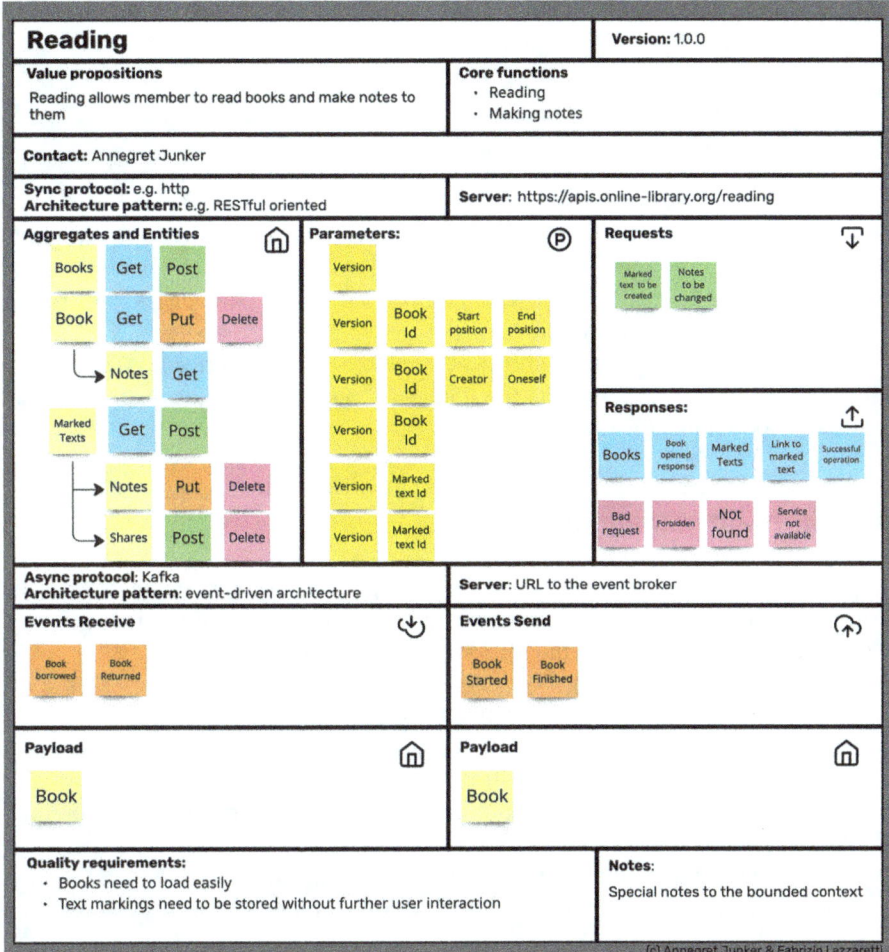

Figure 7-18 API Product Canvas: Reading

to READING. The event `Book finished` was already identified during the DDD process to change the status from READING to FINISHED.

The events handled in reading are a mirror of lending when it comes to sending and receiving.[11]

The service consumes the events

- `Book borrowed` and
- `Book returned`

---

[11] The entire definition can be found in GitHub https://github.com/Apress/Crafting-Great-APIs-with-Domain-Driven-Design/blob/main/Chapter-7/Reading/Reading.aas.yaml

produced by `Lending`.
The service produces the events

- `Book started` and
- `Book finished`

that are consumed by `Lending`.

When a member accesses the media, the media is loaded from `Inventory Management` according to the relevant format. The format provides corresponding functionalities to navigate and provided embedded media like images. The reader provides the respective clients to access the media.

With that, the core functionalities of the library are described. Additionally, functions are used to differentiate the library from other providers. Next, we want to discuss appointment management.

Appointment Management

Appointment Management should allow the negotiation of appointments between members, as well as between parents and volunteers. It can be a customized solution. The solution needs to support the following functionalities:

- Allow appointment templates for members and teaching.
- Allow the definition of time slots of volunteers.
- Create, update, and cancel appointments between members.
- Create, update, and cancel appointments between children (parents for their children) and volunteers.
- Synchronize with the individual calendars of volunteers.
- Notify via email members with a download of appointments.

The functionalities are summarized in the Bounded Context Canvas (Figure 7-19).

To understand the provided API better, we mention some products that offer such solutions. One of these can, for example, be customized later. We can check those APIs and generalize them for the purpose of the online library.

- BookingTime[12]
- ServiceNow[13]
- OnSched[14]

---

[12] https://service.bookingtime.com/apidoc/module

[13] https://www.servicenow.com/docs/bundle/xanadu-api-reference/page/integrate/inbound-rest/concept/appointment-api.html

[14] https://api.onsched.com/index.html

Appointment Management

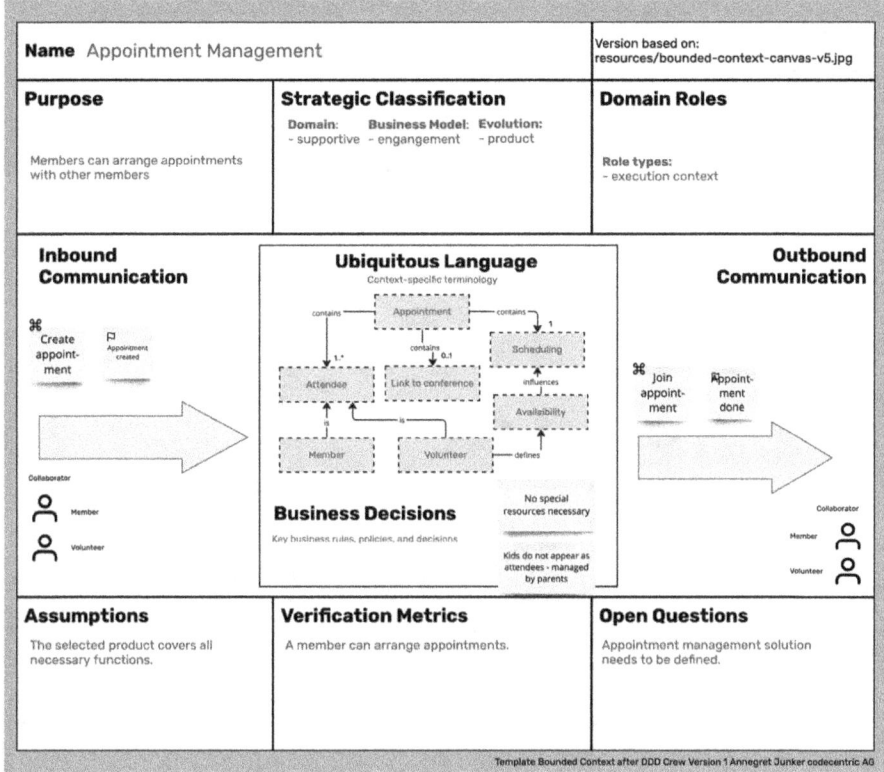

Figure 7-19 Bounded context: Appointment Management

- SAP Appointment Booking[15]
- And many more

The associated API Product Canvas is shown in Figure 7-20.

What we want to do is to arrange an appointment that can be made with a `post` request. An appointment can be changed using a `put`. It includes the change of attendees or change in scheduling. To cancel an appointment, a `delete` can be used.

Volunteers can be set up as calendar owners, where their availability can be checked. The availability of members or volunteers can be checked with a `get` filtered by start and end dates. Volunteers can use a remote conference tool to read to children. A link to a remote meeting can be integrated into the appointment for an easy-to-use appointment.

---

[15] https://help.sap.com/docs/SAP_FIELD_SERVICE_MANAGEMENT/fsm_ai/appointment-booking.html

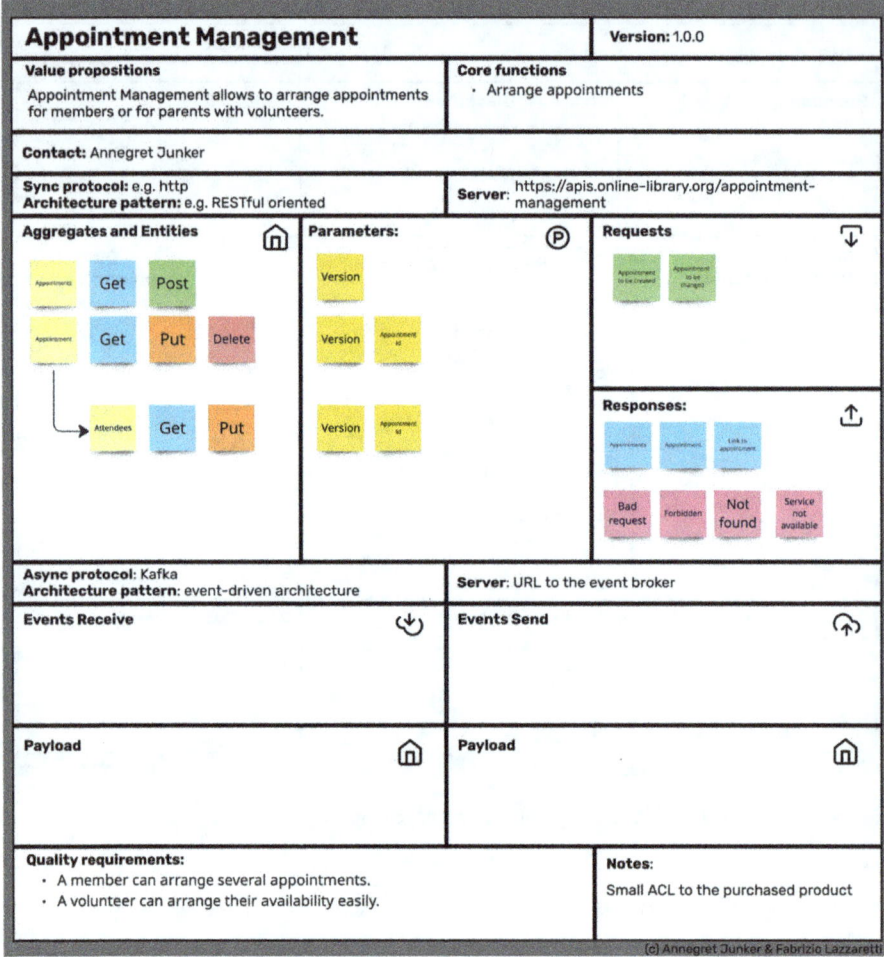

Figure 7-20 API Product Canvas: Appointment Management

Usually when an appointment is created, resources can also be booked. Resources like rooms or technical equipment are not necessary for an online library.

The technical solution is shown in the Architecture Communication Canvas (Figure 7-21).

When looking at the aforementioned Appointment Management products APIs, the ACL can be quite small. For most of them, the API is just a one-to-one layer that allows for changing the Appointment Management product. This is mainly just a mapping from one API call to another that changes the path and sometimes the data serialization structure. The layer acts as a thin facade to make a product substitution simpler.

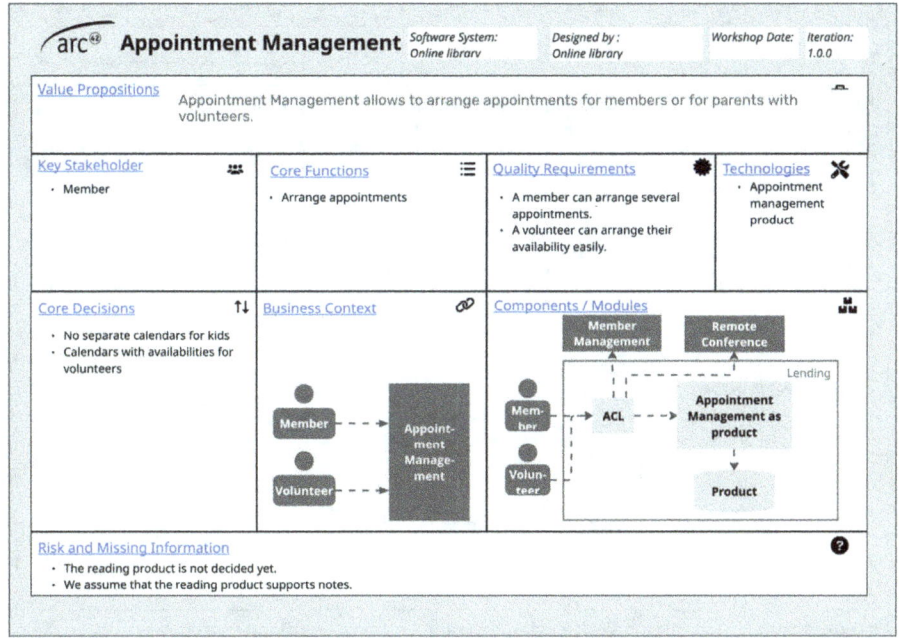

Figure 7-21 Archictecture communication canvas: Appointment Management

The appointment allows, besides appointments between members, appointments between volunteers and parents for their children.

Let us look at `Teaching and reading to kids` in the next section.

Teaching and Reading to Kids

A volunteer has an appointment with children arranged in advance by their parents. The appointment can be attended remotely or in person. For remote appointments, a link can be used to join the conference. The bounded context is presented in Figure 7-22.

A volunteer can join a conference remotely via tool like Zoom[16] or Teams[17] with the parents and the children to whom they were assigned.

Usually, those remote conference tools provide easy-to-integrate APIs, so they are directly integrated into the website, for example, the library portal. Thus, an integration into the backend of the online library is not necessary. The links to the conference tool can be integrated into the appointment created earlier.

---

[16] https://developers.zoom.us/docs/api/rest/reference/zoom-api/methods/#overview

[17] https://developer.microsoft.com/en-us/microsoft-teams

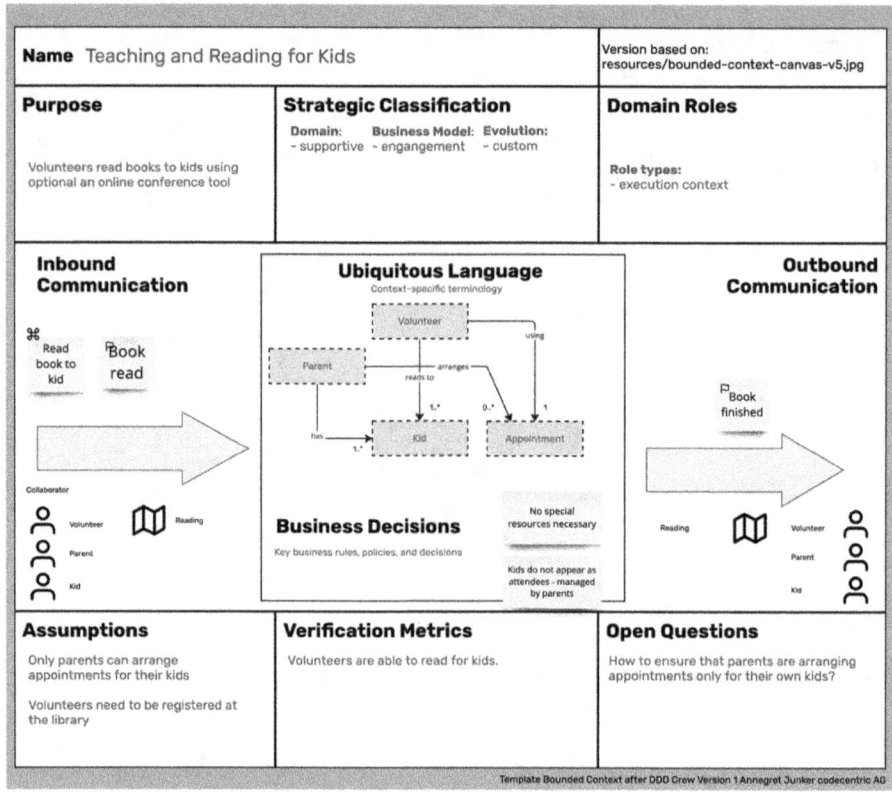

Figure 7-22 Bounded context: Teaching and reading to kids

When the appointment is attended remotely, the volunteer uses the reading application to read the book to the children using the conference tool's screen-sharing capabilities.

The technical solution is shown in Figure 7-23.

The technical solution shows that the bounded context is technically a virtual bounded context that uses two others – the conference tool and the Reading bounded context. A specific implementation is not necessary.

Member Management

Member management is mostly based on a standard identity management system. Standard identity management systems provide authentication and authorization capabilities; see the Bounded Context Canvas in Figure 7-24.

Most big Identity and Access Management (IAM) products support OpenID Connect (OIDC) as an extension standard of open authorization (OAuth) 2.0 [17].

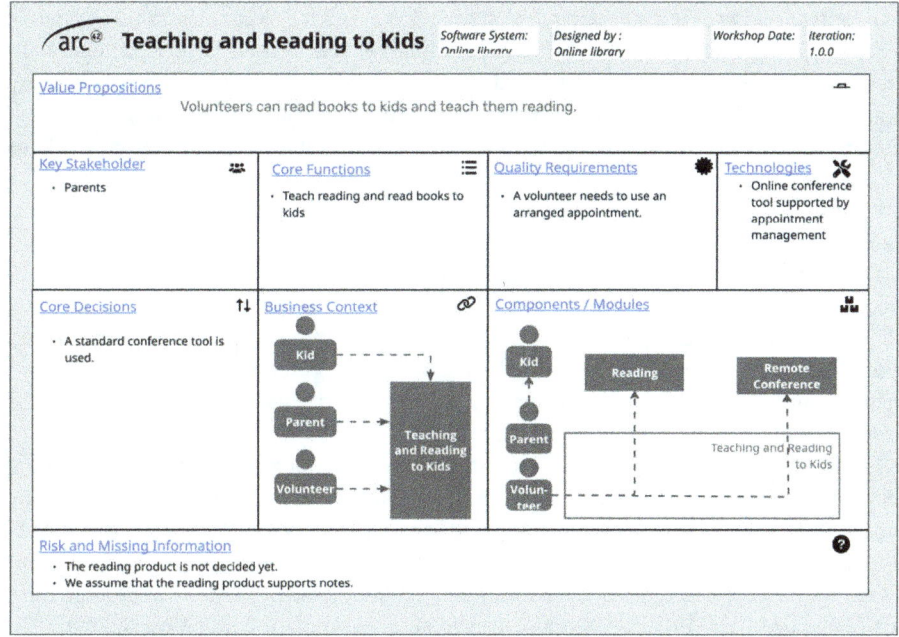

Figure 7-23 Architecture Communication Canvas: Teaching and reading to kids

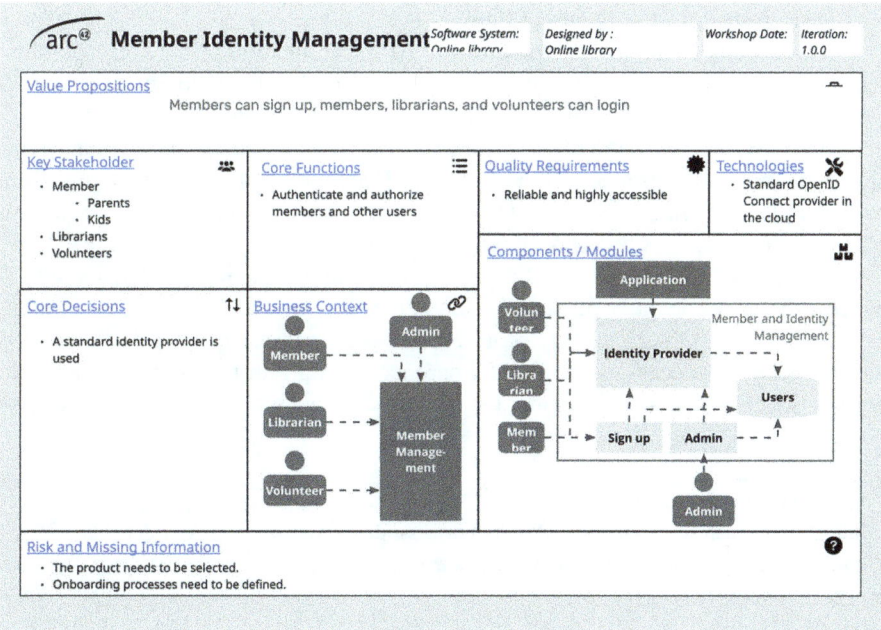

Figure 7-24 Bounded context: Member Management

All large cloud providers have such systems: AWS IAM,[18] Azure Entra ID (formerly Active Directory),[19] or Google Cloud Identity and Access Management.[20] In addition, open-source (e.g., Keycloak[21]) and enterprise products (e.g., Auth0 by Okta[22]) are available as well. For that reason, an OIDC-based solution will be used.

First, we need to differentiate between authentication and authorization. Authentication is the process of proving a person's identity or another application, for example, by a certificate or a shared secret. Authorization grants access to a particular entity or function [18].

The authentication of a library member is done by user name and password. A second secret can also be given, for example, in a two-factor authentication process. Those processes are the responsibility of the identity management system. Which processes are used and how to apply them can be configured at the identity system.[23] The identity provider issues an identity token, which can then be provided to the application requesting the identity check.

Authorization is the process by which an authenticated user (a user with a confirmed identity) can access certain entities in the library. Authorization grants access to a certain entity using a confirmed identity. For the online library, this means that members are allowed to access books, create notes and individual reviews, and share them. Librarians are allowed to manage catalog entries and to write generally valid reviews. The identity provider issues an access token containing the claims a user has access to, whereas the claims for the online library are the corresponding resources like catalog entries or books. The requested application can check these claims.

The simplified flow is shown in Figure 7-25.

The necessary claims are defined in the API specification in the corresponding security block of the requested endpoint. For example, the endpoint to change a catalog entry requires the claim `catalog:write`; see Listing 7-9.

Librarians must have the claim `catalog:write` when they want to change catalog entries (section "Catalog Management"). Even more, an accessing application, for example, `Purchase` (section "Inventory Management") must also present that claim when accessing `Catalog Management`.

Member Management makes it possible to sign up for the library. The identity is checked by a web identity process, which is the responsibility of the identity management system. Other proofing requirements are imaginable could be integrated into the identity process. Additionally, parents can sign up for their kids. Kids do not get the claims to search for books or to create reviews, but they are created as

---

[18] https://aws.amazon.com/iam/

[19] https://www.microsoft.com/en-us/security/business/identity-access/microsoft-entra-id

[20] https://cloud.google.com/security/products/iam

[21] https://www.keycloak.org/

[22] https://auth0.com/

[23] E.g., for Microsoft Entra ID: https://learn.microsoft.com/en-us/entra/identity/authentication/concept-mfa-howitworks.

Member Management

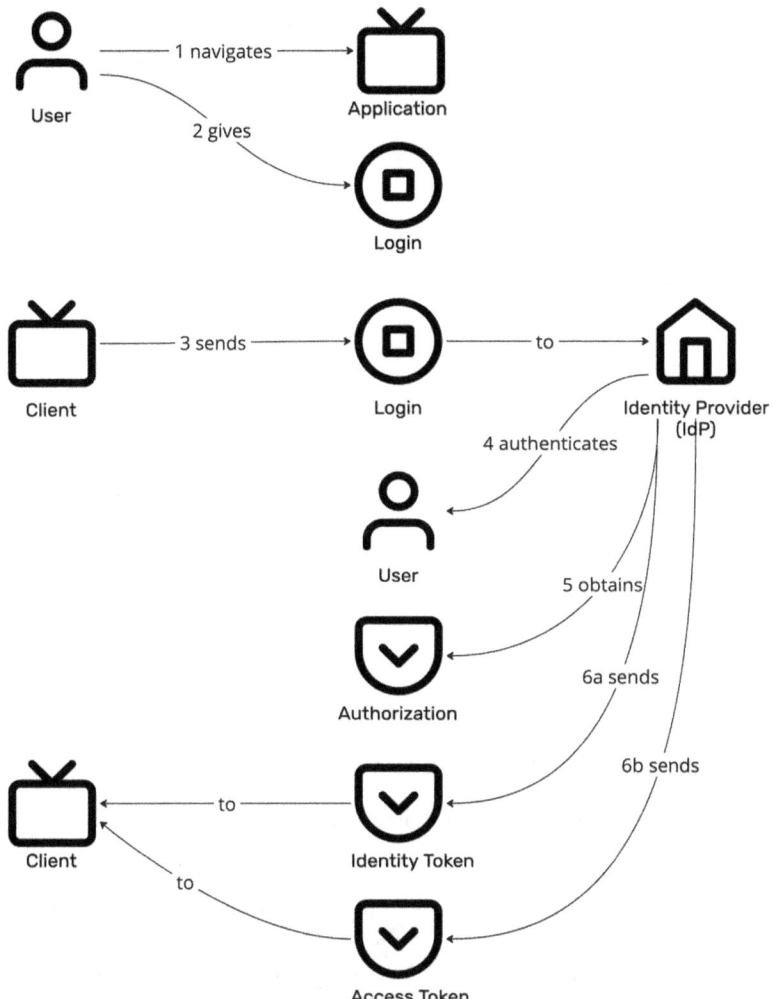

Figure 7-25 OpenID connect flow

```
1  /catalog-entries/{catalogEntryId}:
2    put:
3      description: Changes a catalog entry with all possible entries
4      operationId: updateCatalogEntryById
5      tags:
6        - Catalog
7      security:
8        - openIdConnect:
9            - catalog:write
```

Listing 7-9 Example of a claims definition in an endpoint

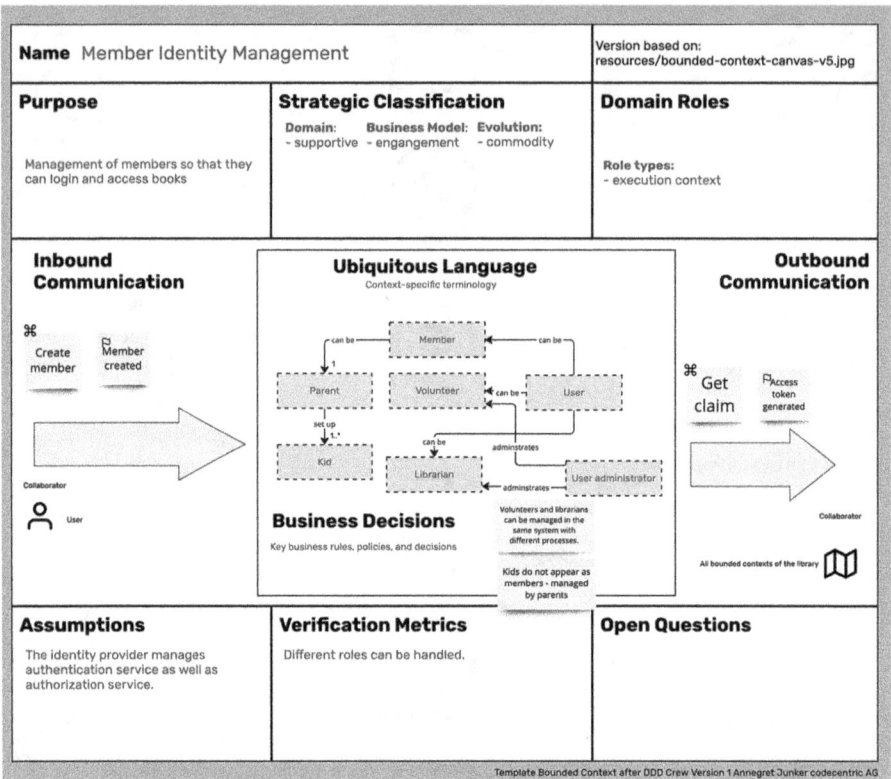

Figure 7-26 Architecture Communication Canvas: Member Management

users in Member Management. However, parents can create reading lists for their children so that they know which books they have already read.

The identity provider can be used to authenticate librarians and volunteers as well. However, the onboarding processes are different. Whereas the onboarding process for members should be as easy as possible, the process for librarians and volunteers requires proof of professional qualifications. Those processes can be handled even outside of the software base. An administrator can set up librarians and volunteers in the identity management system. So far, the identity management system is more than just a Member Management system (Figure 7-26).

Among the stated purposes of the online library, it is used as a member and identity management system.

We described all APIs of all bounded contexts of the library. Let us see what we have learned from the API design.

Conclusion

In this chapter, we discussed the bounded contexts of the online library. We learned that the original approach needed to be adapted in the following points:

- `Inventory Management`
  Besides the event `Book purchased`, `Inventory Management` delivers data synchronously on author details, including author portrait and book details, including cover image. Digital books are provided via standard interfaces defined by the corresponding file format.
- `Catalog Management`
  This bounded context also produces the events `Catalog entry updated` and `Catalog entry deleted` consumed by `Catalog search` and `Lending`.
- `Catalog Search`
  The produced event of the `Catalog search` is renamed from `Book found` to `Book selected` to emphasize the user interaction. The event is still client-side, handled through the client's choreography, as designed in the original design.
- `Lending`
  This bounded context is enhanced by the submodules `Sharing` and `Rating`. Sharing items belonging to `Reading` are no longer part of the submodule. The submodule provides functions to share ratings, reviews, and reading lists. `Rating` is a submodule of `Lending` as well. Integrating those submodules to one bounded context allows for easier integration and minimizing requests. `Lending` no longer consumes member data changes from `Member Management`. It allows for the provisioning of member data or, better yet, user data via the identity provider's standard mechanism.
- `Reading`
  This bounded context produces additionally the `Book started` event. Notes are shared inside the `Reading` context without a generalized `Sharing`. `Reading` accesses digital books as standard interfaces from inventory management.
- `Appointment Management`
  This bounded context is a standard product with a small facade. An invitation is accessed by a volunteer reading to children and by a remote conference tool.
- `Teaching and Reading to Kids`
  This bounded context exists only virtually. An implementation is not necessary.
- `Remote Conference Tool` As a standard tool, it is integrated into the library frontend via a standard API.
- `Member Management`
  This bounded context is implemented using a standard identity and access management. It allows simple onboarding processes for members. The identity management system is also used for librarians and volunteers, whereas the onboarding processes are more complex.

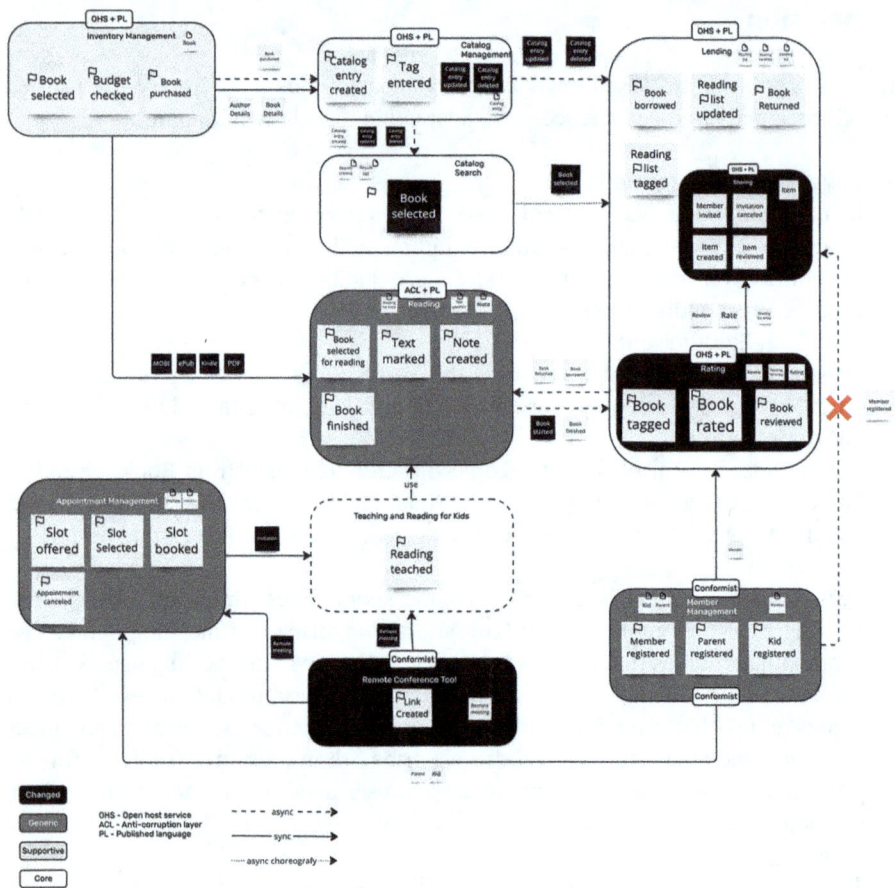

Figure 7-27 Changed context map

The changed context map documents those decisions during API design (Figure 7-27).

The changed context map leads to a simplified team structure:

- A team is responsible for inventory management and providing the reading component.
- A team is responsible for catalog management, search, and lending.
- An integration team integrates commodities like appointment management.

The team structure is shown in Figure 7-28.

In more detail, we will discuss the relations between architecture and organization in Chapter 9, "Collaborative Design and Agility."

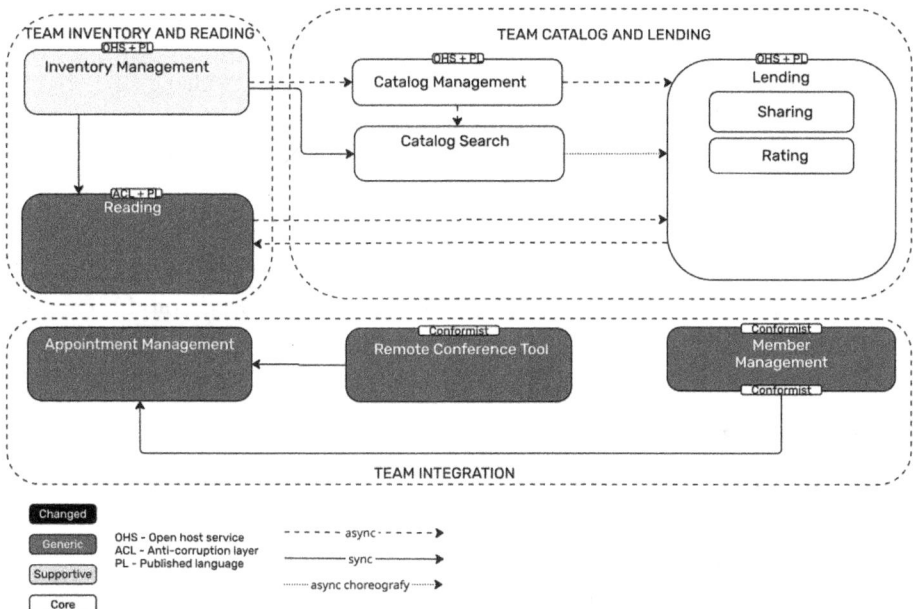

Figure 7-28 Team structure

The API design brought new knowledge about system structure and even deeper insights. The API Product Canvas is helpful in designing APIs. It can be used to steer necessary discussion and as a result used as documentation of the APIs of the bounded context. It is based on the gathered knowledge of Domain Storytelling and Event Storming (Chapter 5, "API Design Supported by Domain-Driven Design") but now brings even more insights. Based on those insights, the context map can be precise, and the team structure can be defined. In this way, we get independent teams. The implementation can be parallelized in client and server implementations, because the DDD process allows an API-First approach. The strategic design part is complete. The presented design of the online library can be stepwise extended following enhanced requirements; see Chapter 10, "Iterative Enhancements". Later in the book, we will discuss an additional example as a brownfield development (Chapter 11, "Brownfield Project").

Implementation can start after the API design step. We will discuss some important notes when starting with the implementation of an API in the following chapter, Chapter 8.

Points to Remember

- A **Bounded Context Canvas** describes a bounded context in its application design. It documents its ubiquitous language.

- An **Architecture Communication Canvas** describes the architectural approach of the associated bounded context up to the component view.
- Designing the APIs leads to architectural decisions that can change the original design of the Event Storming.
- Using the conformist pattern [19], the associated bounded context does not need an additional API design (section "Member Management.")
- Using the ACL pattern [19], the API design can be printed on the given API of the selected product.
- Using the changed context map, the team structures of the entire application can be defined. It shows that team structures deeply depend on the intended architecture.

Review Questions

7.1 What kind of canvas would you use to discuss a bounded context?

(a) Bounded Context Canvas
(b) Architecture Communication Canvas
(c) Business Model Canvas
(d) API design canvas

7.2 What kind of canvas would you use to discuss the architecture of a bounded context?

(a) Bounded Context Canvas
(b) Architecture Communication Canvas
(c) Business Model Canvas
(d) API design canvas

7.3 What kind of map would you use to define a team structure?

(a) Wardley map
(b) Architecture Communication Canvas
(c) Business Model Canvas
(d) Context map

7.4 Which language must be used to define an API?

(a) Technical language defined in a standard
(b) Domain-specific language defined in a standard
(c) Ubiquitous language as published language of the bounded context
(d) A new API language

7.5 Is it helpful to change the overarching service design when designing APIs?

(a) Yes
(b) No
(c) Yes, but only in the insurance domain
(d) No, but in exceptional cases, it can be done in a PoC

7.6 What kind of canvas is helpful when designing APIs for a bounded context?

(a) Architecture Communication Canvas
(b) Bounded Context Canvas
(c) API Product Canvas
(d) Business Model Canvas

7.7 Can an API be implemented after it is designed?

(a) Yes
(b) No
(c) Yes, but all APIs of an application must be designed first
(d) No, the by the bounded context used APIs need to be implemented first

7.8 Can an API be changed after it is designed?

(a) Yes, but only by the product owner
(b) No
(c) Yes, but good versioning needs to be applied
(d) No, changes are only allowed for legal reasons

References

1. Zalando Restful API and Event Guidelines (2024). Source at https://github.com/zalando/restful-api-guidelines. [Online]. Available: https://opensource.zalando.com/restful-api-guidelines/. Visited on 24 Aug 2024
2. Group D (2024) The bounded context canvas. [Online]. Available: https://github.com/ddd-crew/bounded-context-canvas. Visited on 28 Dec 2024
3. Junker A (2024) Bounded context canvas. [Online] Available: https://miro.com/miroverse/bounded-context-canvas-emplate/. Visited on 28 Dec 2024
4. Tune N (2020) Bounded context canvas. Visited on 28 Dec 2024
5. Starke G, Roos P (2024) Architecture communication canvas. [Online] Available: https://canvas.arc42.org/architecture-communication-canvas. Visited on 28 Dec 2024
6. Hruschka P, Starke G (2024) Arc42. [Online] Available: https://arc42.org/. Visited on 28 Dec 2024
7. C4 model (2024). [Online] Available: https://c4model.com/. Visited on 28 Dec 2024

8. Barbugli T (2023) [bug] inheritance with typescript using allof generates code without inheritance #15074. [Online] Available: https://github.com/OpenAPITools/openapi-generator/issues/15074. Visited on 30 Dec 2024
9. Vincent-Ch (2019) [java] "allof" does not support inheritance #3172. [Online] Available: https://github.com/OpenAPITools/openapi-generator/issues/3172. Visited on 30 Dec 2024
10. Open API Server - Implementation Using Openapi Generator. [Online] Available: https://www.baeldung.com/java-openapi-generator-server. Visited on 30 Dec 2024
11. Fowler M (2011) Cqrs. [Online] Available: https://martinfowler.com/bliki/CQRS.html. Visited on 30 Dec 2024
12. Richardson C (2019) Microservices Patterns: With Examples in Java. Manning Publications, Shelter Island, 11 pp. Includes bibliographical references and index. - Description based on print version record. ISBN: 978-16-17294-54-9
13. Sturgeon P (2024) Api design basics: Pagination. [Online] Available: https://apisyouwonthate.com/blog/api-design-basics-pagination/. Visited on 30 Dec 2024
14. Jwt (2024). [Online] Available: https://jwt.io/. Visited on 06 Jan 2025
15. Jones M, Bradley J, Sakimura N (2015) Json web token (JWT). [Online] Available: https://datatracker.ietf.org/doc/html/rfc7519/#section-4.1. Visited on 18 Jan 2025
16. Jacobson I, Christrson M, Jonsson P, Övergaard G (1992) Object-Oriented Software Engineering: A Use Case Driven Approach. Addison-Wesley, Reading. ISBN: 02-0154-435-0
17. What Is Oidc? (2025). [Online] Available: https://www.microsoft.com/en-us/security/business/security-101/what-is-openid-connect-oidc. Visited on 11 Jan 2025
18. Krebs B (2025) The openid connect handbook. [Online] Available: https://auth0.com/resources/ebooks/the-openid-connect-handbook. Visited on 11 Jan 2025
19. Evans E (2004) Domain-Driven Design: Tackling Complexity in the Heart of Software. Addison-Wesley, Reading

Part III
Enabling Transformation

How do we enable Domain-Driven API Design?

Developer Experience and API Implementation 8

Having discussed the definition of an API in Chapter 6 and practically designed it in Chapter 7 for the online library, we now discuss what is important when implementing the API. Some points we will discuss will also focus on the end of the design phase. However, they are closely related to implementation problems and are, therefore, covered in this chapter.

Versioning of APIs

APIs must be versioned. A version indicates a state in an API specification. Two different versions indicate a change in the API. Usually, versions are indicated by a so-called semantic versioning, or "SemVer" for short [1].

A semantic version number is represented by three numbers: MAJOR, MINOR, and PATCH [1]:

1. **MAJOR**
   A version change indicates incompatible API changes. Those changes are even called breaking changes because they break the current implementation. This could be changes in types or mandatory properties.
2. **MINOR**
   A version increase indicates additional functionality. However, the functions do not break the current implementation. They are backward compatible.
3. **PATCH**
   A number change indicates backward-compatible bug fixes. This can represent fixes like descriptions, examples, or typos.

A typical semantic version might be 2.0.1. When the more significant number is increased, the lower one is reset to zero, for example, causing a breaking change in 2.0.10 to result in 3.0.0. Additional prereleases can be represented as labels added to the patch version, for example, 1.0.0-alpha001 [1]. The version of an API specification in OpenAPI or AsyncAPI is given in the info field "version."

We would like to discuss different versioning concepts of API specifications because it represents a contract between the provider and consumer of the API. Both sides must rely on the contract.

Calendar versioning

Besides semantic versioning, other approaches exist. One variant is versioning based on the date of release, known as calendar versioning (CalVer) [2]. The release date indicates the version, for example, year and month. The Ubuntu project has used this versioning type since its launch, with the first release, 4.10, in October 2004 [2].

Furthermore, the version number must indicate a sequence of changes and the changes' significance.

When we speak about versioning of APIs, we want to allow different versions of the API to be available. This allows clients to invoke the deprecated or the latest version. A so-called big bang, where all consumers of an API and the provider itself must switch the version simultaneously, is avoided. Also, consider if a new version is the correct solution to the business problem.

What Is a Breaking Change?

Before diving into when to use breaking changes and how to manage them, let us briefly recap what a breaking change is.

In general, we have two types: semantic and syntactic changes.

Semantic changes   The semantic description changes the meaning. A semantic change can happen in a comment or description of an API without touching the specification in code.
Syntactic changes   A syntatic change is in the specification itself. Such a change can be validated by tools.

Semantic changes need to be reviewed by humans (or AIs with reasoning skills). Syntactic changes can be tested at least partially automatically. Therefore, it also makes a difference in which schema type the data are sent. We will look at JSON, Apache Avro, and Protobuf separately because different schemas have different schema evolution properties [3].

We use our book example from section "Data Formats and Their Schemas." First, we look at the JSON example.

```
{
    "isbn13": "9783161484100",
    "category": "NON_FICTION",
    "tags": ["DDD", "Modeling"],
    "rating": 4.5
}
```

Listing 8-1 Reduced book example in JSON

JSON

Here, we used the JSON example with fewer fields relevant to this example (Listing 8-1 with JSON Schema Listing 8-2).

In this example, we can add new fields as desired. This would not be breaking. However, removing a field that is required by a schema (e.g., isbn13) is a breaking change as the consumer/client of the message expects the field to be here; renaming a field results in the same problem.

Adding new fields can work without a breaking change when the schema has the property additionalProperties set to true, or the consumer still allows for the parsing of unknown fields.

The tags can be used quite freely. However, if the maximum length of 50 were extended, this would be a breaking change. Reducing it would be possible if the change came from the server on a REST API, and as a result, the object would only be used to send books but not receive them as this can then lead to still sending 50 elements by an old client.

If the tags are represented as strings, new tags can be added without a problem; the category is a fixed enum with only two properties. Adding another one would be breaking for clients consuming this object. As they may have a fixed logic, they are only able to handle these two values.

Avro

The main criteria for breaking and nonbreaking changes are the same; however, there are a few more nonbreaking changes in Apache Avro. Field names can be changed, and aliases can still be used to be compatible with the old version [4].

In addition, default values can cover the case where a field is not set. However, the default field does not make the field optional [4].

Protobuf

Protobuf has, like Apache Avro, the capability to rename fields without breaking and the capability to set default values to make interaction between new and old code smoother [5].

Having looked at what a breaking change is and how we can overcome it, let us now look at which changes are not compatible at all and which are backward, forward, or even fully compatible.

```json
{
  "$id": "https://example.com/book",
  "$schema": "https://json-schema.org/draft/2020-12/schema",
  "title": "Book",
  "description": "A schema representing a simplified book",
  "type": "object",
  "properties": {
    "isbn13": {
      "type": "string",
      "pattern": "^[0-9]{13}$",
      "description": "The 13-digit ISBN"
    },
    "category": {
      "type": "string",
      "enum": ["NON_FICTION", "FICTION"],
      "description": "The main category of the book"
    },
    "tags": {
      "type": "array",
      "items": {
        "type": "string",
        "minLength": 1,
        "maxLength": 50
      },
      "uniqueItems": true,
      "maxItems": 20,
      "description": "List of tags associated with the book (top 20)"
    },
    "rating": {
      "type": "number",
      "minimum": 0,
      "maximum": 5,
      "multipleOf" : 0.1,
      "description":
          "The average rating of the book (0 to 5 stars, with one decimal place)"
    }
  },
  "required": [
    "isbn13",
    "category",
    "tags"
  ],
  "additionalProperties": false
}
```

Listing 8-2 Reduced book JSON Schema

> **Automate breaking change detection with syntactic changes**

In section "Testing," we will look at how we can automate testing for breaking changes.

Backward, Forward, and Full Compatibility

Understanding if we need backward, forward, or full compatibility is essential. This decision depends on the use case. Depending on the answer to that question, a change can be breaking or not. Let us look at the types:

Backward compatibility In backward compatibility, data that were produced with an older schema can be consumed. In this mode, new consumers can read old data [3]. This is often used when a message broker is involved, as the message on the broker can be older, and the consumer needs to be updated. This is also important if a server still wants to accept requests from old clients (request data only, no response).

Forward compatibility In forward compatibility, consumers can read data using the defined schema with newer versions than the one defined at build time [3].
This is useful when old clients want to read server responses. In a scenario where the server gets updated immediately, clients need time to update.

When diving into this topic, it is always good to first think of some potential use cases, play that through in a test, and check if it works. Once the first version is created and used, it will be much more challenging or impossible to make nonbreaking changes if these concepts were not understood beforehand and care was not taken to implement them.

Consider If a New Version Number Is the Solution

Before diving into versioning strategies, first for synchronous APIs and then asynchronous APIs, we want to understand if a new version number is needed.

Many teams tend to do all types of updates with a new version of an endpoint. However, most changes that result in a new API have also changed business requirements.

Therefore, it is crucial to consider whether a new version of the same endpoint or a new one would be better. The endpoint or event name should also change if the business requirement has changed. However, if it is unnecessary to rename the endpoint, we need to version the endpoint or event. Therefore, we will dive into versioning strategies for synchronous APIs next.

Synchronous APIs

We will only consider REST APIs in this section. In REST, the version of the contract can be given in different parts of a request. The client can provide the expected version in a property, for example, in the header, or the server side might expect the client to call the correct version, for example, by providing the version in the URL.

Endpoint Versioning

A version in an endpoint might look like this:
https://online-library.org/catalog-management/v1/catalogs

The URL part "v1" means the major version 1 of the API specification. This concept is reliable because the client cannot invoke a wrong version. However, it is often unreliable as well because, with a new major version, all infrastructure components providing the API, like API gateways, firewalls, or even proxy servers, need to be updated at once if no preview concept was implemented to deliver multiple versions at the same time [6].

Header/Parameter Versioning

Another variant is that the client provides the requested version as a header variable or query parameter. We already formulated those in the examples of the online library in Chapter 6.

The versioning strategy has the benefit that the new version does not require changes along the request route, for example, changes in a gateway configuration or firewall [6].

This is, of course, only the case in default configurations; sometimes, special configurations should be different in different versions. However, the benefit of header versioning is that it keeps it small by design and not as focused on endpoint versioning.

Media-Type Versioning

A third type of versioning is so-called media-type versioning. It uses the content versioning capabilities of HTTP [7]. The version is formulated based on the media type of the request body and the response. The client does not need but can formulate the version directly in the request. The client can pass an Accept: header to negotiate a preferred response; see the example in Listing 8-3. This strategy can be used for endpoints or request/response objects, however, not for a whole API. It is a strategy that is quite advanced and not recommended to start with [6, 9, 10]. However, if implemented correctly, it can give way more flexibility as a consumer needs only to update an endpoint if something has changed. It can be applied nicely for many consumers with significant and advanced APIs.

```
GET /greeting HTTP/1.1
Accept: application/vnd.example.resource+json; version=2

HTTP/1.1 200 OK
Content-Type: application/vnd.example.resource+json; version=2.1.3
```

Listing 8-3 Example of media-type versioning with content negotiation [8]

> **Example of media-type versioning**
>
> The Adidas API Guidelines require that all APIs must implement HTTP content negotiation [8].

Keep the Flexibility

It is essential to formulate the API specification to be as flexible as possible with respect to supporting versions. We can make some suggestions on how to do so, as follows.

- **Extensible enumerations**
  Formulating string variables as enumerators is preferable because the requests are stable, and wrong information is avoided. However, if the enumerator needs to be enhanced, a new version of the API is necessary. It is a backward-compatible change, but the API specification needs to be changed. When other values should be allowed in an enumerator, "x-extensible-enum" can be used, which enables the formulation of an enumerator that allows additional entries [11, 12]. Formulating those allows additional entries in an enumeration without creating a new API version.
- **The response is always an object**
  A search result often needs to be delivered as an array because multiple result entries can be found. A JSON Schema allows formulating a direct response with an array. However, if later additional information is necessary, for example, who searched or when the search was done, a breaking change is required because the response must now be an object containing the additional information, in addition to the array containing the result set. To avoid those breaking changes, even a simple array should be wrapped in an object [11, 12].
- **Backward or forward comparability**
  Think about the use case and where an update will happen first, at the producer or consumer.

Asynchronous APIs

Versioning events and messages that go over a middleware, for example, an event broker, is even more difficult because, unlike direct peer-to-peer communication, there is no single point where the change can be made: Produced events can be read after days, weeks, months, or even years by new consumers and need to be there for compatibility (section "Eventing").

Therefore, it is preferable to avoid the necessity of versioning at all [13].

Avoiding Breaking Changes

When we look at events, we can avoid versioning. A business change should result in a new event and not in an updated one, as discussed in section "Consider If a New Version Number Is the Solution". Events should also be small and should not have too much data. Breaking versions should be minimized when this is strictly considered [14].

Double Writing

With a completely new version of an event, the producer must write both versions so that the consumers can handle either the old or the new ones. Both messages should be sent over different channels. Hence, the later shutdown of the old version of the event will be more straightforward to control. First, it can be ensured that nobody reads the events from the old channel. Then access to the channel can be removed before finally removing the channel. This strategy is called double writing [14] and is the strategy recommended by the authors. There can be variants of this; for example, either the producer writes both versions or the producer writes a single version and a stream processor converts the event to the other one.

Multiple Versions in the Same Channel

In this approach, different versions will be written with the same event identifier to the same stream. Consumers must read all messages and filter the ones they are interested in. So it is very costly because the consumer reads messages that are not intended for it [14].

In this case, it is very important to check if the use case should be implemented with forward or backward comparability. Depending on the selection, the producer or consumer needs to update it first. If none of the cases can be guaranteed, full compatibility needs to be guaranteed.

Conclusion

Events can be versioned, but this needs to be done carefully. We can discuss whether a new version is necessary or whether we can work with enhanced events or additional ones, but in any case, handling events and their versions can be difficult. A schema registry helps to handle the schema versions of events and guarantees compatibility.

Schema Registry

A schema registry is just a central location for storing different schema definitions. A consumer can access information on which schemas are available using a schema registry. A schema registry can be used for payload schemas for synchronous and asynchronous calls and responses (as well as headers or key schemas, but they are used less often). However, it is more usual for asynchronous communication, where time can pass between when an event or message is sent and when it is consumed. This also results in multiple versions that need to be read from a persistent storage, for example, a broker, at the same time.

Besides schema management, a schema registry provides different functions [15, 16]:

- It can be used to enforce the use of a particular schema in combination with a broker or proxy.
- It can be a repository where different schemas are registered, validated, evolved, and checked for interoperability.
- It allows for the management of schemas using compatibility rules.
- It allows for data validation against registered schemas on the producer and consumer sides.
- It provides schema identifiers with short references instead of using full schema definitions.

A schema registry's functionality is shown in Figure 8-1.

Usually, a schema registry follows a particular hierarchy, for example, a schema group with different schemas and a schema with various versions [15, 16]. As discussed, the schemas can be defined as JSON Schema, Apache Avro, or Protobuf schemas, for example, and be registered (section "Data Formats and Their Schemas"). All those schemas can be handled in a schema registry [16].

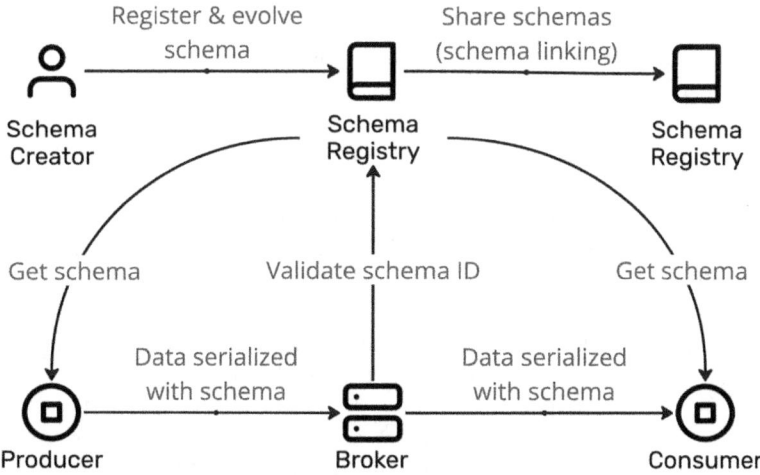

Figure 8-1 Functions of a schema registry

The use of a schema registry helps to avoid data inconsistency by agreeing on the schema at runtime, as the API definition does at design time. It prevents incompatible data formats between producers and consumers by validating schemas against the registry, for example, via a proxy or broker or directly by the producer and consumer. A schema registry supports versioning, allowing different schema versions simultaneously [16]. Between new versions, compatibility can be enforced, and the typical types of compatibility are as follows:

- **BACKWARD**
  For full backward compatibility, deleting fields and adding optional fields are allowed. The current schema is validated against the previous version. Consumers need to upgrade first [17–19].
- **BACKWARD_TRANSITIVE, BACKWARD_ALL**
  This means the same changes as the backward compatibility, but the data are checked against all previous versions [17–19].
- **FORWARD**
  Forward compatibility allows adding fields and deleting optional fields. It is checked against the last version, and producers must be upgraded first [17–19].
- **FORWARD_TRANSITIVE, FORWARD_ALL**
  The schema registry checks a new version against all previous versions [17–19].
- **FULL, ALL**
  To reach full compatibility, only adding and deleting optional fields is allowed. It is checked against the last version. The producer or the consumer can be upgraded first [17–19].
- **FULL_TRANSISTIVE, FULL_ALL**
  This kind of compatibility checks a new version against all previous versions [17–19].
- **NONE**
  The schema registry allows all version changes, and compatibility checks are disabled. The changes affect the upgrade order [17–19].

These rules will affect validity checks for syntax and semantic rules [18]. Those are principal rules. However, one needs to be aware that the compatibility rules differ for Apache Avro, Protobuf [20], and JSON Schema [21], as briefly discussed in section "What Is a Breaking Change?." Please refer to the aforementioned sources for more detailed information.

The principal functions of schema registry providers, such as Confluent,[1] Apicurio,[2] Microsoft Azure,[3] and AWS,[4] are comparable. However, there are slight differences. The documentation needs to be studied carefully.

---

[1] https://www.confluent.io/product/confluent-platform/data-compatibility/

[2] https://www.apicur.io/registry/docs/apicurio-registry/3.0.x/index.html

[3] https://learn.microsoft.com/en-us/azure/event-hubs/schema-registry-overview

[4] https://docs.aws.amazon.com/glue/latest/dg/schema-registry.html

```
1   asyncapi: 3.0.0
2   info:
3     title: Example with Schema Registry
4     version: 0.1.0
5   servers:
6     kafkaServer:
7       host: kafkacluster:9092
8       protocol: kafka
9       bindings:
10        kafka:
11          schemaRegistryUrl: 'http://localhost:8080/apis/registry/'
12          schemaRegistryVendor: 'apicurio'
13          bindingVersion: '0.5.0'
14    ...
15  components:
16    messages:
17      userSignedUp:
18        bindings:
19          kafka:
20            bindingVersion: '0.5.0'
21            schemaIdLocation: 'payload'
22            schemaIdPayloadEncoding: 'apicurio-new'
23            schemaLookupStrategy: 'TopicIdStrategy'
24        payload:
25          schemaFormat: 'application/vnd.apache.avro+yaml;version=1.9.0'
26          schema:
27            ....
```

Listing 8-4 Referencing schemas in a registry in AsyncAPI

In AsyncAPI, a link to the schema registry can be formulated in the server part (Listing 8-4 [22]). In addition, the location of the schema identifier and the lookup strategy can be defined in the binding part of the message. However, explicit references to schemas are not possible; only a hint of a schema registry is used. It would be possible to refer to a schema by a payload reference, which the authors do not recommend. This makes the API specification less self-contained [23, 24]. Referring to the schema version number used in the API, it is also only possible with a hint (`schemaIdLocation`). AsyncAPI does not directly support multiple schema versions, only as binding to a schema registry [25].

Using a schema registry allows consumers of an asynchronous API to be informed about new versions during runtime. That reduces downtimes of services and increases availability.

Knowing different versions of APIs and events can exist, we must test the compatibility of the API with the consumer and producer. The following section discusses how APIs can be tested on different levels.

Testing

An API needs to be tested. A well-documented API is only usable if it behaves as described. Therefore, we will look at testing.

Types of Testing

Testing in software development can be split into different categories. Here, we take the approach of *Thwaites* and split it by the feedback loop (Figure 8-2). What can be tested faster should be tested faster; everything more on the right side needs more time and is more expensive [26, 27].

In the following subsections, we will discuss the different test types. Telemetry is not a part of testing but is essential and will be discussed in section "Analytics and Monitoring." In section "Continuous Integration", we will look at how the testing mechanisms can be integrated into a continuous integration (CI) process.

Static Code Tests and Linting

The fastest feedback is given by a linter. A linter is a static code check that instantly spots problems with the code. The code is only statically checked, and no compilation of it is needed. This makes the feedback loop really fast.

A linter should be used in API design to spot, for example, YAML syntax and specification standard errors. It uses the schema of OpenAPI or AsyncAPI to check the file against the standard. Some linters check more advanced rules. For small projects, basic checks are mainly sufficient, but for multiple teams with their own rules (e.g., naming), a customizable linter with its own stylesheets should be considered.

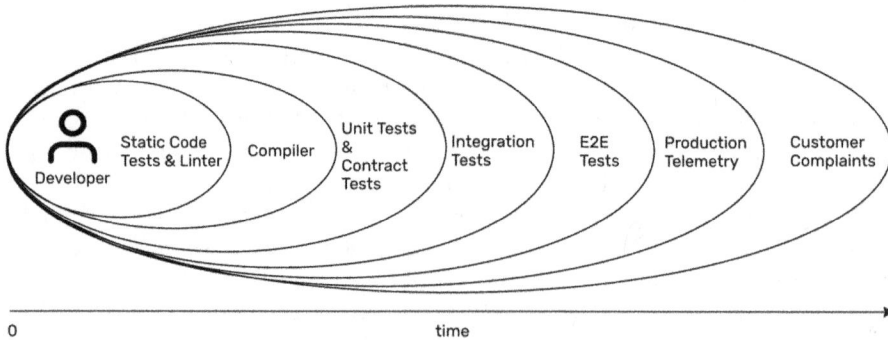

Figure 8-2 Testing feedback loop by Thwaites, extended [26, 27]

The code samples use the following linters:[5]

- A OpenAPI and AsyncAPI linter
- A JSON Schema linter
- An Apache Avro linter
- A XML linter
- A Protobuf linter

Along with these linters, some were already presented in section "Tooling for Linting."

In addition to classical linters, API definitions and data schemas should be checked for breaking changes. There are various tools for OpenAPI, for example, Spectral from Stoplight.[6] For AsyncAPI, there is even an official tool.[7]

Schemas can and should also be checked for breaking changes. A schema registry can take over this role (section "Schema Registry"). There are also full-fledged API registries that can combine all these version checks for APIs and data schemas. For example, Apicurio can act as OpenAPI, AsyncAPI, GraphQL, Apache Avro, Protobuf, JSON Schema, Kafka Connect schema, WSDL, and XSD schema registry. It can then also check for changes when somebody tries to upload a new version [28, 29].

> **Semantic vs. syntactic changes**

These tools can help find syntactic breaking changes. However, semantic changes that were, for example, just described in the description need to be manually checked. An example was presented in section "APIs Without Documentation and Unexpected Behavior": The middle name was misused to represent the second family name. This is for a simple tool, not a problem. It is a semantic change that will break the existing business logic. But it needs to be tested by a human for that. We looked at breaking changes previously in section "What Is a Breaking Change?."

Compiler

Compiler checks are executed when the compiler is running. These are needed to build the code anyway.

A compiler can be used to check errors in definitions. This can involve, for example, wrong annotations, type mismatches, and ambiguity.

---

[5] https://github.com/Apress/Crafting-Great-APIs-with-Domain-Driven-Design/blob/main/docker-compose.yaml and continuous integration (CI) automated with https://github.com/Apress/Crafting-Great-APIs-with-Domain-Driven-Design/blob/main/.github/workflows/validate.yml

[6] https://stoplight.io/open-source/spectral

[7] AsyncDiff is an official library to get information about changes and which are breaking: https://github.com/asyncapi/diff.

A compiler run should be executed multiple times at each commit. It is usually completed in less than a minute, but depending on the product, it can take up to hours.

For API consumers, a compiler can help quickly if a client calls a deprecated API. This is helpful after an update of an API to get informed on which endpoints need to be changed.

Unit Tests

Unit tests can test small parts of code or a unit. They should be designed to be able to run in isolation and, therefore, be great for spotting problems in code and reproducing problems in units [30]. The isolation of the tests also allows the test runs to be parallelized.

This is what is mainly used in concepts like Test-Driven Development (TDD), as shown in Figure 8-4.

In the scope of APIs, the remote APIs cannot be tested well with units, as they extend the scope of a classical unit; these will be tested with contract tests (section "Contract Testing") or integration tests (section "Integration Tests").

However, parts of operations and, sometimes, the operations themselves can be tested, for example, the creation of a thumbnail from the main image.

Besides the testing with unit tests for new features or requirements, unit tests should also be written when a bug is found (on a unit level). First, a test is written that fails and reproduces the bug. Then, the bug is fixed, and the bug fix results in a successful test.

Unit tests should run at least at every commit level to see if the Git commit was successful. A testing library normally implements unit tests; some popular examples are JUnit for Java, xUnit for C#, and pytest for Python.

Contract Testing

A special kind of test are contract tests. Contract testing can be used to test the interface specification as the contract between API provider and consumer. It makes it possible to test the contract earlier than in an expensive integration test (section "Integration Tests").

Meszaros first described comparable approaches called the test double [31]. The idea is that a mock or a stub replaces a remote dependent. In this way, a system can be tested as a whole, but the efforts of complete integration testing are avoided.

Contract testing is built on this idea. A consumer is not entirely interested in the working server. It is interested in the server's behavior as can be expected based on the API specification [32]. Therefore, the consumer can formulate a test.

The consumer formulates a unit test when accessing the API. The API provider builds the consumer's test in its continuous integration and continuous delivery/deployment (CI/CD) flow. The test can run as often as a unit test, and a successful test guarantees that the system will behave as expected by consumers after each change [33].

The API provider integrates the consumer's test in its workflow (e.g., in a CI pipeline; see section "Continuous Integration"). The flow is oriented to the flow

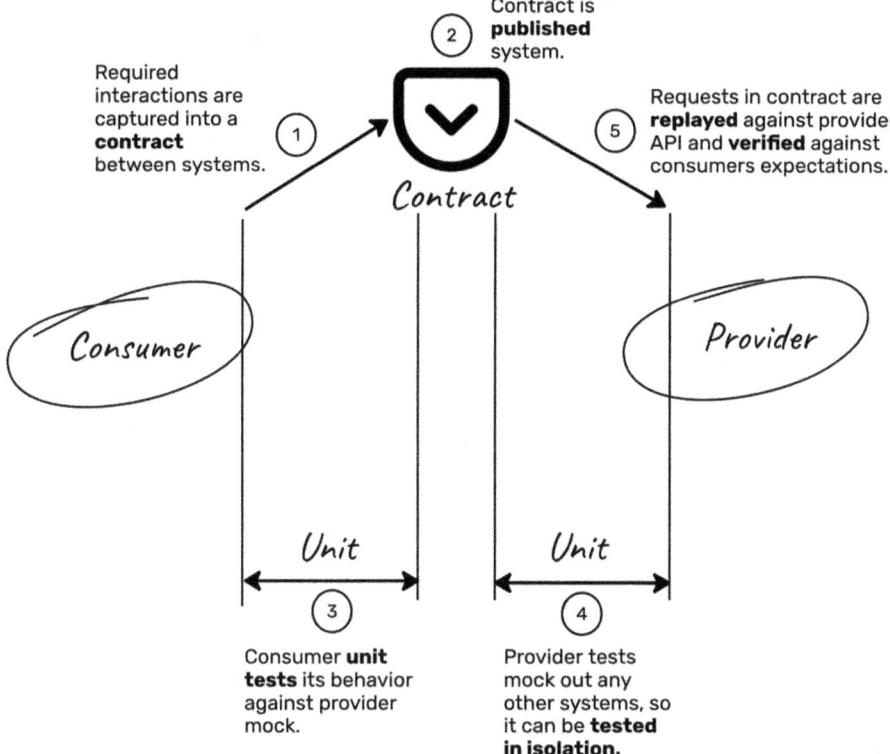

Figure 8-3 Contract testing

presented by *Fellows* [33, 34]. However, it stresses the preferred API-First approach (Figure 8-3).

The process runs as follows:

1. Interactions are captured in an API specification. The specification is created collaboratively by experts from the provider and the consumer sides (Chapter 5).
2. The contract is published. It can be published using a management system or just by sending it to the affected parties (section "API Management Platform").
3. Consumers can unit-test their system by mocking the provider.
4. Providers can unit-test their system by mocking the consumer.
5. Providers test the contract requests and verify the behavior against the defined unit test provided by consumers.

A contract test is a great way to guarantee that changes on the provider side do not disturb the expected behavior on the client side. Ultimately, it combines the idea of unit tests and the test double.

Moreover, it makes it possible to check for breaking changes, because changes can be tested. Even if the contract is changed, the expected change might not be changed so that a consumer is not forced to change its implementation. For example, adding an endpoint does not force a consumer to change its implementation, when the new endpoint is not necessary for the consumer. More critical might be a situation where a property is changed in a request. The consumer test can show if the corresponding consumer is affected by the change.

The tests are formulated as typical unit tests. This means that the consumer test provides a certain request and formulates the expected response. So the behavior can be tested on the server side without an up-and-running consumer.

The same approach can be used for events because the test double idea works for events as well [34, 35]. Pact is an example of a contract testing tool that allows for contract testing of synchronous and asynchronous APIs.[8]

Using contracting testing closes the gap between unit and integration tests. We will discuss integration tests in the following subsection.

Integration Tests

Integration tests are bigger than unit tests and can test much more. These should be used carefully, as they normally are hard to maintain and take a long time to execute. Each additional test normally extends the runtime, as they can normally not run in parallel because of the extra components that are needed. These tests usually need to start components (e.g., in a Docker or Open Container Initiative (OCI) container) or use instances of existing test deployments (e.g., use an already existing test database) [30, 36]. They can also be used to reproduce bugs as they were reported and retest them to ensure they have been resolved and no longer appear. However, they should be used carefully as they have a slow feedback loop. The authors have seen systems that need a couple of hours to execute integration tests, which, to reduce costs, run only once each night on a single development stage. Getting feedback and seeing why a problem happened is complicated when the loop runs so long. A team needs a whole day to determine which change causes the test to fail.

It would be much better if it were possible to keep the integration tests set up smaller so they run in at least on each pull request of a developer so that the problem is addressable. An additional solution can be splitting the integration tests into subparts that run per pull request and a subset that only runs once daily.

Here it is also important not to be tempted to test things that are given by a framework. For example, the call of a controller and deserialization of a data type come from the box and normally work without issues. In that case, the test should simply check the logic inside the controller. Of course, the HTTP call can also be tested. However, this will make the test run a lot longer and make the feedback loop larger. In addition, such tests are harder to maintain, as there is more logic in them (e.g., serialization, deserialization) and harder to find a problem when they fail.

---

[8] https://docs.pact.io/

Along with the classical integration tests that verify functionality, additional integration tests can be created:

Load and performance tests Load and performance tests can verify how APIs are performing in general (speed, latency, concurrency) and under load [37]. These tests can be very important in tracking key performance indicators (KPIs) over time and verifying that performance does not degrade over time and that changes do not harm these KPIs [36]. Here, tools like Gatling or Apache JMeter exist to execute such tests.

Reliability testing Testing the components for reliability is also crucial. This can happen with recovery tests in classical scenarios. In distributed applications, it is even more important. There, the practice was extended into chaos engineering, which has set up practices and toolings to improve reliability; this is a very interesting topic that we will, however, not be able to cover here [37].

E2E Tests

End-to-end (e2e) tests are bigger than integration tests. They are closest to real-world scenarios. Here, the application, or part of the application, needs to be deployed in a test setup. Then the parts can be tested together. Whole flows and processes can be tested. In addition, the behavior of the system can be tested. These tests also often include UI tests. The tradeoff is that these tests are pretty slow, and when an error causes failure, it is not always easy to find the problem [30]. Such a test can, for example, test "purchase book." If the test fails, it needs to be investigated whether there is a UI problem, API problem, deployment problem, business logic problem, or database problem.

When using REST APIs in the browser, this is often only tested with End-to-end (e2e) tests as these tests are the only ones that should cross bounded contexts. These tests will verify that the bounded contexts can integrate with each other.

It is also often the case that e2e tests are executed fully or partially manually. Sometimes, these are also the final acceptance tests.

If the tests are automated, tools like Selenium or Cypress can be used for web browser–based e2e tests.

In addition to the functional verification tests on e2e tests, other tests are normally seen, in addition to the extended concepts presented in integration tests (load, performance, and reliability tests):

Security tests Security tests should be implemented on all levels. However, API-based security tests are mainly seen at an e2e level. Here, APIs are tested for vulnerability, for example, with penetration tests against the OWASP top 10 [38]. There exists far more security tests; this topic will, however, not be discussed in this book.

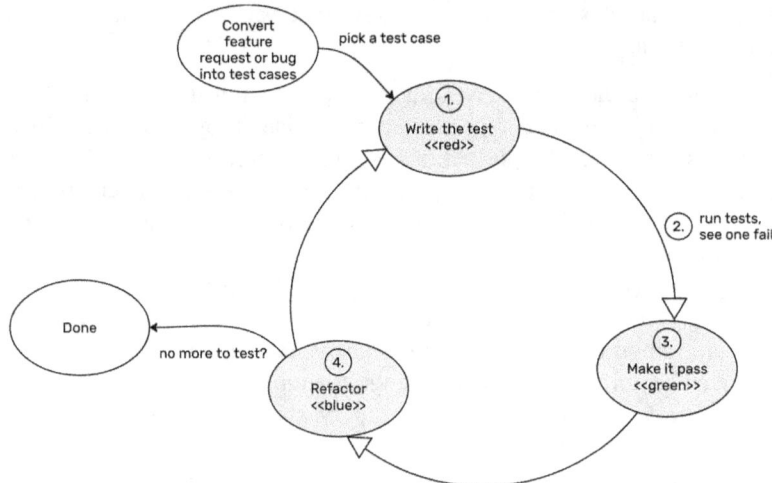

Figure 8-4 The circle of Test-Driven Development (TDD), also called Red-Green-Refactor (or Red-Green-Blue) [40]

Concepts

These types of tests can be used inside different development approaches. Next, we will discuss some interesting concepts.

Test-Driven Development

TDD is a concept focusing precisely on this: The developer first implements tests and then carries out the implementation until the tests are green (Figure 8-4). The concept works as follows [39, 40]:

1. Add a test.
2. Run all tests and see if the new one fails.
3. Change the code so that the test goes to green (do not add more).
4. Run all tests until they all go to green (successful).
5. Refactor if necessary.

This can be applied to API testing as well. We first write the API and our feature request and then add tests. These tests can then be implemented. After each test the implementation can be adapted.

TDD emphasizes an "outside-in" view: The interface/API is in focus, the implementation is second. This implements the API-First approach on a unit-test level. The contract is defined before implementation.

Behavior-Driven Testing

Behavior-driven development (BDD) tries to close the gap between business experts and IT specialists because the tests can be formulated as business requirements. It supports building a shared understanding of what needs to be executed and tested, the precise goal of the DDD workshops in Chapter 5, "API Design Supported by Domain-Driven Design." In addition, BDD also tries to add documentation as part of testing and reduce feedback loops [41, 42].

BDD derives its roots from DDD, TDD, and Agile and is therefore a perfect approach to testing APIs defined by the proposed DDD process and communicating low-level requirements.

The goal is not to have business requirements that then get translated to test code but to have a written statement that supports as documentation and as a runnable test. This is done using a common language that is easily understood by all parties but also reduces ambiguity, like a programming language. The language should be seen as an executable specification that can be used as acceptance tests, guidelines for developers when implementing the code (after the text, as in TDD), and regression tests as the test ensures that the implementation will not be broken afterward [42,43]. The language needs to use the ubiquitous language defined in the DDD process.

With BDD, the TDD approach gets extended, so the first written tests are now specifications defined by the business and developers together. These will then be as in TDD. The developer not only starts implementing the code necessary to fulfill the test but also ensures that no information is lost between the business experts and the developers [42]. In the authors' experience, this approach is not enough; that is why this book was written; however, this would also be a helpful approach, and the methods can be combined. Mainly, TDD adds low-level specifications and reduces friction between testers and developers. The ubiquitous language of the DDD process needs to be used to be consistent in the code and test. We will discuss this in Chapter 9, "Collaborative Design and Agility."

> **More on that later**
>
> In section "Behavior-Driven Design," we will look at how we can bring BDD into the process of designing APIs in a collaborative way.

In practice, BDD can be implemented with a framework like Cucumber or JBehave [42].

Conclude Testing

We presented the different layers of tests with different feedback loops. In addition, concepts like TDD and BDD were presented to help build the software right, not just the right software. These concepts also worked perfectly with the API-First concept.

This should be seen as a starting point to look into the area as it is important to use testing, do it on the right layer, and include the API focus there as well.

Continuous Integration

Continuous integration (CI) is a technique to integrate software frequently, at least daily. This integration is part of testing and building a project with the latest changes from team members. The goal is to reduce integration problems and find mistakes faster [44, 45].

In practice, this involves all developers building their code locally, pushing it to a version control system (usually a couple of times a day), and then, the code gets built and tested by an agent. All tests must pass, and the build must work (sometimes without any warnings). This is shown in Figure 8-6, steps 4, 8, and 11.

A broken pipeline or a failed build must be the team's highest priority to ensure the software works [44]. These tests can involve all steps from section "Testing," but do normally include at least everything up to integration tests. Sometimes, e2e tests are not covered, but they can be.

CI does not necessarily need to be automated by definition. However, it is typically automated because this approach enhances reliability, increases the frequency of integration, speeds up the process, boosts confidence, improves resilience, provides transparency, reduces costs, and ensures compliance. The primary reason for automating CI is that it encourages more frequent integrations. Furthermore, the code itself serves as documentation, clearly outlining the process by eliminating the risk of divergence between documented processes and actual practices, a common issue with traditional documentation. There is no chance of outdated documentation, as the code should serve as the state of what happens. Additionally, automation helps to minimize errors that can occur when a person performs repetitive tasks such as deploying an application [44–46].

Git Flow

The general standard to use as a version control system (VCS) is Git. Therefore, we will only look at Git and use some Git-specific naming. Mostly, the approach gets combined with a simplified git-flow approach, visualized in Figure 8-6 [47]. The approach has core features that each developer uses for each bug or each feature and its own branch, which should be quickly merged into a main or master branch (Figure 8-5). The original git-flow approach has a great system to support hotfixes and multiple released versions; however, to get started and for most projects, a simplified approach is more practicable (e.g., GitHub flow[9]) [47].

---

[9] https://docs.github.com/en/get-started/using-github/github-flow

Continuous Integration

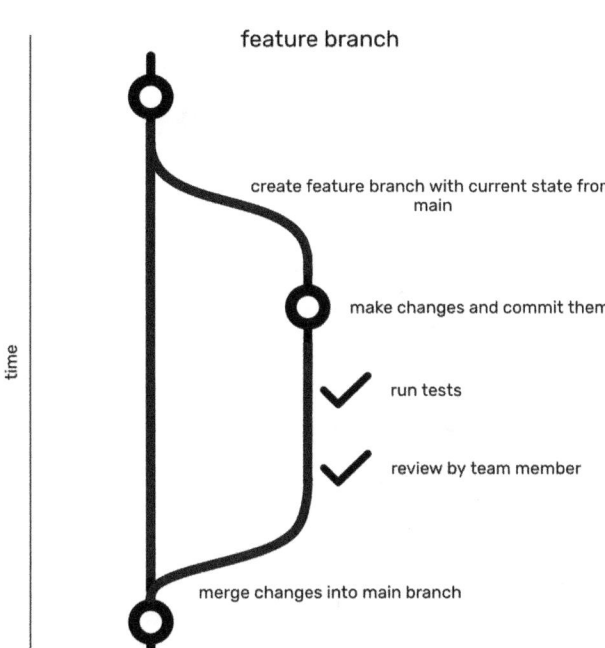

Figure 8-5 Simplified git-flow

Usually, the CI runs on the main branch on every commit on this branch [45]. Features and bug fixes are tested when a pull request (PR) gets created (step 7 in Figure 8-6).

Continuous Deployment

A closely related term is CI/CD. Continuous integration and continuous delivery/deployment combines continuous integration (CI) with continuous deployment (CD). With continuous deployment (CD), release to production is done frequently. The goal is to bring each commit to the main branch to production as well. This will, of course, only be done if the CI runs successfully and maybe even more checks are passed [45]. The combined process is visualized in Figure 8-6.

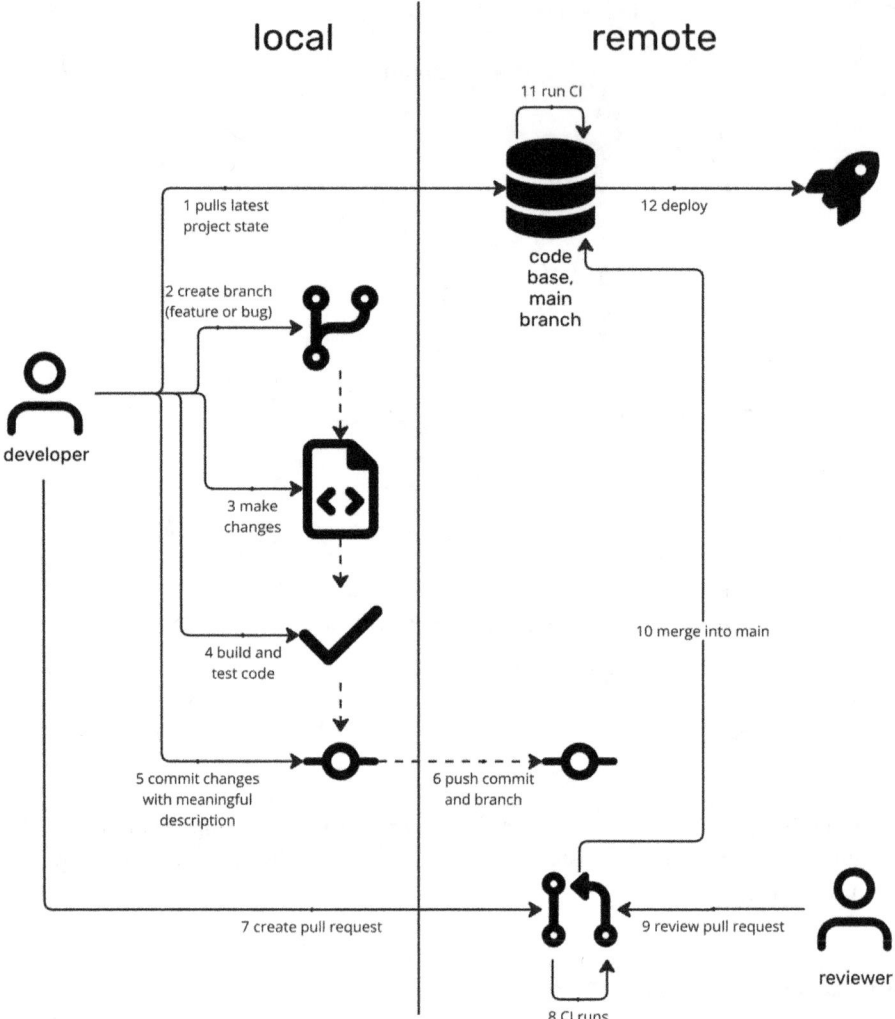

Figure 8-6 Continuous integration and continuous deployment combined with a simplified gitflow approach

Benefits and Use of CI/CD with APIs

The discussed approaches allow for better quality by reducing MTTR by faster and more reliable deployments and rollbacks, as well as a better understanding of the processes, as they are implemented, not just written, in plain text [44, 48].

The approach is also great for APIs: It helps with fast and small releases, allowing one to see progress and work faster and collaboratively with other teams. Small parts of an API, for example, one endpoint after another, can be released and already integrated by the other team. Problems can then be identified faster, and the time to market is shorter.

One must also consider how to implement big features or big new endpoints that do not allow a fast release. Here are two approaches that can help:

Keystone interfaces In this approach, a feature is first implemented in the backend; the business logic gets implemented piece by piece and tested with unit codes. The code is, in addition, tested by the CI and other tests; library parts can also be reused in the code by other changes. However, the endpoint does not exist to call it. This part will be built last so that a feature will get exposed only when it is finished and fully tested [45, 49, 50].

Feature flags These flags activate or deactivate code by checking a flag. This can normally be decided by the environment, allowing them to test the feature in a test environment but not plan a productive release when it is not ready for production. However, a feature flag should also then be tested in the deactivated position in a test so that it is clear that nothing is broken when the feature is deactivated.[10]

> **Are keystone interfaces a Code-First approach?**

Some can think that keystone interfaces are a Code-First approach, as you will implement the code before the API is visible. But actually, it is not. We start with API-First, as we first design the interface. Just in the implementation step, we then jump from the interface to the logic part behind it before we expose the interface. A controller can also be implemented before its function is implemented. However, when we use TDD (section "Test-Driven Development"), we will not be able to conduct a green test until the logic behind it exists. Therefore, we can start with a controller; however, we will not be able to finish it until we implement the parts behind it. These can be tested and implemented, and then already go through the PR review process, be merged, and potentially go to production before the API/controller layer is finished.

Conclusion

In this section we looked at CI/CD concepts and showcased notions of how to implement APIs by following an integrate daily approach. In the next section, we will look at API management and see how we can combine CI/CD with API management.

---

[10] Something like feature flags can be simply implemented by a library, for example, by the open-source project https://openfeature.dev/.

Publishing APIs

In this section, we will focus on the work that needs to be done to publish APIs and to discover them.

We start with the classic API Management platform, then turn to lean alternatives such as API Gateways, and wrap up the section with API discovery and ApiOps.

API Management Platform

Classic API Management platforms include an API Marketplace and API Gateways. They are centralized and often managed by a single team that makes APIs available to external users. Some companies use them for every API exposed by a bounded context, whereas others use them only for public APIs.

We will look at the components in API Management platforms and their roles. They can often become a bottleneck and slow the process of publishing APIs. An alternative approach is ApiOps; we will look at this process, too [51]. We recommend only using a classic API Management platform for public APIs and using direct access or a service mesh for internal APIs with separate discovery tools. Service mesh will be discussed in section "Service Mesh". Section "API Discovery," will describe internal developer portals.

> **API Management is more than a platform**
>
> The term API Management contains not only the platform for APIs but also the associated processes, governance, and organizational mindset with APIs [52, 53].

API Gateway

An API Gateway is a component that delivers the API to a consumer. It is a proxy that can provide security, monitoring, and routing functionality [53].

Let us discuss the key benefits of an API Gateway:

Security An API Gateway can act as a single, clean endpoint to all APIs in the company and provide access to them with a centralized security model [53]. Authentication can be managed centrally for external partners and customers, and authentication with central policies can be provided.[11]

Monitoring As a central component, the API Gateway can act as a good overview of the APIs that are used and by whom. Scaling and throttling can be implemented here as well; see section "Analytics and Monitoring," for more details.

---

[11] The gateway than acts then as a policy decision point (PDP).

In addition, rate limiting, quota management, and metrics can also be added to the API Gateway, as the metrics are there. Those features even support a billing solution.

Routing This component is responsible for routing to the correct endpoints. Examples of this use case are as follows:

- **Combine different APIs** The component can also combine different internal APIs into a single API (if the logic is simple, alternatives will be discussed in section "Publishing APIs.")
- **Version management** This component provides multiple API versions at the same time and manage compatibility.
- **Routing** A gateway is in the end also a router that sends data packages from one place to another. Therefore, it can be used for, for example, Open Systems Interconnection (OSI) Layer 7 (e.g., HTTP) and Layer 4 (e.g., TCP) routing.[12]

Although an API Gateway can combine all these capabilities, it should be carefully considered if all of them should be used. This centralizes a system and can lead to bottlenecks. Combining different APIs into a single one is convenient; however, would this not mean that the initial API was poorly customized? If necessary, it should be done by the team developing the corresponding bounded context, as otherwise, it is in a different team, managed and released in other cycles. In addition, this also means that the API of the team should not be used by other teams but only by externals over the gateway (as the API itself is not nicely cut).

API Marketplace

An API Gateway often comes with an API marketplace. A marketplace can be used to discover published APIs. This will be discussed in section "API Discovery." In addition to this, it can also help to onboard new users with self-registration and user management for external users; billing and rate limiting can also be integrated.

Sometimes it makes sense to separate the external API marketplace from an internal developer portal (section "API Discovery"), as the external one has fewer details than the internal one. However, this differentiation can make the process harder and hinder teams from exposing the API in all available places.

Backends for Frontends (BFF)

As discussed in section "API Management Platform," an API Gateway can also act as a single entry point and change and simplify APIs from different microservices into one. But when this action becomes too complicated, a Backends for frontends (BFF) should be considered [54].

---

[12] Even though an API gateway can be used for things like network zone routing that can be done with a simple Layer 4 router, it should typically not be used for this as a single purpose as there are better and cheaper solutions.

> **Warning: The need for a BFF or a gateway can be the source of a bad API**
>
> Be careful before considering a BFF or a gateway. Ask yourself why it is needed and what it solves. The need for a BFF or a gateway can indicate that the API was not designed, cut, or implemented properly. If this is the problem, iterating on the design would be a better solution.

The Backends for frontends (BFF) is a commonly seen pattern, where for each frontend type, for example, a web frontend and a mobile app, a separate backend is built. This backend should not contain business logic but only help to gather all information from different APIs to serve them specifically for a frontend, for example, minimize client-server communication with fewer calls or bundle requests to different APIs to reduce latency. This backend is normally maintained by the team that builds the frontend.

It is crucial not to put business logic into it to ensure the domain separation is not lost. To deliver a better user experience, logic and behavior should only be contained in a BFF [55, 56].

Service Mesh

Service meshes address the challenges of enormous service deployments and the need for secure, resilient, fault-tolerant, and observable communication between the services that come with a microservice architecture. A service mesh controls all service interactions by adding a dedicated layer. It is often implemented with small, transparent service proxies between the services [57].

Service meshes are often deployed in a cloud native setup, (e.g., on a Kubernetes cluster) [57]. Some popular products in the Kubernetes space are Istio, Linkerd, or Cilium.

In practice, the following areas can be addressed with a service mesh:

Traffic control   Discovery of services, route traffic, including OSI Layer 7 routing (e.g., by HTTP headers) for, for example, canary deployments, and load balancing between instances [57].

Security   Block and allow traffic between service instances and to the outside. Authenticate and authorize service-to-service communication and encrypt traffic transparent to the application with mTLS. It can also add rate limits and so on [57]. This is an enabler for zero trust.

Analytics   Track and correlate interactions between services and generate traces.

Resilience   Service meshes can inject faults for testing and act as a circuit breaker [57].

Service mesh is mainly a concept used with microservices; it can, however, also work with other big deployments of services [58]. A typical pattern is to deploy a "sidecar" (also known as sidecar pattern [54]) next to each service that handles the

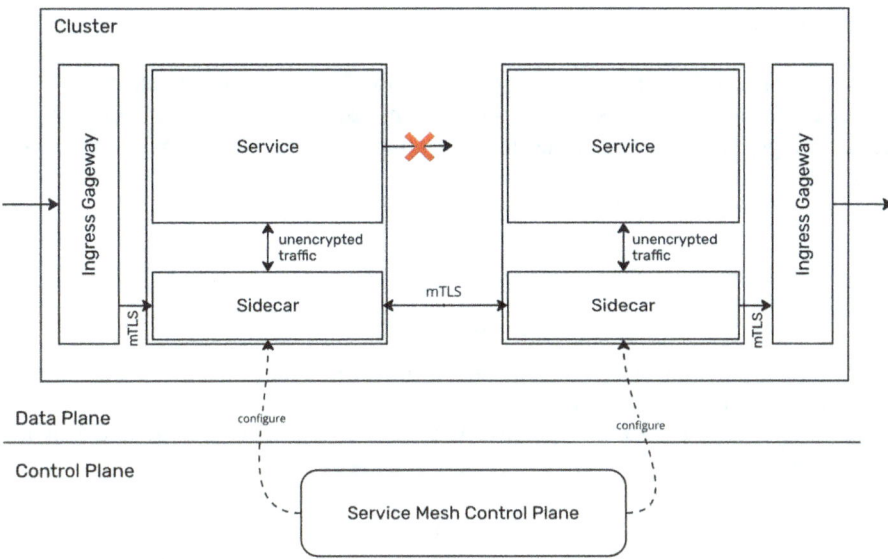

Figure 8-7 Service mesh with sidecar in a cluster deployment

service mesh functionality for each service [58]. This can include routing, retrying failed requests, and terminating TLS connections (Figure 8-7).

Be aware that a service mesh can bring some significant advantages; however, it is not free and needs know-how and resources to be deployed correctly. In addition, such a classic service mesh with sidecars uses a lot of resources. Newer versions do not use a sidecar but concepts like ambient mesh (in Istio) or Extended Berkeley Packet Filter (eBPF) (in Cilium) to overcome these disadvantages [59–62].

API Discovery

API discovery is the process of finding an API. This process is usually done by developers or architects who can search for a product, but it can also come from a business. External product APIs can often be found in an API marketplace. However, in addition to that, a developer portal can often be used internally to help developers acquire a deeper view into all internal APIs available (as well as components and libraries).

Along with human discovery, discovery can also be automated. There are automated tools to find undocumented APIs in a company. On the other hand, registries can help systems find and update APIs to the latest version. Schema registries like Apicurio, as discussed in section "Schema Registry," can be used for that.

ApiOps as a Process

ApiOps describes the process of bringing API design, implementation, and run closer to each other, and its validation is automated. It does what DevOps does for code. The goal is to make it easier to make changes and deploy them in an iterative way [51, 53, 63].

The process has the following steps [51]:

1. Store API definitions in version control (e.g., Git).
2. Check API compliance, with linters and other static code testing tools (section "Static Code Tests and Linting").
3. Run API tests against contract (section "Contract Testing").
4. Store declarative API configurations in version control.
5. Reconcile the desired state of the API, so that the API is in a state that the repository has.

The process works perfectly with API-First. Be aware that this list focuses only on the main points. These steps also need to be combined with what was discussed in section "Continuous Integration." Depending on the regulatory environment of the API, sometimes approval needs to be given by a human or by a business and IT board.

Developer Experience When Integrating an API

An important aspect to consider when implementing APIs is the developer's experience with the integration process of consumers. APIs are used by developers. Therefore, we need to look at their experience. This enhances the quality of the user experience in the area of flexibility (section "Flexibility"). As discussed, this can be done by adding observability, user error protection, self-descriptiveness, self-contentedness, and understandability to the API specification.

This can be done by a good API design, as discussed in Chapter 5, "API Design Supported by Domain-Driven Design," but also by some implementation factors, which we will discuss here.

An API should be seen as a product: API as a Product (AaaP). It goes through all processes, like a product, and is handled as a product that should be sold. With this aspect the developers are the customers, and they should have the best possible experience [53].

The publication and discovery of APIs were already discussed in section "Publishing APIs." Now we will focus on the steps an integrator wants to have when an API is found and the implementation process starts.

Sandbox

A very important factor is to be able to test an API. There is nothing more frightening than to need to start working with an API that you do not know on the basis of productive data. How should one try to test an API function and at the same time try not to make changes to a productive system?

Therefore, it is very important to give developers access to test or sandbox systems where they can test the API. The access should be simple, and they should be able to observe the data. Even though access should be simple, it is very important that the API behave as similarly as possible to the production system. Thus, different authentication mechanisms make it harder to know how it will behave in production and should be avoided.

A very elegant implementation of a sandbox can also be used in an OpenAPI, so that the developer can test API calls directly in the UI representation of the OpenAPI.

The author considers a great example to be the implementation of Stripe.[13] Stripe provides in their documentation an API key that can be used to test their API without the need to generate an API key. However, if you want a real test environment, you can sign up and get a test environment. Keys can be generated in their API.

Observability

Getting insights into an API is mainly important for the team that provides it. This will be discussed in section "Analytics and Monitoring." However, the caller also wants to know what happens and obtain insights.

The first and simplest thing is a status page, where the integrator can check if the service should work as expected, normally indicated by a green check mark if everything is okay. The next step is to also provide historical data and explain incidents that happened. This helps consumers of APIs to know when something does not work. The error does not need to be on their side. Historical data help to pinpoint problems.

An excellent example is the status page of AWS.[14] It shows the uptime per service and API, per region and day. In addition, open issues are shown, and historical events and planned changes are visible. There is also an option to be notified about an incident.

If such a service were built up, it would be important to isolate it from the other services, as the website should be reachable to get in touch with customers if there is a major incident. A good option for that is to also buy this dashboard as a service and get the first simple tests as well as a service offering. A very simple solution is

---

[13] Stripe is an online payment-processing and credit card–processing platform; stripe.com.
[14] https://health.aws.amazon.com/health/status

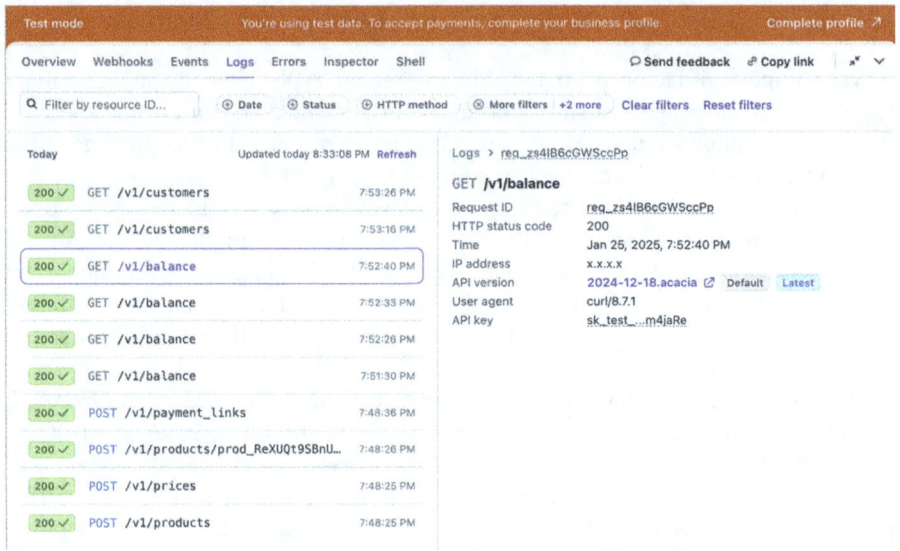

Figure 8-8 Screenshot from Stripe dashboard (https://dashboard.stripe.com) showing logs of all API requests; IP address was anonymized

updown.io.[15] These services regularly check if endpoints are reachable and notify the provider if they are down.

Even better would be a dashboard, where customers can see what they have called and what came back. An example is shown in Figure 8-8.

Ability to Get Help

It is also very important to find help and be able to get support when you are stuck. Sometimes the documentation is not enough, and the observability does not help, either.

Providing an email where support can be requested is very important.

Software Development Kits (SDKs)

Sometimes it can also help to build SDKs around an API. This can make it easier to integrate an API into an application for a developer.

Certain products make sense to use an SDK, for example, metrics.

---

[15] https://updown.io/

Sometimes they can facilitate integrating an API into a framework or they can facilitate the proper use of an API by some client-side logic, for example, by combining certain API calls together or reauthenticating if needed.

SDKs can sometimes also help to hide breaking API changes.

However, bear in mind that building a good SDK in multiple programming languages is a huge effort and complicates testing. Each programming language has its own styles that need to be respected. If the SDK is not implemented with great care, it may be better to not implement an API and only provide a good API.

Consequently, we also show here an example of Stripe. Stripe provides multiple SDKs to facilitate handling with their API.[16]

Other use cases for SDKs are when API calls are combined with client logic. For example, Google Maps[17] uses an SDK that facilitates the integration and integrates a view on the client side.

Common Standards

Whether or not SDKs are used, consistency and use of standards help users to navigate around APIs quickly and to use tooling support. Therefore, adopting standards and being consistent with APIs make sense. In section "New Requirements and Challenges in the Cloud," we will review some important cloud best practices and refer to further sources.

If an SDK is provided for an API, it should follow the same semantic versioning as the API (see "Versioning of APIs" section).

Use a schema registry to be able to get schemas over an API, as mentioned in section "Schema Registry."

Other best practices and standards were discussed previously in Chapter 6, "Interface Definitions."

Data Mesh Concept

In section "Different Scenarios for AsyncAPI," we briefly touched on the term "data mesh." Now we wish to take a deeper look at the concept.

To understand a data mesh, it is easy to start with the past. In the past, centralized databases existed with centralized data schemas. With the shift to microservices and agile working teams, this has changed to a looser coupled organism. However, in many organizations, data analytics and processing of data from different areas happen in a central team. The data are centralized into a data warehouse. A data warehouse requires structured data, which are hard to get with different teams having their own data models. Therefore, data lakes were implemented which pulls raw data from all sources together. The problem is, then, however, that these data

---

[16] https://docs.stripe.com/sdks

[17] https://developers.google.com/maps/documentation/javascript

end up as a data swamp. As the data model of the source team will change and the data in the "lake" will change as well, without the know-how of that central data team. This team needs, then, a lot of resources to again understand the data and clean the data to finally be able to use them.

The data mesh concept tries to bring the shift from microservices to the analytics space by shifting the responsibility to the operational teams: Each team that processes business logic in a domain should also be required to provide analytical data through APIs to interested parties.

With that shift, data should be seen as first-class citizens and as products the team is serving: the Data Products.

Consumers who need analytical data now should directly request it from the team that implements the business logic.

For that new approach, centralized discovery for these Data Products and APIs is needed, as well as the infrastructure to provide them. In addition, a federated governance structure should exist [3, 64, 65].

After reducing the cost, the approach also has the benefit of speed, as the new Data Products provided directly by the team can be consumed faster, as there is no data team in between.

Analytics and Monitoring

To provide a great API during runtime, insights into the deployed environment are needed for the implementation team. Therefore, quality attributes need to be measured, and errors must be tracked and analyzed. Fostering metrics, logs, and traces is the preferred way to do that. Of course, here we improve the quality perspective observability [53]. This section extends what we have already discussed about monitoring for the consumer in section "Observability."

Before we dive into the capabilities for which analytics, tracing, and monitoring can be used, we will briefly describe the terms:

Metrics These are a measurement of quantity, for example, the amount of API calls per hour.

Logging Logs are entries on a time axis. They help in reproducing problems and events.

Tracing A trace captures a program run or API call, including many substeps. This can help, for example, see how long an API call needs to execute.

Error Discovery

In section "Testing," we presented the testing feedback loop in Figure 8-2. Production telemetry helps in obtaining the insights that were not possible to obtain in the faster testing loops. However, production telemetry helps find errors before users see them.

Logging and tracing can be used to record, report, and finally eliminate errors. The error reporting can happen at three levels [53]:

Caller level/end user When the user needs to report an error, the feedback loop is bigger, and the experience is lowered; this should be mitigated.
Gateway level Errors are reported by the gateway, which can be part of an API management solution (section "API Management Platform").
Service level Errors are reported directly by the service that implements the API.

Usage Insights into APIs

After problems are found, it also helps to know if an API will be used at all or how much. Tracking all requests can also give insights into the use of APIs and can give hints as to what should be extended or optimized.

The data can then also be used to formulate objectives and key resultss (OKRs) or KPIs and measure a team or product by its APIs [53].

Billing with Metrics

Metrics, for example, of the number of API calls, are perfect for billing an API. Such a concept can be used to build the API as a Product (AaaP) and to sell it like a product as well.

Analyzing Performance

Tracing can also help to optimize the performance of an implementation. However, this will not be discussed in greater detail in this book.

Conclusion

We looked at some important aspects that should be considered when an API is implemented. This chapter should be considered a starting point for methods and concepts; however, more in-depth research will be needed to apply them.

Points to Remember

- Be careful with breaking changes and check whether they constitute API changes or a new business requirement representing a new endpoint.
- Different levels of testing have different feedback times and the tests should therefore test things as fast as possible.

- Contract testing can speed up development and bring assurance to the contract as an API.
- TDD and BDD are concepts for writing tests that create a value add to an API-First strategy.
- CI brings agility and keystone interfaces and feature flags were presented as a way to deploy new features faster without long development cycles and uncertainty.
- API Gateways are helpful for integrating APIs with externals but should not be overused.
- Developers are the customers of APIs. They need to understand what happens when an API is called and have insights about the semantics.
- API Gateway and, optionally, BFF can be used for public-facing APIs. Internally, a service mesh can be facilitated.
- The data mesh concept addresses key problems with classic data analytics and brings the concept of Data Products to product teams.
- Logging, metrics, and tracing bring insights into an API for the development team and help with problems like errors, performance, and billing.

Review Questions

8.1 Are breaking changes always avoidable?

(a) Yes
(b) No
(c) Breaking changes only appear with business changes.
(d) Breaking changes should be preferred over nonbreaking changes.

8.2 What tests should be performed when implementing an API?

(a) Only contract tests
(b) Only unit tests
(c) APIs should be tested only manually
(d) Tests should be carefully considered taking into account their time intensity and the level of feedback.

8.3 What concepts can be used to specify tests?

(a) BDD and TDD
(b) Domain Storytelling
(c) Event Storming
(d) context mapping

8.4 What functions can be fulfilled by an API gateway?

(a) Only routing
(b) Only billing
(c) Routing, authorization, and rate limiting, for example
(d) Only authentication

8.5 When should the use of a service mesh be considered?

(a) For public-facing APIs
(b) For internal APIs
(c) Only in combination with a BFF
(d) To secure an API gateway

References

1. Preston-Werner T (2023) Semantic versioning 2.0.0 [Online] Available: https://semver.org/. Visited on 02 Feb 2025
2. Calendar Versioning (2019). [Online] Available: https://calver.org/. Visited on 04 Feb 2025
3. Bellemare A (2023) Building an Event-Driven Data Mesh: Patterns for Designing and Building Event-Driven Architectures, 1st edn. O'Reilly, Beijing, 1 p. ISBN: 10-9812-757-9
4. Specification |apache avro (2022). [Online] Available: https://avro.apache.org/docs/1.11.1/specification/. Visited on 24 Aug 2024
5. Language guide (proto 3) |protocol buffers documentation (2025). Source: https://github.com/protocolbuffers/protocolbuffers.github.io/blob/main/content/programming-guides/proto3.md. [Online]. Available: https://protobuf.dev/programming-guides/proto3/. Visited on 09 Feb 2025
6. Parks Y (2023) Api versinioning: Url vs header vs media type versioning. [Online] Available: https://www.lonti.com/blog/api-versioning-url-vs-header-vs-media-type-versioning. Visited on 02 Feb 2025
7. Gupta L (2023) Content negotiation in a rest API. [Online]. Available: https://restfulapi.net/content-negotiation/. Visited on 02 Feb 2025
8. Content negotiation |API guidelines, adidas (2024). [Online] Available: https://adidas.gitbook.io/api-guidelines/rest-api-guidelines/message/content-negotiation. Visited on 08 Feb 2025
9. Fränkel N (2023) API versioning. [Online] Available: https://medium.com/apache-apisix/api-versioning-f8c662588768. Visited on 08 Feb 2025
10. Fielding RT, Nottingham M, Reschke J (2022) HTTP Semantics, RFC 9110. https://doi.org/10.17487/rfc9110. [Online] Available: https://www.rfc-editor.org/info/rfc9110. Visited on 29 July 2024
11. Zalando restful api and event guidelines (2024). Source at https://github.com/zalando/restful-api-guidelines. [Online] Available: https://opensource.zalando.com/restful-api-guidelines/. Visited on 24 Aug 2024
12. Bundesbahnen S (2024) Best practices. [Online] Available: https://schweizerischebundesbahnen.github.io/api-principles/restful/best-practices/. Visited on 02 Feb 2025
13. Dudycz O (2020) How to (not) do the events versioning?. [Online] Available: https://event-driven.io/en/how_to_do_event_versioning/. Visited on 02 Feb 2025
14. Young G (2017) Versioning in event-sourced systems, self-published. [Online] Available: https://leanpub.com/esversioning. Visited on 02 July 2025

15. Pelluru S, Chhabria A, West R, Indrasiri K (2025) Schema registry in azure event hubs. [Online] Available: https://learn.microsoft.com/en-us/azure/event-hubs/schema-registry-concepts. Visited on 02 Feb 2025
16. Confluent (2025) Schema registry for confluent platform. [Online] Available: https://docs.confluent.io/platform/current/schema-registry/index.html. Visited on 02 Feb 2025
17. Confluent (2025) Schema evolution and compatibility for schema registry on confluent cloud. [Online]. Available: https://docs.confluent.io/cloud/current/sr/fundamentals/schema-evolution.html. Visited on 02 Feb 2025
18. Apicurio Registry User Guide (2024). [Online]. Available: https://docs.redhat.com/en/documentation/red_hat_build_of_apicurio_registry/2.6/html/apicurio_registry_user_guide/intro-to-registry-rules_registry. Visited on 04 Feb 2025
19. Aws glue schema registry (2025). [Online]. Available: https://docs.aws.amazon.com/glue/latest/dg/schema-registry.html#schema-registry-compatibility. Visited on 07 Feb 2025
20. Yokota R (2021) Understanding protobuf compatibility. [Online] Available: https://yokota.blog/2021/08/26/understanding-protobuf-compatibility/. Visited on 02 Feb 2025
21. Yokota R (2021) Understanding json schema compatibility. [Online] Available: https://yokota.blog/2021/03/29/understanding-json-schema-compatibility/. Visited on 02 Feb 2025
22. Nema A, Gupta S, Goyal A, Pahti A (2024) Message validation. [Online]. Available: https://www.asyncapi.com/docs/guides/message-validation. Visited on 07 Feb 2025
23. Hagen N (2022) Asyncapi and apicurio for asynchronous APIS. [Online]. Available: https://www.asyncapi.com/blog/asyncapi-and-apicurio-for-asynchronous-apis. Visited on 08 Feb 2025
24. Hagen N (2021) How to master schemas and asyncapis with registries. Visited on 08 Feb 2025
25. Davis M, Lane D, Urbańczyk M, Méndez F, Gornicki L, Goodman N (2021) Introduce schema versioning #697. [Online] Available: https://github.com/asyncapi/spec/issues/697. Visited on 07 Feb 2025
26. Thwaites M (2024) Production comes first - an outside-in approach to building microservices. [Online] Available: https://www.infoq.com/presentations/microservices-outside-in/. Visited on 02 Jan 2025
27. Wiggers S-J (2024) Production comes first - an outside-in approach to building microservices by martin thwaites. [Online] Available: https://www.infoq.com/news/2024/04/qcon-london-outsidein-testing/. Visited on 02 Jan 2025
28. Registry lapicurio (2024) Source: https://github.com/Apicurio/apicurio.github.io/blob/3c945cd5d9402e4f31a9865c0d3c31f9513868b3/pages/reg_Red_Hat. [Online] Available: https://www.apicur.io/registry/. Visited on 19 Jan 2025
29. Apicurio registry content rules (2024) Source: https://github.com/Apicurio/apicurio.github.io/blob/main/registry/docs/apicurio-registry/3.0.x/getting-started/assembly-intro-to-registry-rules.html Red Hat. [Online] Available: https://www.apicur.io/registry/docs/apicurio-registry/3.0.x/getting-started/assembly-intro-to-registry-rules.html. Visited on 19 Jan 2025
30. Cruz M, Prescott L (2024) Contract Testing in Action. Manning Publications Co. LLC, Shelter Island, p 305. ISBN: 978-16-33437-24-1
31. Meszaros G (2011) xUnit Test Patterns: Refactoring Test Code, 1st edn. Addison-Wesley, Reading. ISBN: 978-01-31495-05-0
32. Fowler M (2011) CQRS. [Online] Available: https://martinfowler.com/bliki/CQRS.html. Visited on 30 Dec 2024
33. Fellows M (2023) What is contract testing and why should i try it?. [Online] Available: https://pactflow.io/blog/what-is-contract-testing/. Visited on 27 Jan 2025
34. Cruz M, Prescott L (2024) Contract Testing in Action: With Pact, Pactflow, and Github Actions. Manning, Shelter Island. ISBN: 978-16-33437-24-1
35. Introduction to message based API testing. [Online]. Available: https://docs.pact.io/implementation_guides/javascript/docs/messages. Visited on 27 Jan 2025
36. Khorikov V (2024) The Art of Unit Testing: With Examples in JavaScript, Osherove R (ed), 3rd edn. Manning Publications, Shelter Island, 1288 pp. Includes Index. ISBN: 16-1729-748-8

37. Rosenthal C (2020) Chaos Engineering, System Resiliency in Practice, Jones N, (ed), 1st edn. O'Reilly Media, Beijing, 1275 pp. Includes bibliographical references and index. - Description based on print version record. ISBN: 10-9819-146-3
38. Owasp Top Ten, OWASP Foundation (2021). [Online]. Available: https://owasp.org/www-project-top-ten/. Visited on 14 Aug 2024
39. Beck K (2015) Test-Driven Development, By Example. A @Kent Beck Signature Book), 20. Printing. Addison-Wesley, Boston, 220 pp. ISBN: 978-03-21146-53-3
40. Fowler M (2023) Test driven development. [Online] Available: https://martinfowler.com/bliki/TestDrivenDevelopment.html. Visited on 02 Jan 2025
41. Behaviour-driven development. The Cucumber Open Source Project (2024). [Online] Available: https://cucumber.io/docs/bdd/. Visited on 05 Jan 2025
42. Smart J (2014) BDD in Action, Behavior-Driven Development for the Whole Software Lifecycle. Manning Publications Co. LLC, New York, 1307 pp. Description based on publisher supplied metadata and other sources. ISBN: 978-16-38353-21-8
43. Thomas P (2023) Is behaviour driven development (bdd) right for api testing?. [Online] Available: https://www.karatelabs.io/learning/is-behaviour-driven-development-bdd-right-for-api-testing. Visited on 05 Jan 2025
44. Duvall PM, Matyas S, Glover A (2013) Continuous Integration: Improving Software Quality and Reducing Risk. A @Martin Fowler Signature Book, 8. print. Addison-Wesley, Upper Saddle River, 283 pp. Literaturverz. S. 273–274. ISBN: 03-2133-638-0
45. Fowler M (2024) Continuous integration. [Online] Available: https://www.martinfowler.com/articles/continuousIntegration.html. Visited on 12 Jan 2025
46. Hohpe G, Danieli M, Landreau J-F, Hashmi T (2021) Cloud Strategy: A Decision-Based Approach to Successful Cloud Migration. Architect Elevator Book Series. leanpub.com
47. Driessen V (2010) A successful git branching model, Revised in 2020. [Online] Available: https://nvie.com/posts/a-successful-git-branching-model/. Visited on 12 Jan 2025
48. Beyer B, Jones C, Petoff J, Murphy NR (eds) (2016) Site Reliability Engineering: How Google Runs Production Systems, 1st edn. O'Reilly, Beijing, 1 p. Description based on publisher supplied metadata and other sources. ISBN: 978-14-91951-18-7
49. Fowler M (2020) Keystone interface. [Online] Available: https://www.martinfowler.com/bliki/KeystoneInterface.html. Visited on 12 Jan 2025
50. Beck K (2012) Extreme Programming Explained: Embrace Change. The XP Series, Andres C (ed.), 2nd edn. 11. print. Addison-Wesley, Boston, 189 pp. Includes bibliographical references and index. ISBN: 978-03-21278-65-4
51. Nwaiwu I (2024) Automating API Delivery: APIOps with OpenAPI, van der Hecht M (ed), 1st edn. Manning Publications, Shelter Island, 1400 pp. Includes bibliographical references. ISBN: 16-3343-878-3
52. De B (2017) API Management: An Architect's Guide to Developing and Managing APIs for Your Organization, 1st edn. Apress, New York, 11 pp. Includes bibliographical references. - Description based on print version record
53. Medjaoui M (2019) Continuous API Management, Making the Right Decisions in an Evolving Landscape, Wilde E, Mitra R, Amundsen M (eds), 1st edn. O'Reilly, Beijing, 267 pp. ISBN: 978-14-92043-55-3
54. Richardson C (2019) Microservices Patterns: With Examples in Java. Manning Publications, Shelter Island, 11 pp. Includes bibliographical references and index. - Description based on print version record. ISBN: 978-16-17294-54-9
55. Newman S (2015) Building Microservices. Designing Fine-Grained Systems, 1st edn. O'Reilly Media, Sebastopol, 1259 pp. ISBN: 978-13-22879-90-1
56. Backends for frontends pattern, source: https://github.com/microsoftdocs/architecture-center/blob/main/docs/patterns/backends-for-frontends.yml, Microsoft (2024). [Online]. Available: https://learn.microsoft.com/en-us/azure/architecture/patterns/backends-for-frontends. Visited on 26 Jan 2025

57. Sharma R (2020) Getting Started with Istio Service Mesh: Manage Microservices in Kubernetes. Springer eBook Collection, Singh A (ed). Apress, Berkeley, 1321129 pp. ISBN: 978-14-84254-58-5
58. Posta CE (2022) Istio in Action, Maloku R, Brewer E (eds). Manning, Shelter Island, 450 pp. ISBN: 978-16-17295-82-9
59. Leggett B, Kohavi Y, Sun L (2024) Maturing istio ambient: Compatibility across various kubernetes providers and CNIS, 2024. [Online]. Available: https://istio.io/latest/blog/2024/inpod-traffic-redirection-ambient/. Visited on 26 Jan 2025
60. Howard J, Jackson EJ, Kohavi Y, Levine I, Pettit J, Sun L (2022) Introducing ambient mesh. Visited on 26 Jan 2025
61. Sidecar or ambient? source: https://github.com/istio/istio.io/blob/release-1.24/content/en/docs/overview/dataplane-modes/index.md (2024). [Online]. Available: https://istio.io/latest/docs/overview/dataplane-modes/. Visited on 26 Jan 2025
62. Rice L (2023) Learning eBPF: Programming the Linux Kernel for Enhanced Observability, Networking, and Security, 1st edn. O'Reilly, Beijing, 217 pp. ISBN: 10-9813-512-1
63. Automate API deployments with apiops, Git hash: 6cfcc34, Microsoft (2024). [Online]. Available: https://learn.microsoft.com/en-us/azure/architecture/example-scenario/devops/automated-api-deployments-apiops. Visited on 17 Jan 2025
64. Dulay H, Mooney S (2023) Streaming Data Mesh: A Model for Optimizing Real-Time Data Services, Mooney S (ed), 1st edn. O'Reilly, Beijing, 1 p. ISBN: 10-9813-069-3
65. Dehghani Z (2022) Data Mesh, Delivering Data-Driven Value at Scale, 1st edn. O'Reilly, Sebastopol, 1403 pp. ISBN: 978-14-92092-34-6

Collaborative Design and Agility   9

A collaborative design process must support us in a VUCA world and tackle complexity (section "Complexity in a Volatility, Uncertainty, Complexity, and Ambiguity (VUCA) World").

We presented a step-by-step process in Chapter 5, "API Design Supported by Domain-Driven Design."

Collaborative Design Process

In Chapter 5, "API Design Supported by Domain-Driven Design," we introduced a design process that allows us to specify APIs and events based on robust and reliable business requirements. The process was introduced by example.

In this section, we want to concentrate on the process called Synergetic Blueprint, its setup, and deliverables [1]. Thus, applying the process to real-world design and development processes becomes possible. The Synergetic Blueprint describes steps to come from a business idea to team-owned bounded contexts with APIs to exchange data between them [1]. The steps are presented using the online library example (Chapter 5, "API Design Supported by Domain-Driven Design"). They are as follows:

- Ideation using sketches
- Business modeling using Business Model Canvas
- Business modeling using Wardley map
- Gathering business requirements using Domain Storytelling and Visual Glossary
- Definition of bounded contexts using Event Storming and Visual Glossary
- Definition of data exchanges and team structures using context map
- Definition of interfaces using API Product Canvas, OpenAPI, and AsyncAPI

Let us start with the ideas and the creation of the business model.

Business Model

In the steps, the business people and technology nerds discuss and sketch ideas more as a game than a serious activity. Those sketches can be first user interface designs, rough workflows, or ideas sketched on a napkin.

A Business Model Canvas can be used to formalize ideas [2]. We already showed the canvas of the online library in Figure 5-1. Those Business Model Canvases are an outstanding way to present a business idea in a compressed fashion and a tool for creating a business plan in a workshop format. Such a workshop can result in the creation of stakeholders' commitment and a shared understanding of the business idea. The workshop result is the Business Model Canvas and the first ideas of business capabilities. In addition to the business model canvas, a capability map should be discussed (Figure 5-10). The capability map then delivers a first impression about generic, supportive, and core capabilities. The impression can be more detailed using a Wardley map [3].

The workshop attendees then discuss the prioritizing of the found capabilities using a Wardley map. An example of a Wardley map is given in section "Wardley Map".

The workshops must be carefully moderated so that ideas and statements are not lost in a majority opinion. The authors propose to engage a seasoned agile trainer or another experienced moderator with no stake in the ideas. This approach makes it possible to moderate in a balanced way without cutting off discussions.

The deliverables of the ideation part are as follows:

- First sketches
- Business Model Canvas
- Capability map
- Wardley map

Those workshops can be held over the course of a couple days. The attendees should include the most critical stakeholders: managers, customers, financiers, and enterprise architects. However, attendance at the workshop should not exceed more than 15 people.

With this groundwork, the business requirements can be gathered in collaborative workshop formats. That starts the strategic design.

Strategic Design

The strategic design of the DDD process leads to the tailoring of the application in bounded contexts. The bounded contexts can be defined using the collaborative approaches of Domain Storytelling and Event Storming.

Domain Storytelling

A Domain Storytelling is a moderated workshop. The attendees must be included when the business experts tell their stories. They need to ask questions and define the Visual Glossary together. Examples of a domain stories and a Visual Glossary are given in Figures 5-12 and 5-17 (section "Domain Storytelling").

To guarantee a successful workshop, the room should be arranged so workshop attendees can see each other and the drawing board equally [4].

The narrated stories should not be prepared. The workshop attendees should see how the stories evolve to create the necessary authenticity.

When one story is told, the workshop attendees collect the used story terms. Moderated, they find the relationships between the terms and give them strong verbs. In this way, they create a Visual Glossary together. When the next story is told, the workshop attendees update the Visual Glossary.

With two or three stories, the stories can be consolidated. First bounded context boundaries can be marked in the Domain Story and in the Visual Glossary.

The step in the process delivers

- Domain stories and
- Visual Glossary.

Workshop attendees should be business experts and technical specialists. It is preferable to include User Experience (UX) and security specialists as well. The 15-attendee limit is to ensure that diverse perspectives can be heard in the workshop discussions. Ideally, developers of the application to be implemented or modernized should be part of it as well. It is not easy to involve creators because the implementation engineers are not named at such an early stage of the development process, but at the very least, the responsible architects should be assigned and take part in the workshop.

Technically the workshops can be conducted in person or remotely, though in-person workshops are preferred to guide the communication among attendees.

The results of the workshops can be collected in different ways:

- A remote whiteboard like Miro[1] or Mural[2]
- Sticky notes for icons and a brownboard to collect stories
- Drawing by code using PlantUML[3] or Mermaid.[4] Additionally, the original webtool egon.io by *Hofer and Schwentner* can be used.[5]

---

[1] https://miro.com/

[2] https://mural.co/

[3] https://github.com/johthor/DomainStory-PlantUML

[4] https://www.mermaidchart.com/

[5] https://egon.io/

Regardless, the moderator should be very familiar with the selected tool. The authors prefer remote whiteboards to document results because they make it easy to use the results as the foundation of later workshops.

For an application of medium complexity, such as the online library, one should calculate two working days of eight hours for those workshops.

Event Storming workshops can be performed using the results of the Domain Storytelling workshops.

Event Storming

Event Storming workshops ideally follow right after Domain Storytelling with the same attendees. At a minimum, the core group of attendees should be the same. Since Event Storming leads to more architectural and technical questions, additional technical specialists can be involved.

We discussed Event Storming in section "Event Storming". One of the examples is shown in Figure 5-27.

The workshop setup depends on the technical foundation. When the moderator utilizes an interactive board, the setup can be demonstrated for Domain Storytelling. In an in-person workshop using moderation materials, the boards should be arranged in front of the attendees so that everyone can see the boards at all times during the workshop. More complex tools like PlantUML[6] and Mermaid[7] should only be used if all workshop attendees are already familiar with them. The authors prefer whiteboards because they allow the workshops to be interrupted; although interruptions of workshops are not preferred, sometimes they are necessary to ensure the availability of all attendees.

Event Storming workshops typically require twice the time of a Domain Storytelling workshop. This means the Event Storming workshop of the online library would take four days while the Domain Storytelling workshop took two days.

The Event Storming workshop delivers the business process based on domain events and a fine-tuned Visual Glossary. The Event Storming workshop result contains the process including

- aggregates and read-only models,
- commands triggering the domain events,
- roles and processes triggering or performing the commands, and
- the bounded contexts around the domain events.

Additionally to Event Storming, a context map can be created in the workshop.

Context Map

As discussed in section "Context Map", a context map shows the dependencies between the found bounded contexts. It is used to discuss the used pattern as

---

[6] https://github.com/johthor/DomainStory-PlantUML

[7] https://www.mermaidchart.com/

conformist, anticorruption layer, and open host service. Those patterns were already defined by *Eric Evans* [5]. Other patterns like CQRS were defined later with the emergence of microservices [6]. We also use context maps to discuss the necessary APIs.

An example is given in Figure 5-40.

Workshop attendees create a context map at the end of the Event Storming workshop. Based on those first bounded contexts and team topologies, the teams can turn their attention to tactical design.

Tactical Design

Based on the results of the strategic design, the teams can discuss the solution design using several tools. We discussed them in Chapter 7, "Defining the Online Library Interfaces." Out of this are created some deliverables:

- Bounded Context Canvas (Figure 7-2),
- Architecture Communication Canvas (Figure 7-3),
- API Product Canvas (Figure 7-4), and
- Architecture Decision Record.

The Architecture Decision Record follows the recommendations of the arc42 group as one part of a full arc42 documentation. It contains the alternatives that were discussed for the decision and the alternative the team decided on. It is a great tool to document architecture decisions of teams in the corresponding team repository. It is more suitable for teams because team members access those day by day.

A template for an Architecture Decision Record is given in the samples.[8] Other templates can be found at *MADR*[9] and *Y-Statements*.[10] The examples also contain a completed Architecture Decision Record (ADR).[11] The ADR is a great tool to document decisions in teams and even overarching decisions like the combination of bounded contexts, as discussed in Chapter 7.

Overview

The presented workshop formats are based on the *Synergetic Blueprint* introduced by *Junker* [1]. An overview of the workshops and the deliverables is shown in Figure 9-1.

---

[8] https://github.com/Apress/Crafting-Great-APIs-with-Domain-Driven-Design/blob/main/Chapter-9/ArchitectureDecisionRecord.md

[9] https://adr.github.io/madr/

[10] https://medium.com/olzzio/y-statements-10eb07b5a177

[11] https://github.com/Apress/Crafting-Great-APIs-with-Domain-Driven-Design/blob/main/Chapter-9/postgre.md

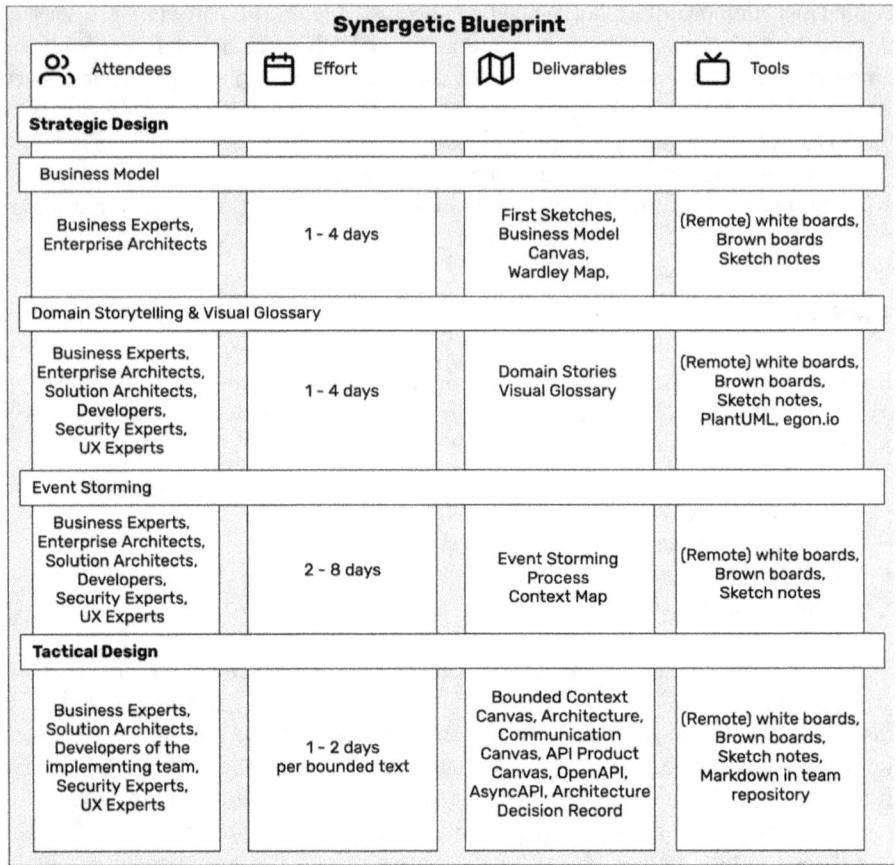

Figure 9-1 Synergetic Blueprint

Each workshop session should follow a standard workshop format with the following features [7]:

- **Check-in**
  All attendees have enough time to get used to the workshop. The authors do it, for example, with a short introduction round and a question about one's favorite ice cream.
- **Check-out**
  At the end of the workshop, the attendees can reflect on the workshop and share their individual results.
- **Small groups**
  A workshop should not exceed twelve people; groups with six people are preferred.

- **Retrospectives**
  A format to obtain feedback to get iteratively better at the workshop format. The retrospective format should be changed, as it can get boring when it is repetitive [7].

Reliable and sustainable architectures can be designed using the flow presented in Figure 9-1. The collaborative approach requires from all attendees a specific mindset, which we will discuss in the following section.

Necessary Mindset for API Design

To design APIs, not only technical skills are necessary. Engineers and developers need skills to visualize and to collaborate. In this section, we will discuss how methodologies can enhance the skills of developers and engineers, so that they are able to design sustainable APIs in a collaborative way.

First, we will discuss how additional methodologies or techniques can enhance the *Synergetic Blueprint* process.

Necessary Mindset

To facilitate a collaborative modeling workshop, one needs to make sure that the group can do the modeling and decide on certain product models that emerge during the workshop [7]. Therefore, it is recommended to do sensemaking exercises. Sensemaking exercises will help the group to create a common understanding from individual beliefs, opinions, concerns, and so forth. A mental model can be tested in the group, for example, the attendees group the application to be built from a commodity product into a full custom-built product. Such a scale makes the unknowns of the group explicit [7].

As was already discussed, workshops are used to create a common understanding among the different stakeholders in a project. Therefore, it is more important that a model be understood than that it is accurate [4, 7]. A more complete model should be preferred over a more accurate model [7].

To create an atmosphere in which ideas are allowed to emerge during the workshops, chaos is to be preferred over a well-structured workshop [7]. This aspect is well supported by Event Storming during the storming phase [8].

As pointed out in connection with Domain Storytelling, Event Storming, and especially a Visual Glossary, all discussions should be visualized. Even discussions in smaller groups, for example, in teams, are worth visualizing [7].

Additional Methodologies in the Process

In the *Synergetic Blueprint* process, we used the methodologies capability map, Business Model Canvas, Wardley mapping, Domain Storytelling, Event Storming, and context map. In this subsection, we want to discuss additional methodologies, which can enhance the *Synergetic Blueprint* process.

Business Process Model Notation

Business Process Model Notation (BPMN) is a formal notation of business processes [9]. The business processes can be modeled using tasks represented as rectangles. Actors and technical systems are usually modeled as swimlanes. An example of a BPMN is given in Figure 9-2.

The given example was modeled with Miro.[12] Models can be developed in BPMN.io[13] as well. The authors would even prefer to draw BPMN in Miro because the models are rougher and therefore more intuitive, which means they are more suitable for workshops.

BPMN is an excellent tool for formalized models. With experience, it is less suited for collaborative modeling. It is challenging because each workshop attendee needs to know the model language, and experts will examine in special classes many details that need to be modeled correctly in BPMN notation. The focus on users gets lost when they are not experts in BPMN. In the authors' experience, those workshops can be narrowed to a dialog of two heads. Even though the necessary APIs can be detected exceptionally well at the swimlane borders (Figure 9-2), BPMN is not recommended for modeling workshops. Sometimes teams use BPMN runtimes that include modelers. In those teams, it makes sense to use a modeler for the process models in team workshops. Detailed modeling of processes is possible because tools can be used to run the processes as simulations. Across different teams detailed modeling is less suitable. Tools to do so include, for example, Camunda,[14] jBPM,[15] and Kogito.[16] However, one must be careful because those tools are powerful and probably too expensive for the problem at hand. They require a high level of expertise and time to create models.

To model business processes collaboratively with experts, Domain Storytelling should be preferred. A great tool to visualize business processes is User story mapping.

User Story Mapping

User story mapping is a great tool for visualizing the dependencies of user stories along a business process. It helps large groups to obtain a common understanding [10].

Along with Domain Storytelling and Event Storming, user story mapping allows teams to know the place of their bounded context with its user stories in the overarching development environment. User interactions are placed on a board from left to right. Below it, user tasks are depicted as a skeleton walking through the application. User tasks can be imagined as steps in a user interface. Now the necessary user stories correspond to the user tasks. Additionally, they are sorted by

---

[12] https://miro.com/

[13] https://bpmn.io/

[14] https://camunda.com/

[15] https://github.com/kiegroup/jbpm

[16] https://kogito.kie.org/

Necessary Mindset for API Design

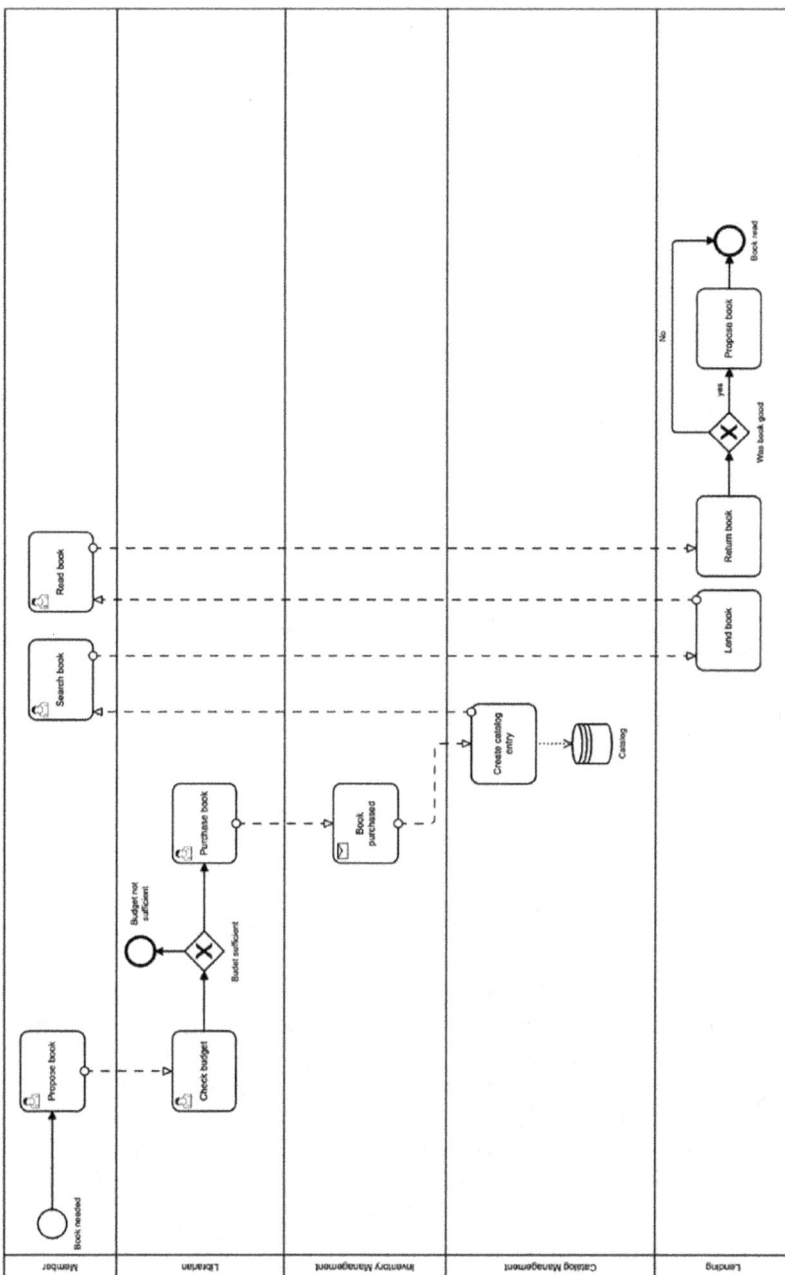

Figure 9-2 Example of BPMN for lending and searching

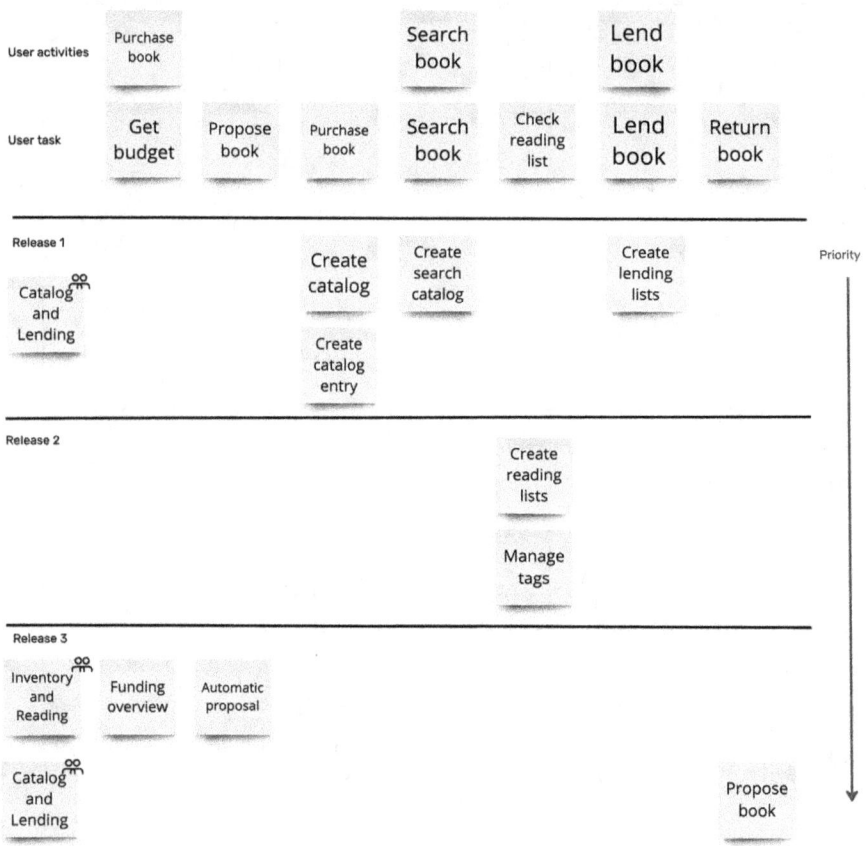

Figure 9-3 Example of user story mapping

their priority below each other. That means two user stories belonging to the same user's task are sorted above each other, whereas the user story given higher priority will be higher than the other.

An example of User story mapping is given in Figure 9-3.

User story mapping is a great tool for planning user stories created from Domain Storytelling and Event Storming to prioritize and visualize team dependencies. It can be used as a visualization tool in overarching team planning and prioritizing workshops.

Behavior-Driven Design

BDD is a methodology that is used to design software applications according to their requested behavior to bridge the gap between business and IT [11], as was discussed in section "Behavior-Driven Testing" in section "Testing." BDD uses conversations around business rules and examples, expressed in a form that can be easily automated, to reduce lost information and misunderstandings. Those conversations can be supported by detailed Event Stormings when discussing invariants [8].

BDD is a good approach to stabilizing the ubiquitous language and the bounded context in a team and to support test cases (section "Behavior-Driven Testing"). However, technical experts need to be involved earlier in the entire process to be sure that business requirements are being understood and the ubiquitous language is well defined.

Example Mapping

To steer discussions about user stories in BDD, Example Mapping can be used. Examples help to explore the problem domain and are a good foundation for design acceptance tests [12].

Example Mapping defines general rules for a user story. For example, the user story "Create catalog entry" has the rule that "only one catalog entry can be created at a time." The user story is written on a sticky note. The rules written on sticky notes are placed below the user story,. Then examples are formulated for this rule. Those rules are placed below the rules. An example might be when a librarian purchases two books in one day, each creating a catalog entry separately. Or a librarian can only create one catalog entry manually using the provided dialog. Formulating those rules steers questions written on sticky notes and placed on sticky notes to the story on the board [12].

The discussion about examples and rules provides instant feedback to the teams and helps them better understand the user stories and business requirements. Example Mapping should be done in teams when refining user stories. It can be done in the API design process to find the best examples and descriptions in the API specification (Figure 9-4).

Conway's Law

The knowledge of team structures and necessary interactions is essential for building successful architectures. A deep dependency exists between architecture and team structures following *Conway's Law* [13].

Conway stated that an organization only produces something that reflects the organization itself [13]. For architecture, we need to organize the implementation teams in a form that reflects the intended architecture, as we discussed with respect to the online library. The team structures must follow the found structures in the context map.

Ideally, a bounded context is implemented by one team. As we saw in Chapter 7, "Defining the Online Library Interfaces," that is not implementable in every case. It might be that, at the beginning of a project, the team sizes are too small or the costs for larger teams are too high. In such cases, teams need to take care of multiple bounded contexts that belong together, or even more than one team can take care of all bounded contexts.

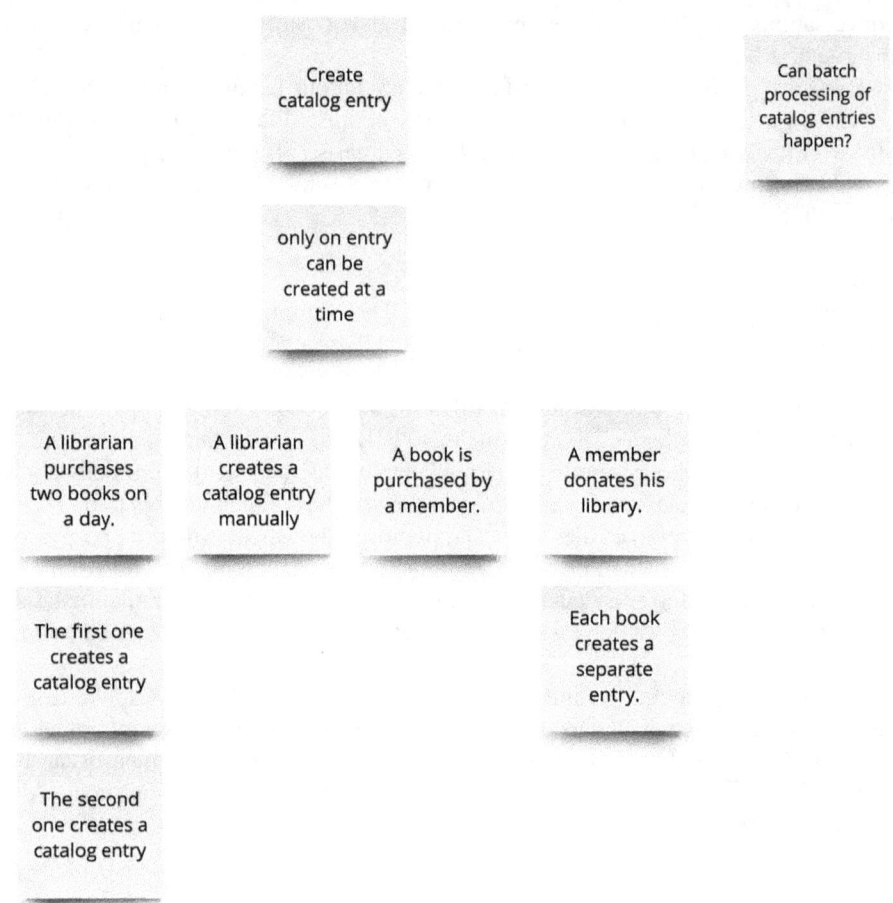

Figure 9-4 Example Mapping

This carries the risk that the bounded contexts will grow together as a monolith. In the case of the online library, the risk is high that catalog management and lending will grow together as a monolith. Teams must have rules to avoid those situations. Teams must take over architecture governance to follow the intended architecture. External governance, for example, from an architectural chapter or guild, cannot take over the role because they are not involved in the teams' day-to-day work.

We must consider team structures when creating APIs.

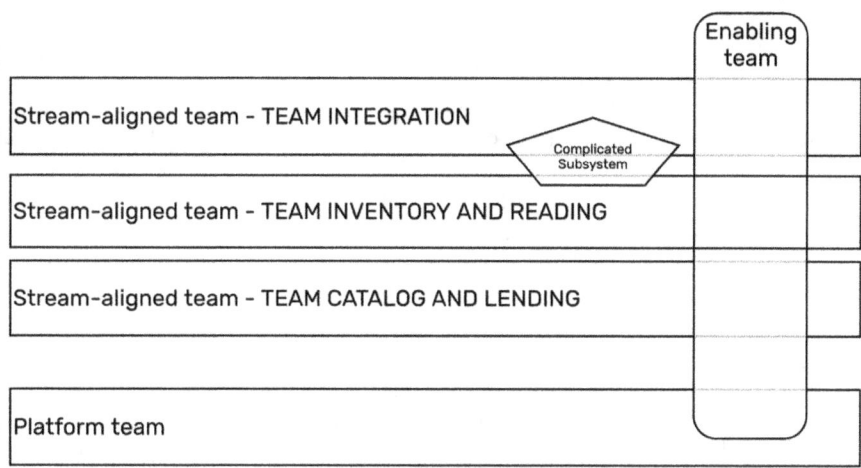

Figure 9-5 Team topologies with the example of the online library

Team Topologies

Bounded context teams usually deliver business functions day by day. They are called stream-aligned teams [14]. However, we need to be aware of different types of teams. Those different types of teams are discussed in team topologies [14].

Team topologies describe ways to organize teams so that the dependencies between them are reduced. There are stream-aligned teams that deliver constant content as teams working on bounded contexts. They use the methodologies of DDD to define and prioritize content. Those teams use services provided by a platform team that provides support to deploy and operate applications [14] (Figure 9-5).

An enabling team and a team for a complicated subsystem can also exist. An enabling team is recommended when new technologies or methodologies are needed [14]. The enabling team enables the stream-aligned teams to use them efficiently and owns the necessary governance processes. Those are necessary, e.g., in a highly regulated environment like health software.

Another type of team is the complicated subsystem. Such a team owns subsystems needed by two or more teams [14]. A typical complicated subsystem is a tariff calculator for insurance. Those teams need precise business knowledge and can facilitate stream-aligned teams with an application that encapsulates such knowledge.

Collaborative Processes in Classical Project Environments

In the preceding subsections, we simply assumed that teams work together in an agile way following the *Agile Manifesto* [15]. Moreover, using Domain-Driven

Design (DDD) leads us to teams with few synchronization points that can act self-responsibly. However, this concept depends highly on company culture. The famous quote "Culture eats strategy for breakfast" attributed to *Peter Drucker* means that a strategy cannot simply stated by management [16]. The strategy needs to be implemented with the people in the company. Even more, the defined strategy needs to be applied on a daily basis.

In large, traditional companies, for example, insurance companies, a hierarchical behavior is evident. Young companies whose strategy largely depends on the internet tend to act more freely and agilely.

Using the self-responsible approach of bounded context teams requires a long and tiresome change process. However, such a process is worth starting because the software developed will be more functional and be of higher quality, even though a large organization will need to adapt to team-oriented agile frameworks like *Scrum* [17]. Frameworks like SAFe [18], Flight Levels [19], Kanban Maturity Model [20], LeSS [21], Scrum@Scale [22], and Spotify model [23] focus on this point. These frameworks support independent teams and teams with their own responsibility. However teams need to be synchronized with respect to higher business targets. In addition, certain governance processes for architecture or user experience rules need to be established.

We will look deeper at the Spotify Model and Flight Levels to become more accustomed to those approaches.

Spotify Model

The Spotify model represents a matrix organization. The rows of the matrix are skill-based and are called chapters. The matrix's columns are stream-aligned teams or, as we prefer to call them, bounded context teams. The chapters take over governance tasks, just as the architecture chapter takes over architecture governance [23]. The bounded context teams deliver customer increments. Even platform teams, enabling teams, and complex subsystem teams can be organized in this structure. A matrix handling one large product is called a tribe. The model is shown in Figure 9-6.

A tribe is led by a trio: product, design and architecture, and business leads. Such a trio can synchronize the different and sometimes contradictory requirements across squads.[17] When multiple tribes are necessary to implement a product, they are organized into an alliance [23].

The Spotify model was developed by Spotify.[18] It represents the culture of the organization. Therefore, it cannot be simply copied to another organization [24]. The culture of the organization needs to be changed step by step to an agile organization. And, the model needs to be adapted step by step to the needs of the organization. It is a long-running process. The Flight Levels model can support such a long-running change process.

---

[17] A squad in software development is a small team that can implement a business functionality independently.

[18] https://newsroom.spotify.com/company-info/

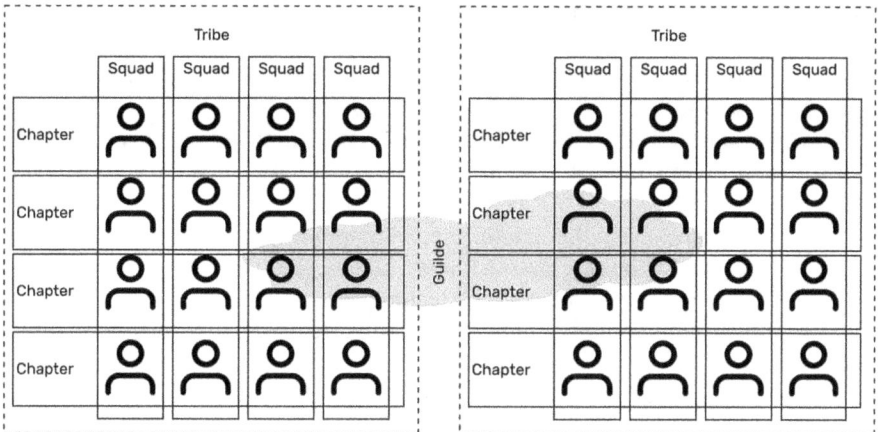

Figure 9-6 Spotify model

Flight Levels

Flight Levels tries to solve the problem of teams being agile; however, the entire business process is not agile (yet). The teams have too many dependencies (tackled by DDD). All teams are agile, but customer deliveries are too slow. Priorities constantly shift (tackled by a Wardley map). Because of missing overarching synchronization, waterfall plans that do not represent reality are made [19].

Flight Levels introduces an organization as three levels of interrelated goals [19]:

- **Level 1: Operations**
  The level of teams and individuals where work gets delivered.
- **Level 2: Coordination**
  The level where teams need to be coordinated.
- **Level 3: Strategy**
  The level at which strategic goals are decided and what guides the teams and sets the company's priorities.

Those Flight Levels can be organized, for example, in a task management system. They can guide squads and tribes in organizing their day-to-day work.

Usually, large organizations use one of those approaches. More often, large organizations still behave more traditionally and hierarchically. This is acceptable as long as the organization uses an organizational approach that supports the architecture and the API design. However, as we will see in the following section, a certain mindset is necessary to design APIs.

Short Feedback Loops and Stable APIs

If we apply just an agile mindset to APIs, we will get unstable APIs, because the responsible team will change an API as they need it.

As a consumer of an API, one expects an API will change slowly and with a long lead time (minimum half a year). An API is a contract, and the contract is not allowed to change.

Both statements, even though they are contradictory, are accurate. As API designers, we need to ensure that APIs are stable. We can achieve that stability we guarantee by applying collaborative techniques throughout the design process. The design process already includes consumers, so providers and consumers design the contract.

When, on a strategic level, business goals change, the design process starts again. Providers and consumers can agree on contract changes collaboratively. The teams can work agilely and independently because the contract foundation on which to work without specific synchronization points. Using such a collaborative approach, consumers, as well as providers, can initiate a contract change. Using versioning (section "Versioning of APIs") different versions can exist in parallel. Such a strategy is possible even when not all consumers are known to the providers.

With those collaborative approaches in mind, stable APIs do not contradict an agile mindset.

Let us see in what follows what mindset is necessary to design APIs.

Conclusion

We saw in this chapter that we can follow the *Synergetic Blueprint* to design and implement APIs.

We can use ideation techniques for the initial sketches. Those ideas are formalized with a Business Model Canvas [2]. Using a capability map, the identified business capabilities can be categorized into core, supportive, and generic business capabilities. This points to a prioritization of capability implementation, which can be fine-tuned by a Wardley map [3].

Using this prioritization, business requirements can be found using Domain Storytelling [4]. Event Storming helps to identify boundaries between contexts and define corresponding bounded contexts [8]. From the Event Storming session, a context map [5] can be defined, which will show team dependencies and synchronous and asynchronous data flows.

The synchronous and asynchronous APIs of a bounded context can be outlined using the API Product Canvas. They are based on the data flows of the context map. Based on the canvas, the outlined APIs can be specified using OpenAPI and AsyncAPI.

Conway's Law requires that the organization follow the architecture if it should produce the intended architecture [13].

Designing APIs is not solely the task of architects. It requires agile and collaborative mindsets. The mindsets are supported by the collaborative modeling techniques presented. However, applying those mindsets in large, traditional organizations is challenging. Scaled agile methodologies like Flight Levels [19] or the Spotify model [23] help to establish a collaborative mindset.

There are several successful projects publishing beautiful APIs. They do not start with the result we see now. The APIs were extended step by step. Let us see how the online library can be extended successfully with different features without breaking the principal architectural ideas.

Points to Remember

- Ideation can be done with sketches.
- Ideas need to be formalized with Business Model Canvas, Wardley map, and capability map.
- Business requirements must be gathered using Domain Storytelling.
- Bounded contexts are defined using Event Storming and context mapping.
- Bounded contexts are documented with a Bounded Context Canvas.
- APIs are outlined using an API Product Canvas.
- APIs are specified using OpenAPI and AsyncAPI.
- The architecture of a bounded context is documented with an architecture communication canvas.
- Successful and sustainable APIs can only be defined in a collaborative and agile atmosphere.
- The architecture needs to define the implementing organization following Conway's Law.

Review Questions

9.1 Do I need to shape ideas to establish a business?

(a) Yes
(b) No
(c) Only for the bank
(d) Only for discussions with friends

9.2 Do I need to collect business requirements for designing APIs?

(a) Yes
(b) No
(c) Only for the bank
(d) Only for discussions with friends

9.3 Do I need context mapping to design elegant APIs?

(a) Yes, for the data flow and the team organization
(b) Yes, but only for the AsyncAPI
(c) Yes, but only for the AsyncAPI
(d) No

9.4 Does a collaborative mindset support API design?

(a) Yes
(b) Yes, but only for managers
(c) Yes, but only for consultant companies
(d) No

9.5 Are collaborative and agile processes possible in large organizations?

(a) Yes
(b) Yes, but only for middle management
(c) Yes, but only for internet companies
(d) no

References

1. Junker A (2025) Mastering domain-driven design. BPB Online. ISBN: 978-93-65892-52-9
2. Strategyzer (2024) The business model canvas. [Online]. Available: https://www.strategyzer.com/library/the-business-model-canvas. Visited on 21 July 2024
3. Wardley S (2016) Exploring the map, chapter 3. [Online]. Available: https://medium.com/wardleymaps/exploring-the-map-ad0266fad59b. Visited on 27 Aug 2024
4. Hofer S, Schwentner H (2022) Domain Storytelling, a Collaborative, Visual, and Agile Way to Build Domain-Driven Software. Pearson International, London. ISBN: 978-01-37458-91-2
5. Evans E (2004) Domain-Driven Design: Tackling Complexity in the Heart of Software. Addison-Wesley, Boston
6. Richardson C (2019) Microservices patterns, With examples in Java. Manning Publications, Shelter Island, p 11. Includes bibliographical references and index. - Description based on print version record. ISBN: 978-16-17294-54-9
7. van Kelle E, Verschatse G, Baas-Schwegler K (2025) Collaborative Software Design. Manning, Shelter Island. ISBN: 978-16-33439-25-2
8. Brandolini A (2024) Event storming. [Online]. Available: https://www.eventstorming.com/. Visited on 30 Jun 2024
9. Pant K, Juric MB (2008) Business Process Driven SOA using BPMN and BPEL. Packt Publishing, Birmingham. ISBN: 978-18-47191-46-5
10. Patton J (2014) User Story Mapping. O'Reilly Media, Sebastopol. ISBN: 978-14-91904-90-9
11. Smart J (2014) BDD in Action, Behavior-Driven Development for the Whole Software Lifecycle. Manning Publications Co. LLC, New York, 1307 pp. Description based on publisher supplied metadata and other sources. ISBN: 9781638353218.
12. Wynne M (2015) Introducing example mapping. [Online] Available: https://cucumber.io/blog/bdd/example-mapping-introduction/. Visited on 05 Feb 2025

13. Conway ME (1968) How do committees invent? Datamation Magazine. [Online] Available: https://www.melconway.com/research/committees.html. Visited on 06 Feb 2025
14. Skelton M, Pais M (2019) Team Topologies: Organizing Business and Technology Teams for Fast Flow. IT Revolution Press, Portland. ISBN: 978-19-42788-81-2
15. Beck K et al (2001) Manifesto for agile software development. [Online] Available: https://agilemanifesto.org/. Visited on 08 Feb 2025
16. Favaro K (2014) Strategy or culture: Which is more important? . [Online] Available: https://www.strategy-business.com/blog/Strategy-or-Culture-Which-Is-More-Important. Visited on 07 Feb 2025
17. What is scrum? (2025). [Online] Available: https://www.scrum.org/resources/what-scrum-module. Visited on 07 Feb 2025
18. Safe(r) 6.0 (2025). [Online] Available: https://scaledagileframework.com/. Visited on 07 Feb 2025
19. What is flight levels? (2025). [Online] Available: https://www.flightlevels.io/what-is-flight-levels/. Visited on 07 Feb 2025
20. Kanban maturity model (2023). [Online] Available: https://www.kanbanmaturitymodel.com/. Visited on 07 Feb 2025
21. Less (2025). [Online] Available: https://less.works/. Visited on 07 Feb 2025
22. The official scrum@scale guide (2021) [Online] Available: https://www.scrumatscale.com/scrum-at-scale-guide/. Visited on 07 Feb 2025
23. Tsonev N (2024) What is the spotify model for scaling agile? . [Online] Available: https://businessmap.io/blog/spotify-model. Visited on 07 Feb 2025
24. Cruth M (2024) Discover the spotify model. [Online] Available: https://www.atlassian.com/agile/agile-at-scale/spotify Visited on 07 Feb 2025

Iterative Enhancements 10

We started with the online library and saw that it needed three different teams. However, the approach changes when the library changes. In this chapter, we will discuss how changes in requirements and new partnerships change the architectural approach.

Adding New Features

We saw in Chapter 5, "API Design Supported by Domain-Driven Design," that APIs could be elegantly designed using the Synergetic Blueprint approach. However, the software does not stay in the same state forever. In contrast, software constantly changes due to changing technical or business contexts. Therefore, we must be prepared to change our software based on business and technical requirements.

First, let us discuss how to manage adding features to our online library. Then we will discuss more significant extensions.

Adding a Small Feature Extension

Small feature extensions seem easy to implement. However, it is a mistake not to take them seriously and to start immediately with the implementation. It is recommended to discuss even small features using Domain Storytelling. We assume that the business priority is given. A Domain Storytelling workshop need not be held in a large group. It can be conducted by implementation teams, where the product owner assumes the role of the business expert.

Let us assume that a member wants to make an appointment with other members using the appointment management feature that is part of the online library. Additionally, the member wishing to make the appointment would like to invite a nonmember to the appointment to promote interest in the online library.

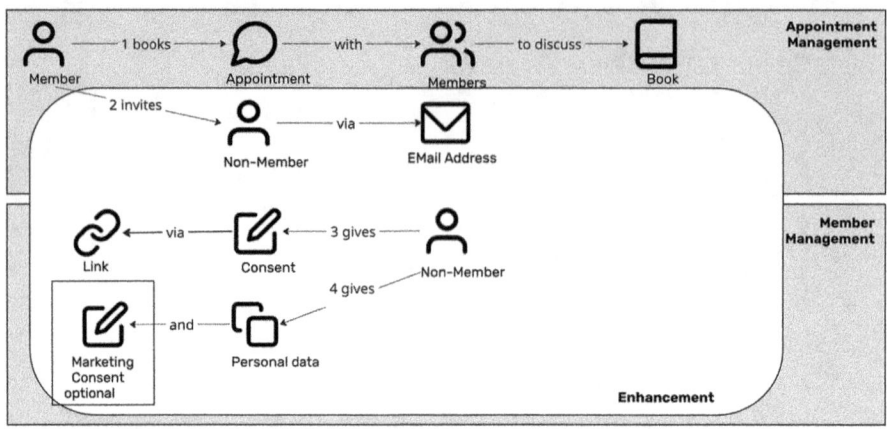

Figure 10-1 Enhancement of appointment management by nonmembers

The story is shown in Figure 10-1. The new parts are marked in white.

1. A member books an appointment with members to discuss a book.
2. The member invites a nonmember via their email address.
3. The nonmember gives her consent to process her data via a link in the invitation email.[1]
4. The nonmember gives her personal data and, optionally, consent to being contacted for marketing purposes by the online library.

This feature is important to the online library because, since even nonmembers can be invited to appointments, the online library is able to attract new members by those invitations.

The domain story shows the changes are less significant than one might think. Appointment Management must be enhanced so that nonmembers can be invited via email. Member Management must be enhanced so that consent can be given via link or at the same time one provides personal data.[2] The other steps have already been implemented for the onboarding process of members. Even so, the last step to becoming a member is left out.

In the authors' experience, implementing such changes without domain storytelling usually leads to more complex solutions that implement more features than initially requested.

---

[1] That is solution was proposed by user experience specialists. Other solutions to giving consent together with personal data are also possible.

[2] Both versions are valid, and user experience specialists must be involved.

Adding New Features

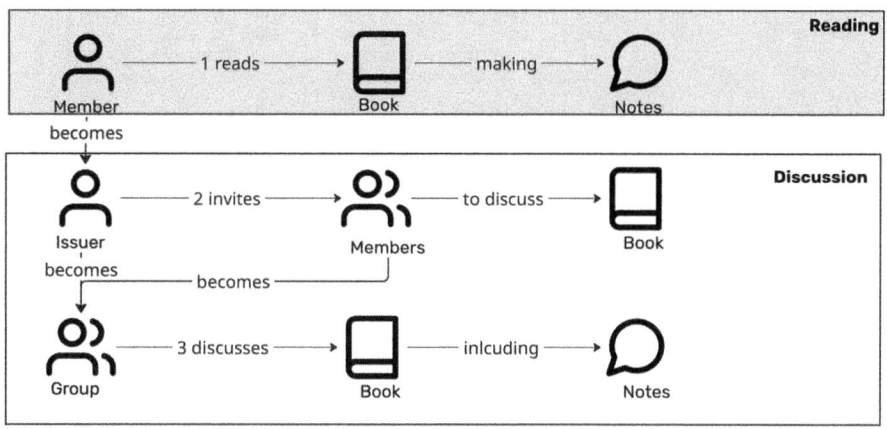

Figure 10-2 Member discussion groups

Enhancement of Online Library by Completely New Features

Another case might be one where a new feature would enhance the online library. Such a feature might be discussion groups.

Discussion groups allow members to be invited to groups discussing a book and have access to the book notes entered by the other group members. The associated domain story is shown in Figure 10-2.

1. A member reads a book and enters notes to the book.
2. An issuer of an invitation (who must already be a member) invites other members to discuss the book.
3. The group (members and issuer) discuss the book, including the notes.

At first glance, the changes are not very difficult to make. However, each attendee of the discussion needs access to the notes of the other attendees. Each member could manually do it. However, that seems tiresome and would not be accepted by the members. Therefore, the notes should be shared when a group is initiated for a particular book. Shares are revoked when the group is closed. This is possible because the discussion group discusses exactly one book. A discussion of the next book will prompt the creation of a new group.

The discussion feature needs a new bounded context because a group, an attendee, and an issuer did not previously exist. The Catalog and Lending teams could implement the bounded context; however, there are overlapping requirements. Nonetheless, the sharing and reviewing features already integrated into Lending are comparable to the sharing notes in a discussion group. Therefore, a new team is established to implement the discussion groups and sharing and reviewing in two new bounded contexts. However, a new team may not be possible depending on the product team size, application size, and budget. However, a new team is principally

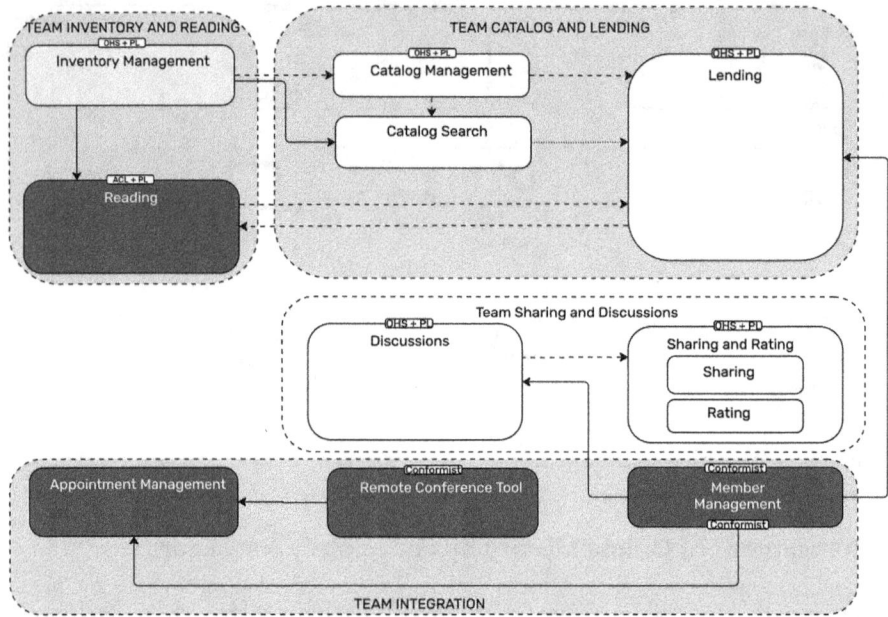

Figure 10-3 Context map with the new sharing and discussions team

because sharing and reviewing were separate modules in lending in the first step and can easily be extracted.

The corresponding context map is shown in Figure 10-3.

Such a significant change requires Domain Storytelling and usually an Event Storming session. The Event Storming session for the sharing and reviewing component was held initially. It can be revisited and used as a foundation to develop the bounded context. Ideally it will only be a "shift and lift" from the Lending component. The change in the organization is a substantial step and will take a while to play out. Following *Conway's Law*, it is worth changing the organization when the architecture requires it.

We saw that the principles of the Synergetic Blueprint process could be applied to feature enhancements of a software product. The following section will discuss how the principles can be applied when integrating external systems.

Integration of External Interfaces

In the previous section, we saw that implementing requirements changed the software application. The changes might be minor, such as an enhancement of a bounded context, or may be more substantial, like adding an entirely new bounded context, including organizational changes.

Integration of External Interfaces

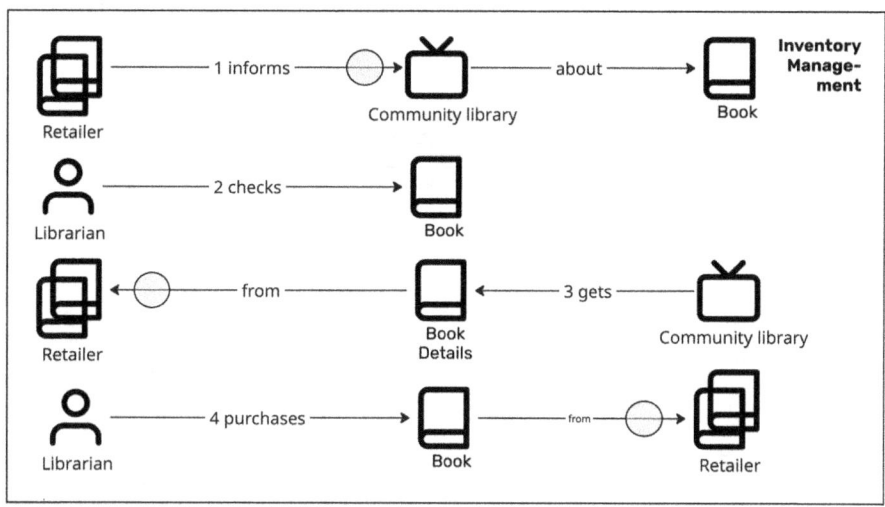

Figure 10-4 Book retailer integrated

However, external changes can even change a software application. That is valid, especially when integrating external APIs.

Assuming we want to integrate a book retailer into the online library, it will help librarians research new books for the library. The retailer provides an interface that the online library can implement. The corresponding domain story is shown in Figure 10-4.

1. The book retailer informs the library community about a new book.
2. The librarian looks into the book.
3. The community library obtains details about the book from the retailer.
4. The librarian purchases the book from the retailer.

The librarian can search for books and purchase them from the retailer in a completely transparent way. They work entirely in the common Inventory Management. The integration is achieved via synchronous interfaces provided by the retailer. The corresponding interfaces are marked with circles in the domain story (Figure 10-4).

The integration is shown in Figure 10-5.

The book retailer provides information about new books via a synchronous post to an interface that is implemented by Inventory Management. Even though the retailer acts as a client in this scenario, the interface is defined by the retailer. The Inventory Management implements this post as a server.

Such a scenario is common to push events over HTTP to partners (Webhook). Events over HTTP avoid setting up a public event broker that both needs to be integrated with and is not product independent. We will discuss this in more detail in the following section (section "Publishing Events to External Partners").

10   Iterative Enhancements

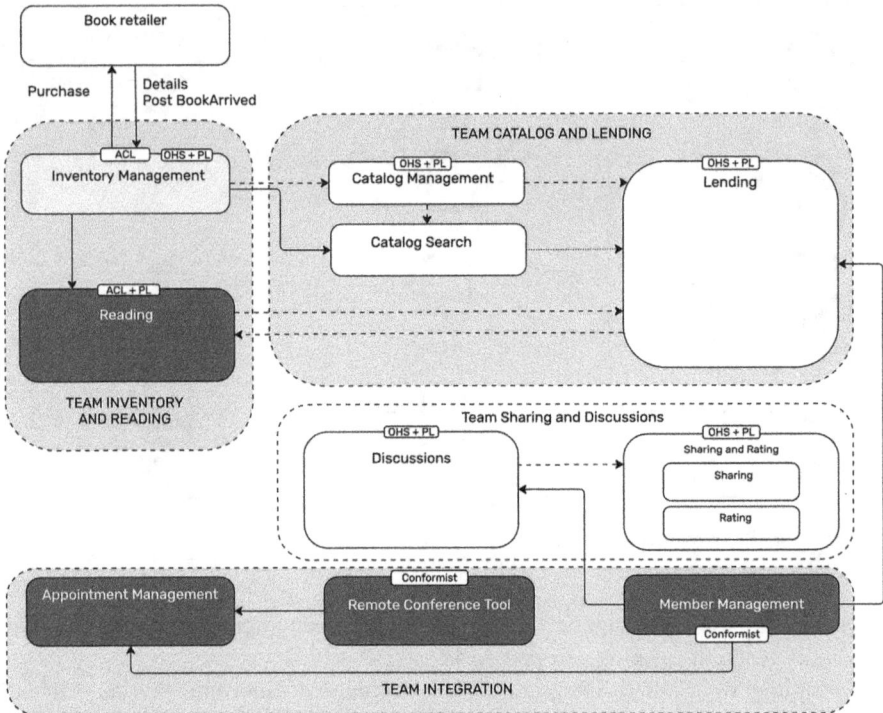

Figure 10-5   Book retailer, integrated – context map

The post about a new book contains a callback link to get the necessary detailed data if the responsible librarian is interested in the new published book. If the book is so interesting that the librarian decides to purchase it, it can even be purchased directly using Inventory Management. The librarian does not need to navigate to another application. The purchase is done in the Inventory Management, which calls the purchasing interface of the retailer.

The interfaces from the Inventory Management to the retailer are decoupled from the Inventory Mangement model by an ACL. Such an ACL takes over three tasks: decoupling the external and internal model, enabling the substitution of the current retailer, or adding new retailers. The ACL works in the end as a standardization component to the outside world.

Standardization

Using APIs allows quite different partners to work collaboratively. Those collaborations have a significant impact on the business. They usually make the partner more successful in the market.

Therefore, standards are recommended when working together because endless synchronizations can be avoided when using the standard. Groups of enterprises are trying to define those standards. In what follows, we list some examples:

- **FHIR**[3]
  FHIR (speak fire) is a standard that defines the interoperability between European health providers.
- **Swift**[4]
  Swift is a global standard for transferring money. Swift defines the financial messages standard (syntax). It provides the network for transmitting messages (proprietary) and a set of software.
- **Open Banking (UK)**[5]
  Open Banking is a standard in the United Kingdom for sharing accounting information.
- **PSD2**[6]
  PSD2 is a revised standard for sharing accounting information in the European Union. It emphasizes the security of end consumers.
- **Open Insurance**[7]
  Open Insurance is a specification for data properties, structures, relationships, and data types. It comes with a technology-agnostic and comprehensible domain model in business terms.
- **Fediverse based on ActivityPub**[8]
  ActivityPub is a decentralized social networking protocol based on the ActivityStream 2.0 protocol data format.[9] It is best known for its implementation in Mastodon.[10] The corresponding API is publicly available.[11] The standard is even adopted by large players like Meta Threads[12] or Flipboard.[13]

Those standardizations are sometimes more principles like PSD2, but can be true APIs like Open Insurance. However, they enable companies to work together based on a generally accepted contract. A specified contract between companies is not necessary.

Those standards only work when the technical base is standardized as well. Therefore, we use almost every time HTTP with REST because the protocol and

---

[3] https://fhir.org/

[4] https://www.swift.com/

[5] https://standards.openbanking.org.uk/

[6] https://www.ecb.europa.eu/press/intro/mip-online/2018/html/1803_revisedpsd.en.html

[7] https://openinsurance.io/standards/

[8] https://activitypub.rocks/

[9] https://www.w3.org/TR/activitystreams-core/

[10] https://docs.joinmastodon.org/

[11] https://docs.joinmastodon.org/client/intro/

[12] https://help.instagram.com/169559812696339

[13] https://about.flipboard.com/inside-flipboard/flipboard-begins-to-federate/

standard are widely used and reliable. Other standards that are not widely used or are only a kind of pseudo-standard without an independent board must be considered carefully. Otherwise, the pseudo-standard may no longer be supported by providers, and a new solution needs to be implemented to break the provider lock-in.

Collaboration Between Partners

More often, you will not find a standard you can use. When integrating, you need to work with externals.

The workshop formats recommended by the Synergetic Blueprint can be used to define common APIs together with externals. By "externals" we mean partners from other companies. Both the internal company and the external company want to integrate with each other. The corresponding business requirements are collected using Domain Storytelling.

As shown in Figure 10-4, the necessary APIs can be marked with circles. Using a Visual Glossary and an API Product Canvas (section "Introduction"), the necessary APIs can be specified in a joint workshop. The needed IDLs can be formulated later in a shared version control system using pull requests to tune the contract. Based on the contract, both partners can implement the agreed interfaces.

Such a collaborative approach is most suitable for working together on a cooperative and trusted base.

Publishing Events to External Partners

As we discussed in previous chapters, EDA is a great approach to building decoupled and sustainable applications. In addition, we discussed in the previous section that it is beneficial to integrate systems with partners.

Thus, it is obvious to consider integrating partners over the internet with event-based API.

Let us imagine a school library that benefits from the community library because the inventory of the community is larger and can be presented online. A student searches for a book and can find it. It is transparent to the student that the book is provided by the community library. The corresponding domain story is shown in Figure 10-6.

1. The student logs into the school library.
2. He searches for a book in the school library.
3. He borrows the book via the school library.

The student is able to search for books in the community library transparently. The corresponding catalog entries are synchronized from the community library to the school library via events. This solution approach is shown in Figure 10-7.

Publishing Events to External Partners

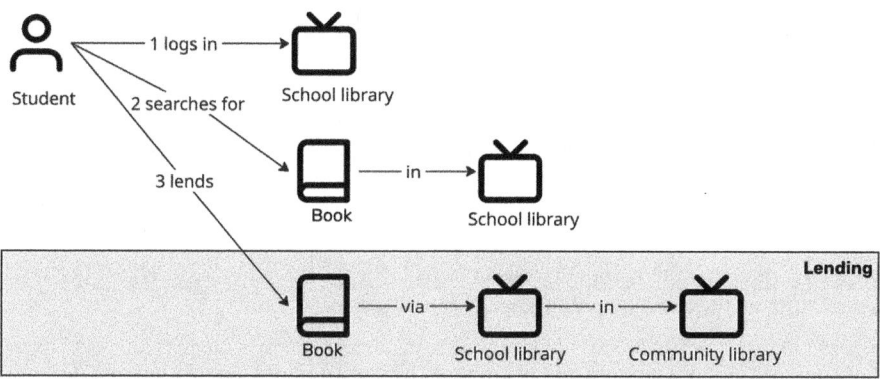

Figure 10-6 School library lending

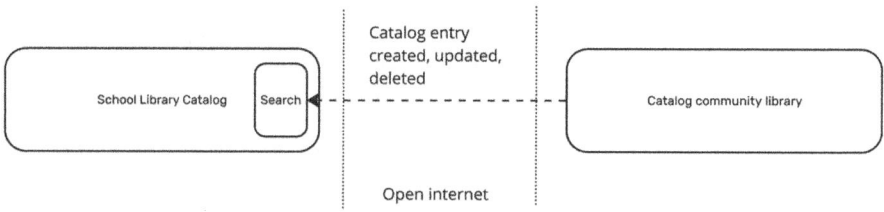

Figure 10-7 Architectural approach to school library

The school library consumes the events `catalogEntryCreated`, `catalogEntryUpdated`, and `catalogEntryDeleted` from the Catalog Management component of the community library. Those events are already implemented supporting the CQRS pattern of Catalog Management and Catalog Search (section "Catalog Management").

The search component of the school library can consume events.

However, as promising as the idea seems, it must be carefully considered. As discussed in the previous section, usually events are distributed over a broker. One of the sides needs to provide such an infrastructure component. Both sides must agree to the broker protocol like Apache Kafka or AMQP they will use. In addition, authentication is not as streamlined as in HTTP. Many more security standards are floating around in the event broker environment.

Publishing External Events

Event brokers like Apache Kafka are not set up to communicate over the public internet with untrusted code. They are not set up to handle too many TCP connections in parallel, which needs to be covered by public services.

Most public API are HTTP-based

Most public API are defined in the HTTP protocol. On the one hand, it is the most commonly used web-based integration mechanism [1]; one factor is certainly that it is the main integration mechanism for browser-based applications. On the other hand, there exists a lot of great tooling and know-how to expose HTTP to the public. A known protocol like HTTP makes it simpler for most integrators to use it, for teams to develop it, and for security officers to approve it. On the other hand, protocols like Apache Kafka do not scale well for millions of users. These protocols have limits as they use long-running TLS connections.[14]

However, data-set-based access control based on dynamic rules, like user properties like age or role in a company, is difficult to apply directly (only via encryption or proxies/routers[15]). The messages need to be tailored carefully using Synergetic Blueprint to apply those security considerations. The cut of messages and channel segregation directly impact security concerns.

A solution approach might be to use proxies to communicate via HTTP. It is shown in Figure 10-8.

Therefore, some message component is needed that can communicate via the internet asynchronously, for example, Webhook via HTTP, WebSockets (section "Communication Styles and Mechanisms").

Alternatives to a fully available broker via the internet are possible, as described in the previous section, by implementing post interfaces defined by the consumer. The solution shown in Figure 10-8 even combines the idea of public events and securing them via communication over HTTP. A really great implementation of this is Webhooks (Listing 6-34). With Webhooks, a user-defined callback can be triggered when an event happens [4]. The publisher can filter the possible callback URL to guarantee security. Webhooks can be specified with OpenAPI 3.1 as "Callbacks."[16]

---

[14] New HTTP servers also use long-running TLS connections. However, these can be terminated by the server at any time and scale well. Protocols like Apache Kafka are designed such that these connections are up all the time, which can be limiting. Therefore, providers often have strict limits that can in such use cases restrict how many new connections can be established in a particular time frame and how many can be parallel open [2, 3].

[15] For example, field-level encryption or a gateway like Conduktor Gateway https://docs.conduktor.io/gateway/, kafka-proxy https://github.com/grepplabs/kafka-proxy, or Zilla https://docs.aklivity.io/zilla/latest/ or a router like Amazon EventBridgehttps://aws.amazon.com/eventbridge/.

[16] https://swagger.io/docs/specification/v3_0/callbacks/

Publishing Events to External Partners 321

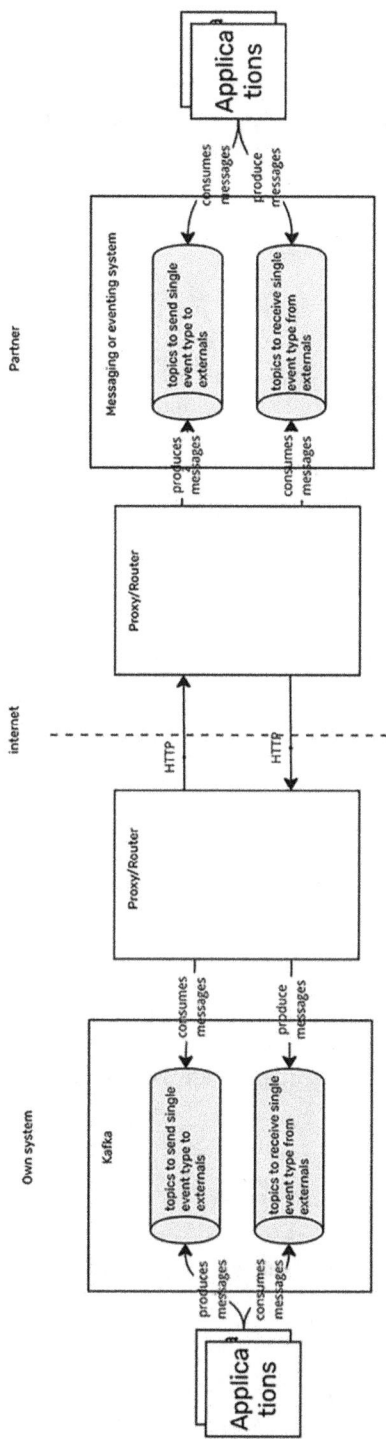

Figure 10-8 Solution approach for public events using proxies

Security Concerns

A data-processing component should not be available via the internet because the data processed cannot be secured against unauthorized access. The access to channels can be secured by broker-provided security models.

Conclusion

We saw that the principles of the Synergetic Blueprint could be applied to extensions, too. Even more, they must be applied to enhance the software after the same principles that were applied when it was initially set up.

The principles of DDD help to tailor and assign enhancements to the suitable bounded contexts.

Even domain events can be enhanced, even though we saw that publishing events via the internet required some consideration.

Points to Remember

- Enhancements of a software application can be done using the principles of the Synergetic Blueprint.
- Enhancements can be done inside of a bounded context or can create new bounded contexts.
- Collaborative work with partners is enhanced by workshop formats and collaborative modeling by the Synergetic Blueprint.
- Publishing events over the internet requires special technical, security, and organizational considerations. Usually, therefore, events are transmitted over HTTP, for example, by Webhook.

Review Questions

10.1 Are enhancements of software applications supported by collaborative modeling?

(a) yes
(b) no

10.2 Can enhancements of a software application create new bounded contexts?

(a) Yes
(b) No

References

1. Madden N (2021) API Security in Action, 1st edn. [Erscheinungsort nicht ermittelbar]: Manning Publications, Shelter Island, 1576 pp. Online Resource; Title from title page (viewed January 3, 2021). ISBN: 978-16-17296-02-4
2. Confluent cloud cluster types | confluent documentation, Confluent (2025). [Online] Available: https://docs.confluent.io/cloud/current/clusters/clustertypes.html. Visited on 12 April 2025
3. Amazon msk quota - amazon managed streaming for apache kafka, Amazon Web Services (2025). [Online] Available: https://docs.aws.amazon.com/msk/latest/developerguide/limits.html. Visited on 12 April 2025
4. Webhooks (2025). [Online] Available: https://developer.atlassian.com/server/jira/platform/webhooks/. Visited on 15 Feb 2025

Brownfield Project   11

In previous chapters, we discussed how APIs could be designed and implemented. We discussed how agile organizational approaches were necessary to design sustainable APIs with a strong business focus. We assumed that our project was started without any previous components. Such a project is called a greenfield project, analogous to greenfield land, where no infrastructure exists yet.

In many cases, however, assuming a greenfield project in software is unrealistic. Some components exist in almost every case. Projects with previous components are called brownfield projects. We cannot design software applications freely. We need to consider the components available and the costs of their change.

We will discuss a brownfield approach using the example of a hypothetical insurance system applying the *Synergetic Blueprint* approach. Let us start with the introduction of an insurance system.

Hypothetical Insurance Legacy System

An insurance company covers risks. To do so, it tries to level out the risks to customers. The probability of an event occurring is more negligible in terms of costs than the premiums paid by the customers.

The math behind this simple statement can be pretty complex.

However, we do not want to concentrate on insurance math; we want to concentrate on an insurance company's example system. To understand the example, a deeper understanding of insurance processes is not necessary. We will explain the terms and relations in their proper place.

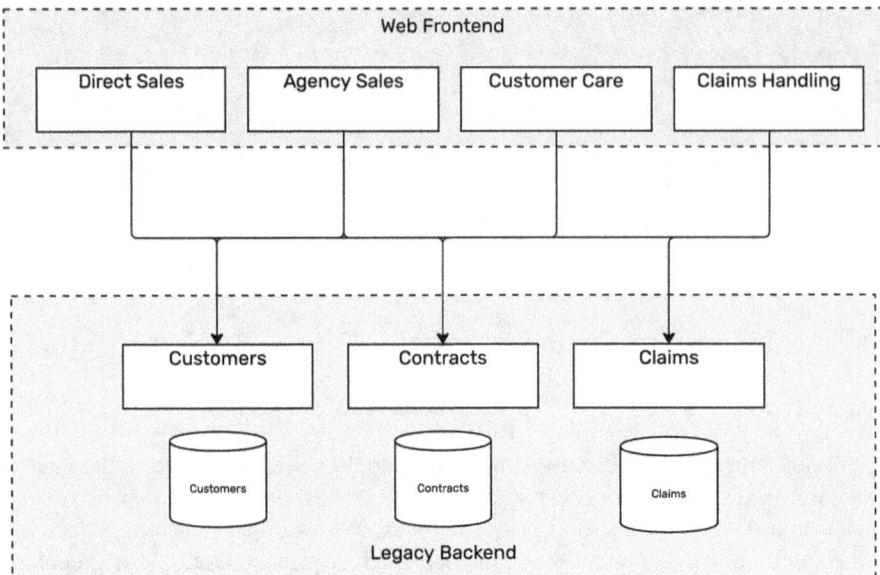

Figure 11-1 Legacy system of an insurance company

Legacy System

An insurance system is based on customers, contracts, and claims. Based on those data, frontends can be built for direct or agency sales, customer care, and claims handling.

A typical legacy system is shown in Figure 11-1.

The frontend applications access the backend databases via corresponding access layers to large databases structured by large data types, which are not differentiated via DDD.

The system faces new challenges because the premium calculation and quoting should be externalized via synchronous interfaces.

Business Analysis in a Brownfield

An external comparison portal for insurances wants to access the rating and quoting directly via REST APIs. The performance requirements are high. If the response to a request is not sent in 0.5 s, the premium or quote is not shown to the end user. Not being directly in the list means loss of revenue. First tests showed that the requirements could not be met using the current architecture. Therefore, an analysis is started on how to make the architecture "future ready."

It might seem that the architecture should be analyzed first. However, it is a better idea to analyze the business first, to find the right moment from a business perspective. Using such a business perspective is more effective because technical changes can be prioritized from a business point of view.

The business analysis starts with the current process. It is recommended to use the same methodologies as in a greenfield approach.

Domain Storytelling of Current Process

Business processes are usually stable, but they change overtime slowly. Those changes are often not detected. In a legacy system, usually the business process of the original creation are documented. Later changes are described, but the overall documentation is not or only slightly updated.

Therefore, it is worth documenting the current process first before the intended process is discussed (Figure 11-2). Such processes are usually called Quote & Buy because a potential customer can buy an insurance policy based on a quote.

The current business process is based on a web application (Figure 11-2):

1. A prospective customer navigates to the insurance homepage.
2. **Quotation** – The prospective customer navigates to the premium calculator.
3. **Quotation** – The prospect gives parameters to calculate the premium on the quotation component.
4. **Quotation** – The premium calculator calculates the premium.
5. **Quotation** – The prospective customer gives his personal information for a quote. For example, the address and some additional items like bicycles for a home insurance quote.
6. **Quotation** – The quotation component calculates the quote.
7. **Quotation** – The prospect accepts the quote.
8. **Quotation** – The prospect gives her banking details.
9. **Quotation** – The prospect sends her application for the quotation.
10. **Contract Manager** – The Contract Manager checks the application.
11. **Contract Manager** – The contracting component sends the contract to the customer.

The domain story already contains the current system borders implemented as frontend applications – Quotation and Contract Manager.

The current process has some difficulties (marked by a flash in the domain story; see Figure 11-2):

- At the moment, prospects can only navigate to the tariff calculator via the homepage. External access is not possible. Some external static links to the tariff calculator are available, but those are seldom used by prospects.
- There is no difference between prospects and applicants, even though an applicant has already accepted a provided quote.
- Applications of potential customers are handled by the quotation component, even though the customer has already given her banking details. To ensure

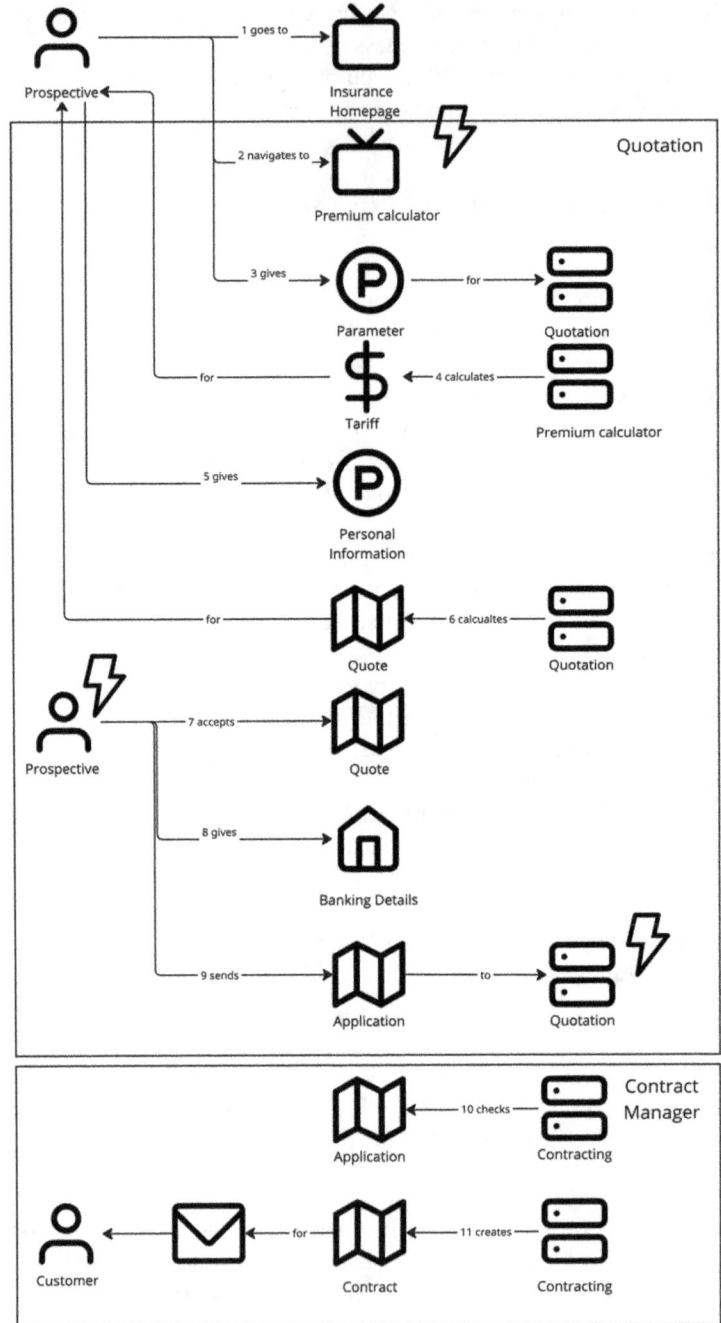

Figure 11-2 Domain story of current sales process

required data usage is minimized, a different component should handle applications so that quotes and applications can be separated.

Those problems are created over a longer period of an existing system. Problems arise over an extended period within an existing system. They are not mistakes; rather, they develop within large systems over years of development.

To fulfill the requirements of a large-scaler portal, the system needs to be modernized. Therefore, we need the intended process as opposed to the current process.

Intended Process

The intended process states that a prospect comes from some kind of portal. That might be a comparison portal or insurance-specific portal. However, the prospective customer wants to receive a tariff as fast as possible and to obtain the contract as fast as possible. The intended process is shown in Figure 11-3.

The process contains a dedicated differentiation between the phases tariff, quote, and application.

1. The prospect goes to the portal.
2. **Premium** – The portal calls the premium calculator.
3. **Premium** – The portal transfers the parameters to the premium calculator via an API.
4. **Premium** – The premium calculator calculates the tariff for the portal.
5. **Quotation** – The prospect enters personal information to obtain a more detailed quote.
6. **Quotation** – The quotation calculates the quote for the prospective customer.
7. **Quotation** – The applicant accepts the quote.
8. **Application Manager** – The applicant gives his banking details.
9. **Application Manager** – The applicant sends his application to the application manager.
10. **Contract Manager** – Contracting checks the application for whether the risk is acceptable to the insurance company.
11. **Contract Manager** – Contracting sends the contract to the prospect.

The intended process tackles the difficulties detected in the current process:

- Access to premium and quoting is open to portals. Portals can be internal and external. The integration can be done via APIs or provided user interface components.
- The process differentiates between prospects, applicants, and customers.
- Quotes, applications, and contracts are handled by different components.

330　　11　Brownfield Project

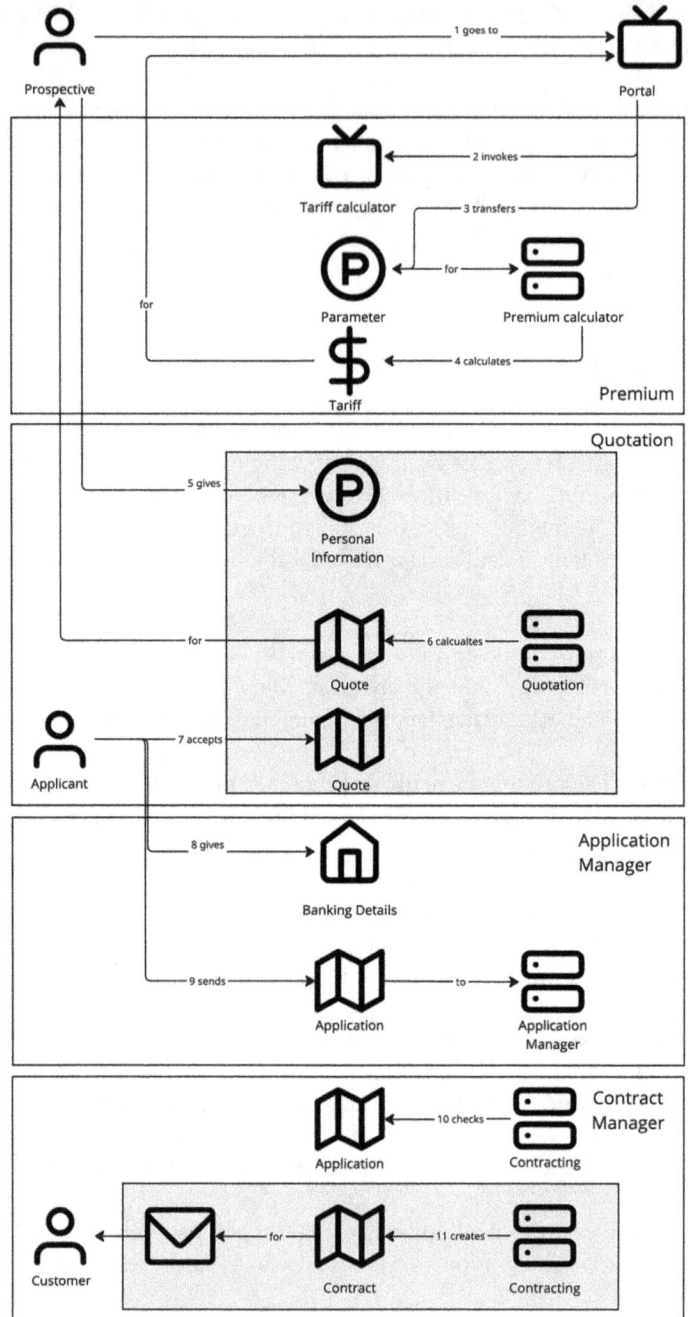

Figure 11-3　Domain story of current sales process

The intended process shows that the system must be more flexible and be tailored along the found bounded contexts. Let us now discuss how we can bring a seasoned system of an insurance company to a modern microservice architecture.

Architectural Approach to Modernization for the Insurance Application

The intended process is outlined within multiple bounded contexts. What's more, the bounded contexts follow an approach that is completely different from that of the existing seasoned systems. The modernization is necessary because frontend applications require different data structures than the existing system, which currently delivers customer, contract, and claim data.

Target Architecture
The target architecture must follow the required structure of the revealed bounded context:

- Premium,
- Quotation,
- Application Manager, and
- Contract Manager.

Additionally, master data are necessary for prospects and customers. Those master data can be externalized to avoid having to deal with doublets in quotations, applications, and contracting.

The external portal provider also requires a revenue overview of the brokered contracts. The sales organization wants to be supported by artificial intelligence, which can provide the next best offers or products based on models trained by the sales of insurance products.

Moreover, external and internal portals require a faster time to market and constant integration and delivery that can only be guaranteed by well-tailored microservices.

The resulting architecture is shown in Figure 11-4.

The architecture vision follows a typical microservice architecture, where single bounded contexts are separate services that provide synchronous endpoints for user interface access. Each of those services requires a separate persistence layer. The services generally need to exchange data via an event broker. However, the bounded context map determines whether the business events can be passed asynchronously or whether a synchronous connection is needed. In our example, the transition between the activities of the user journey requires the following events: `premiumCalculated`, `quoteCreatedUpdated`, `applicationSent`, and `contractCreated` or even `applicationNotAccepted`.

To determine which parts of the architecture should be changed first, we need two analyses:

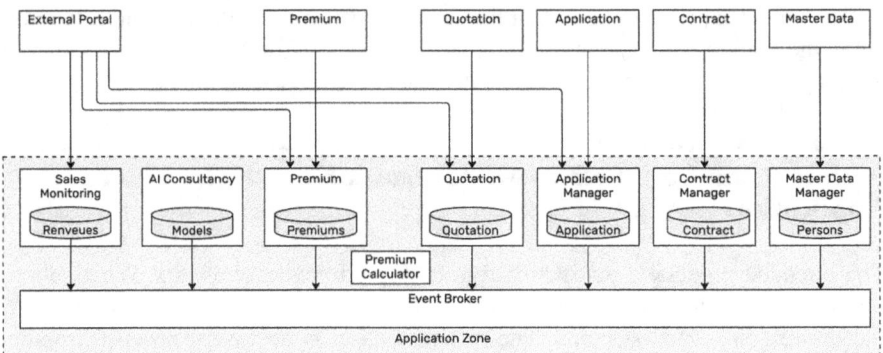

Figure 11-4 Target architecture

1. Analysis of business criticality
2. Analysis of technical criticality

Business Analysis Using Wardley Map Method

Business criticality can be analyzed using a Wardley map, as we did earlier in section "Qualifying Business Ideas." As we learned, we can use a Wardley map to prioritize tasks to be done in a development or modernization project. We search for the users of our application: potential private insurance customers, operators of external portals, and insurance agents.

We apply to each user group the needs they wish to fulfill with software:

Customer Customers want fast and reliable contracting without waiting for long processes that take days.

Insurance agents Insurance agents also want fast contracting and an overview of their revenues.

Operators of external portals The operators of external portals want an overview of their brokerage. However, they are not interested in fast contracting. If the insurance is too slow, it will not appear in the brokered list of tariffs.

In the next step, as described in section "Qualifying Business Ideas," the capabilities necessary to fulfill users' needs are sorted along a visible-to-invisible scale.

Mapping is the last step of a Wardley map. The mapping sorts the identified capabilities into genesis, custom-built, product, and commodity. The result is shown in Figure 11-5.

The Wardley map shows that the components

- AI consultant,
- Premium,

Architectural Approach to Modernization for the Insurance Application 333

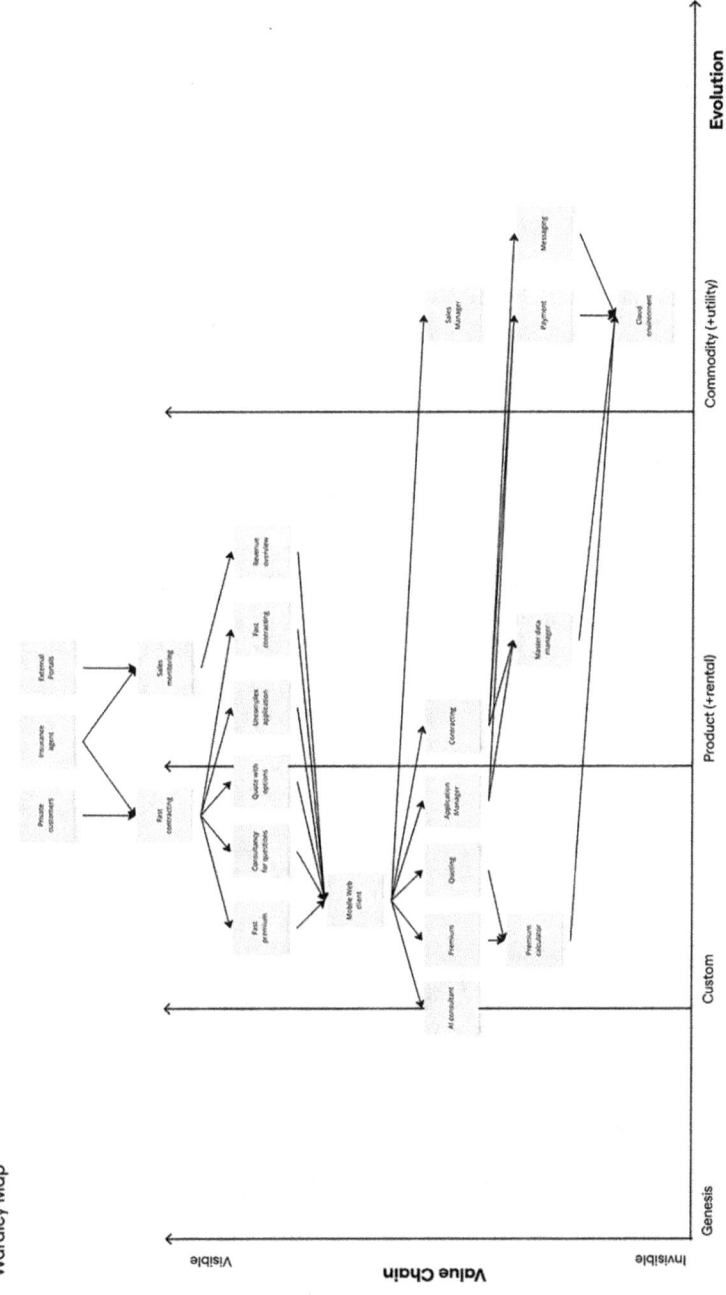

Figure 11-5 Wardley map

- Quoting, and
- Application Manager

get the highest priority because they need to be custom-built as core capabilities. In a modernization environment, we need to fulfill even technical prioritization. Because the AI consultant can be implemented independently of the other modernization activities, it gets a lower priority than premium and quoting, as discussed in more detail in the following section.

As was already explained, the boundaries between them and the corresponding domain events are clear, and an additional Event Storming session is not necessary to define the bounded contexts. However, the Quotation and Premium components both use the premium calculator component. When the premium calculator is called, the event flow is unclear. Therefore, it is worth performing an Event Storming session for the premium calculation with premium calculator, premium, and quotation teams. The result of the Event Storming session is shown in Figure 11-6.

It shows that the premium calculator is called from the Premium component as well as from the Quotation component. The premium calculation itself must be fast during a user interaction. Therefore, it needs to be implemented as a pure calculator that can be called synchronously and stateless. The necessary parameters collected during the Premium need to be provided again during Quotation. Therefore, the Premium component needs to provide the premium to the quote, including parameters, asynchronously.

The resulting context map is shown in Figure 11-7.

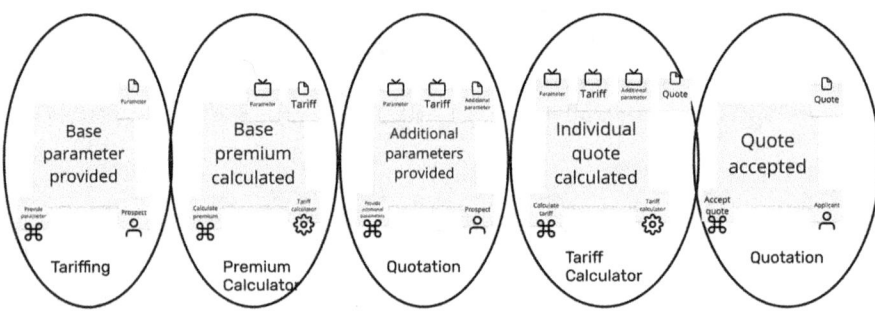

Figure 11-6 Event Storming of premium calculator

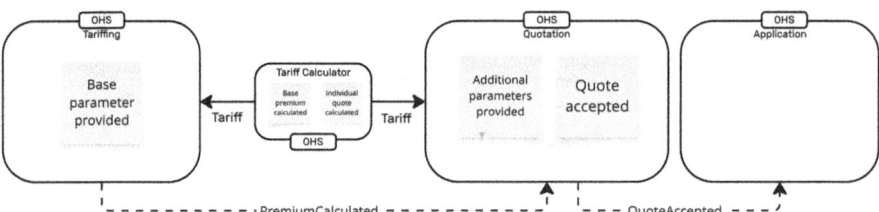

Figure 11-7 Context map premium calculator

The context map corresponds to the target architecture (Figure 11-4).

As we already discussed, the team responsible for the tariff calculator acts as a complicated subsystem team (section "Team Topologies") because the service is used by two teams that need to follow the model of the tariff calculator when giving parameters. The other services for Premium, Quotation, and Application Manager can be implemented as open host services, for which one team is responsible per service.

We defined the target architecture and even a target team cut. The following section will discuss the transition from the current architecture to the target architecture.

Architectural Transition

The transition describes the steps from the current legacy architecture to the new microservice-oriented architecture. The transition architecture can be defined based on the "greenfield approach." As a first step, a technical analysis is necessary to underscore the prioritization based on the business analysis in the previous section.

Technical Analysis

The technical analysis starts with a crude analysis of the current architecture, as we have already shown. The methods of C4 analysis can be used [1]. Ultimately, it is an architectural review process whereby methods such as Lightweight Approach for Software Reviews (LASR) [2] or architecture tradeoff analysis method (ATAM) [3] can be used. Tool-supported code analysis shows dependencies and model incorrectnesses that need to be corrected [4]. During the technical analysis, prioritization for technical modernizations evolves, for example, when a database needs to be substituted because the software is no longer supported. The business and technical prioritization together define the steps of the modernization.

However, analysis of the models shows usually huge differences. The models of the legacy system and target systems differ – usually completely [4]. A large database drives the legacy system models, whereas the models of the target system follow a DDD approach.

Therefore, it is recommended to start with a decoupling layer between the legacy world and the to-be-modernized world to prevent spoilers from the legacy model from entering into the domain-driven model.

Decoupling Layer

To enable an ACL between the legacy layer and a future modernization layer, adapters [5] must translate the legacy model to the domain-driven model [6].

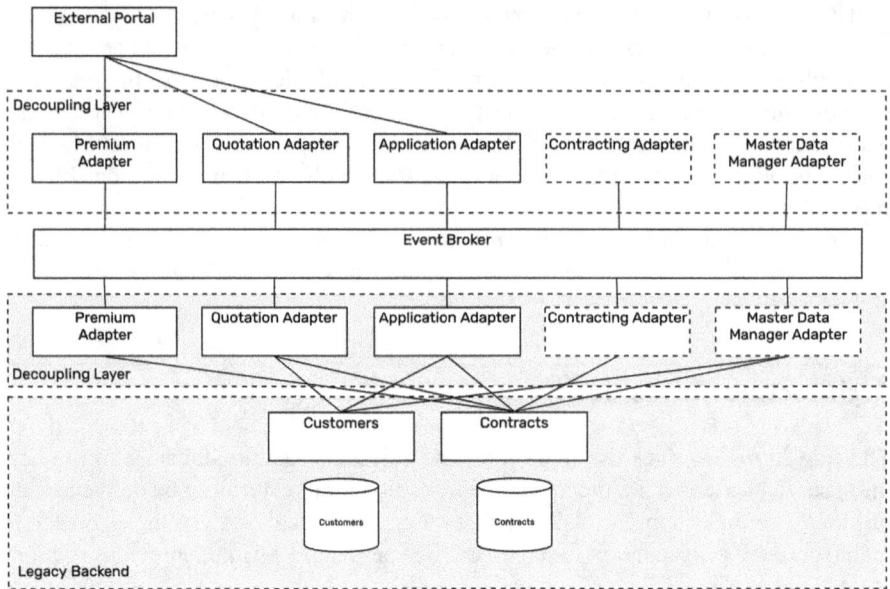

Figure 11-8 Decoupling layer

The adapters watch for changes in the legacy system, which still owns the data. They capture the data changes and publish them to an event broker. Such a pattern is called Change Data Capture (CDC) [7].

The legacy model is huge and affects multiple bounded contexts. Therefore, the data of the legacy system needs to be provided to several bounded contexts. To allow, on the one hand, for capturing changes in the legacy model and, on the other hand, not to fall in a tight coupling between bounded contexts, the accesses need to be decoupled, which is solved by an event broker. Those events are designed to synchronize data between the legacy and the modern world. Usually, those events are not suited for use in an event-driven architecture. The described pattern is called data liberation [8].[1] The domain events of the microservices need to be implemented later in the modernization layer.

The adapters create events consumed by corresponding adapters that provide synchronous REST APIs following the domain-driven model. Those REST interfaces can be designed in a modern style and can be used to decouple modern user interfaces from the monolithic structure of the legacy model. The approach is shown in Figure 11-8.

The modernization should start with the Premium, Quotation, and Application. These components are essential to deliver the premiums and quotes for large-scale

---

[1] There are different approaches to implementing data liberation. Refer to *Bellemare* [8] for detailed information.

portals. Other components like Contracting and Master Data Management do not need to be modernized in this first step (marked with dashes in Figure 11-8). This can be done in a second and third step.

The adapters only map a legacy model to a domain-driven model. Usually, they do not even have a persistence layer (only something like a cache). The data ownership is still in the legacy system. However, the adapters providing interfaces to the outside already represent the DDD of a modern architecture approach. The legacy layer is reduced step by step as in the strangler fig pattern [9].

Modernization Layer

The second step represents the transition from the legacy system's data ownership to the modernization layer's ownership. In the modernization layer, each bounded context implements its model, which is defined by the domain-driven methods Domain Storytelling and Visual Glossary [10, 11]. The REST APIs of the original adapters in the modernization layer (Figure 11-8) are not changed so that external consumers must not change their implementation.

The associated approach is shown in Figure 11-9.

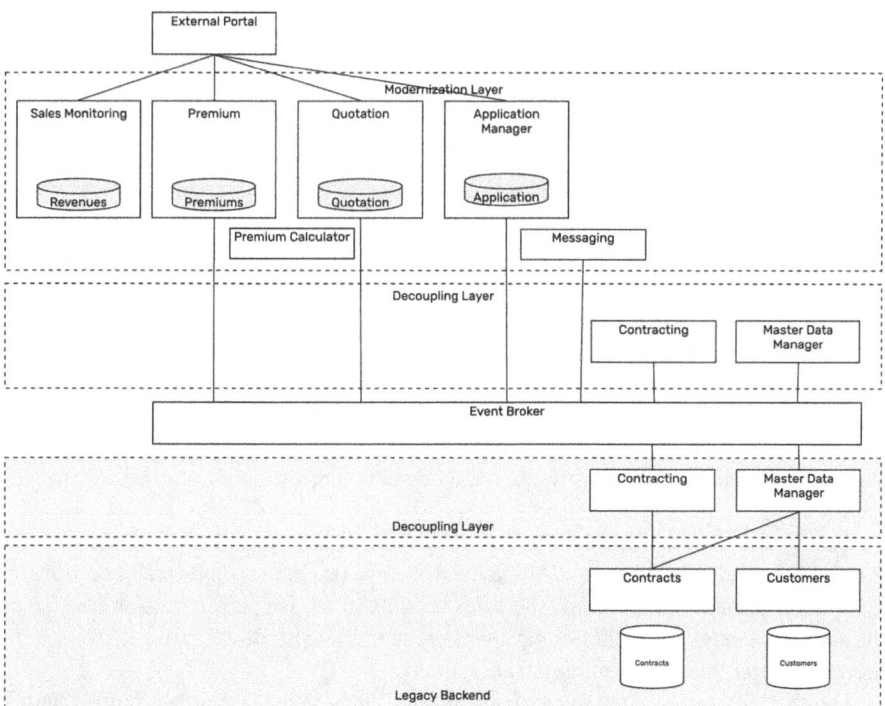

Figure 11-9 Modernization layer

The bounded context services produce and consume domain events that were defined during the Domain Storytelling and Event Storming workshops. Those events are different from the data synchronization events of the decoupling layer.

The modernization layer contains new and transferred services to fulfill the requirements of the external portal's operators. A new service is established to provide sales information like revenues and another one is set up to handle messages to customers. The Premium, Quotation, and Application Manager components are transferred from the legacy to the modernization layer. Other data, such as contracts and customer data, remain in the legacy layer. Such a strategy to transfer the functionality of the legacy monolith to the modernization layer step by step is called the strangler fig pattern, as described by *Fowler* [9]. The legacy layer is weakened by the modernization layer, like a host tree is weakened by the strangler fig.

The transferred services can contain code parts developed before the transition. The technical analysis shows which parts must be refactored thoroughly and which can remain with slight modifications. However, it is recommended to create a new code basis for the services transferred in new repositories to allow the teams to work independently, for example, proprietary pull requests with team rules, separate code rules, and checks, as well as suitable CI/CD pipelines [4].

The code parts in the legacy layer can be switched off step by step as mentioned in the strangler fig pattern [9].

Developed Architecture

The architecture of the modernization layer can be developed further step by step. The developed architecture is shown in Figure 11-10.

The transitioned services Contracting and Master Data Management are available in the modernization layer. Even though the modernization layer services have taken over data ownership from the legacy layer, Contracts and Customers are still available in the legacy layer. Contracts with long lifecycles, like life insurance policies, might have lifecycles of fifty years or even more. Migrating those data and the associated customers is expensive and does not support the modernization and business processes in the modernization layer. Therefore, those data will remain in the legacy backend. They are migrated on demand, for example, when a claims request is initiated based on those contracts. The corresponding event of the Claims Request bounded context initiates an automatic migration of the corresponding contract.

The modernization layer even contains completely new services. One is the Payment module that serves as ACL to different payment providers. The other is the AI component that contains the sales models to advise private customers during the sales processes. No other component preceded it. Therefore, it exists only in the modernization layer without any transition.

The modernization layer even makes it possible to provide independent frontends per bounded context as components to larger portals, marked in Figure 11-10 as Prospect, Applicant, Customer Portal, as Agency Portal, and as Broker Portal.

Conclusion

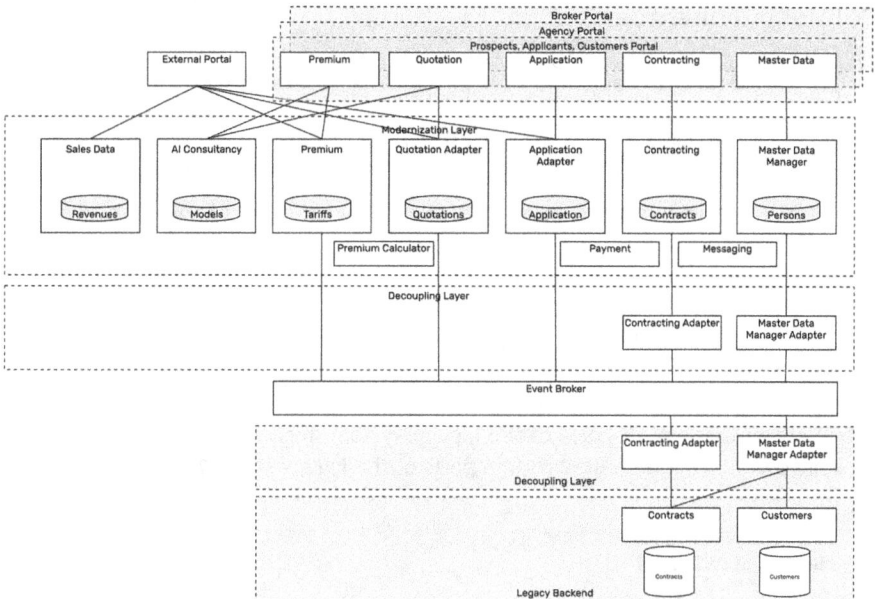

Figure 11-10 Developed architecture

The modern microservice architecture in the modernization layer allows a higher flexibility to fulfill business requirements.

Conclusion

In this chapter, we observed that the principles of DDD can be effectively applied to modernization projects. API design plays an essential role during a modernization project because it is the first tool to deliver the necessary domain-driven model. Events collect the changes to the legacy model in the change capture approach, whereas synchronous APIs provide access to external collaboration partners. The event and API design follows the same principles as in a "greenfield" approach, followed by a transition architecture to arrive at the target architecture. A "brownfield" project must be driven by business, like every other software development project.

We presented a large modernization project in an insurance company. However, the demonstrated process can be applied in smaller modernization projects as well, where not all steps of the Synergetic Blueprint would likely need to be fulfilled. We even used a shortcut in our large example: We used Event Storming only in an unclear situation.

Those shortcuts can be used in certain situations, and we will discuss them in the following chapter.

Points to Remember

- Use a "greenfield" approach with the Synergetic Blueprint to define the target architecture of the legacy system.
- Skip Event Storming when the transitions from one bounded context to another are clear.
- Prioritize the necessary steps of modernization using the Wardley map method and technical analysis.
- Decouple the legacy model from a domain-driven model represented by APIs using an event broker - a data liberation pattern.
- Collect changed data from the legacy system and publish those changes to the event broker.
- Modernizations should always be done in several steps.
- Be aware that not all parts of the legacy system must be modernized. Business consideration should allow certain parts of the legacy system to remain.

Review Questions

11.1 Can the target architecture of a modernization project be designed by the same principles as a "greenfield" project?

(a) Yes
(b) No

11.2 Can certain steps of the Synergetic Blueprint be skipped?

(a) Yes, when the events and transitions are clear
(b) No
(c) Yes, but the business has to agree
(d) Yes, but the project management has to agree

11.3 How can a legacy model be decoupled from the modern model?

(a) No decoupling necessary
(b) Only using synchronous APIs
(c) Using an event broker, CDC, and data liberation design pattern
(d) Decoupling can only be done by separate services

11.4 When modernizing a system, should all parts of the system be modernized simultaneously?

(a) No
(b) Yes
(c) Yes, but only in large systems

References

1. C4 model. [Online] Available: https://c4model.com/. Visited on 28 Dec 2024
2. Lasr - lightweight approach for software reviews (2025). [Online] Available: https://www.embarc.de/themen/lasr-reviews/. Visited on 14 Feb 2025
3. Architecture tradeoff analysis method collection (2018). [Online] Available: https://insights.sei.cmu.edu/library/architecturetradeoff-analysis-method-collection/. Visited on 14 Feb 2025
4. Lilienthal C, Schwentner H (2023) Domain-Driven Transformation. dpunkt.verlag. ISBN: 978-38-64908-84-2
5. Gamma E, Helm R, Johnsson R, Vlissides J (1994) Design Patterns: Elements of Reusable Object-Oriented Software. Addison Wesley, Boston. ISBN: 978-02-01633-61-0
6. Junker A (2023) Event-getriebene intergrationsarchitekturen. [Online] Available: https://www.informatikaktuell.de/entwicklung/methoden/eventgetriebeneintegrationsarchitekturen.html. Visited on 27 Dec 2024
7. Rocha O (2022) Practical Event-Driven Microservices Architecture, Building Sustainable and Highly Scalable Event-Driven Microservices, Filipe H (ed). Apress L. P., Berkeley, 1449 pp. Description based on publisher supplied metadata and other sources. ISBN: 978-14-84274-68-2
8. Bellemare A (2020) Building Event-Driven Microservices, Leveraging Organizational Data at Scale, 1st edn. O'Reilly, Beijing, 1304 pp. ISBN: 978-14-92057-86-4
9. Fowler M (2024) Strangler fig. [Online] Available: https://martinfowler.com/bliki/StranglerFigApplication.html. Visited on 10 Feb 2025
10. Hofer S, Schwentner H (2022) Domain Storytelling, a Collaborative, Visual, and Agile Way to Build Domain-Driven Software. Pearson International, Mississauga. ISBN: 978-01-37458-91-2
11. Zörner S (2015) Softwarearchitekturen dokumentieren und kommunizieren, Entwürfe, Entscheidungen und Lösungen nachvollziehbar und wirkungsvoll festhalten, 2., überarbeitete und erweiterte Auflage. Hanser, München, 277 pp. Literaturverz. S. [269]-272. ISBN: 978-34-46443-48-8

Shortcuts in the Process   12

In Chapter 9, "Collaborative Design and Agility," we saw how the Synergetic Blueprint supported the design of APIs. However, we already saw some shortcuts in the modernization process (Chapter 11) and extensions (Chapter 10). In this chapter, we wish to discuss in detail where such shortcuts are allowed in the entire process.

To provide an overview on possible shortcuts, we enhanced the Synergetic Blueprint of Chapter 9 (Figure 12-1).

The first part discusses the use of a North Star like a guide on the ocean.

North Star

When creating extensions to an existing software application or smaller brownfield projects, a complete business analysis using Business Model Canvas, capability map, and, probably, a Wardley map is usually not necessary. However, the decision on this issue needs some guidelines to support the intended business.

A vision statement can serve as such a guideline. As the name suggests, the text conveys a vision containing the most essential product outcomes.

A North Star helps to formulate a vision statement containing the following points [1, 2]:

- Customer value,
- Vision and strategy, and
- Leading indicator of success.

A vision statement must be actionable and understandable to nontechnical partners. It must have measurable metrics that are not vanity metrics [1].

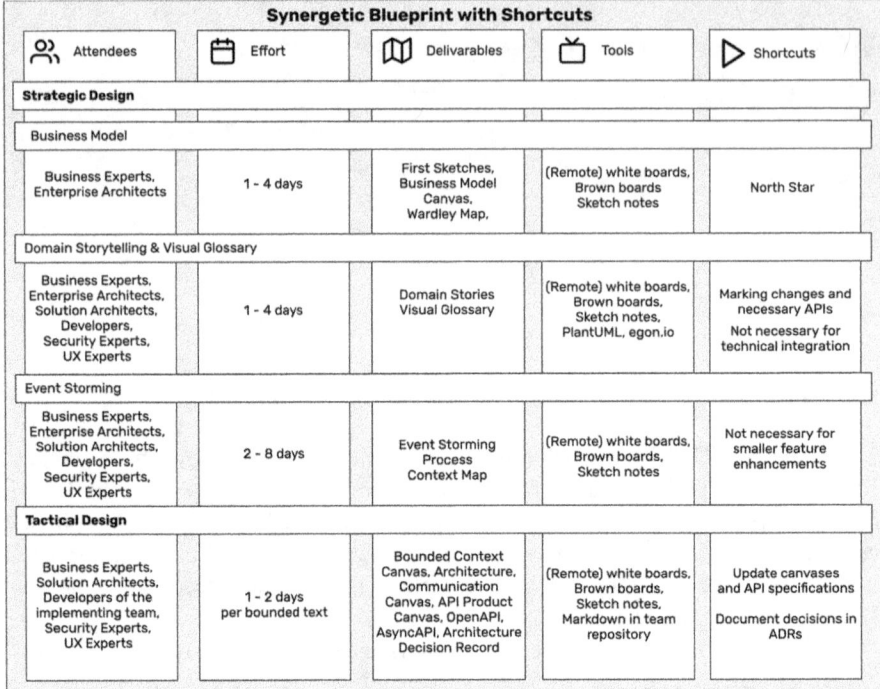

Figure 12-1 Synergetic Blueprint including shortcuts

Modernization Vision Statement

In the case of insurance modernization, we can formulate such a vision statement: "Our sales journeys are understandable and fast. They can be used by end users and by large-scale portal operators. Premium, quotation, and contracting are performed without media breaks and unnecessary waiting times. The tedious work of clerks is completely automated to serve end customers with fast contracting and clerks with interesting and challenging work. The entire sales process can be monitored to observe successful contracting and can be provided seamlessly to partners."

Such a vision statement serves as a North Star, the star that once guided sailors over the oceans. Especially in modernization programs, it is helpful to align all activities. The business processes are not changed. Therefore, changes in the Business Model Canvas and the capability map are not expected. However, customers' and prospects' expectations regarding response times and the quality of the software will change. Therefore, a Wardley map together with a vision statement help to align all partners.

Definition of REST APIs Directly After Domain Storytelling

In Chapter 10, we saw that we could define APIs without Event Storming. We can omit this step when the bounded contexts are clear and well defined. For example, we can leave out context mapping and Event Storming in projects with smaller feature enhancements. However, the Visual Glossary should be adapted in every case, because it serves as the foundation for formulating the API.

The specification of the API can then be done using the API Product Canvas.

Definition of Events Directly from Event Storming

The definition of events can be formulated using Event Storming when extensions or changes are made from a purely technical point of view. We saw a suitable example in section "Architectural Approach to Modernization for the Insurance Application". Complex technical changes should be supported by Event Storming, even though Domain Storytelling is not necessary in every case. Even if domain storytelling is left out, the Visual Glossary must be adapted to reflect the changes in the domain model.

Tactical Design

Even if certain steps of the business analysis and the bounded context tailoring can be left out, but the tactical design cannot. The necessary documents,

- Bounded Context Canvas,
- Architecture Communication Canvas,
- API Product Canvas,
- OpenAPI specification, and
- AsyncAPI specification,

must be updated due to the necessary changes. It includes an update of the Visual Glossary. Architectural decisions must be documented in ADRs.

Conclusion

The Synergetic Blueprint process is flexible and can be adapted to the needs of "greenfield," "brownfield," and extension projects. Not all steps are necessary in all cases. However, careful consideration must be given to the decision to leave out a particular step. Valuable information might be lost when a step is omitted.

Points to Remember

- Business Model Canvas and capability map can be left out in modernization projects when the business process has not be completely overhauled. A Wardley map should be used in every case. To align all product and project partners, a North Star should be used.
- Domain Storytelling can be left out for purely technical changes. The model must be adapted in the Visual Glossary.
- In smaller feature enhancements, Event Storming can be left out. The APIs can be defined using Domain Storytelling and the API Product Canvas.
- A tactical design must be carried out in every case.

Review Questions

12.1 Does a North Star help to align partners?

(a) No
(b) Yes

12.2 Can I skip Domain Storytelling when implementing technical enhancements?

(a) No
(b) Yes

References

1. Kouzmanoff A, Bashir I, Cutler I, Clark T (2024) The North Star Playbook. Amplitude, San Francisco. ISBN: 978-19-39623-12-6
2. Ellis S (2017) What is a north star metric?. [Online] Available: https://medium.com/growthhackers/what-is-a-north-star-metric-b31a8512923f. Visited on 16 Feb 2025

APIs and Events in a Serverless World 13

Before we wrap up the book on how to avoid errors and present examples of beautiful APIs, we would like to look at the differences when it comes to designing APIs compared to events in a serverless world. We come up against new challenges that emerge with scalable cloud infrastructures and what needs to be considered in these scenarios related to APIs. Those considerations are necessary not only in the cloud; it emerges with changes from classic Three-Tier Architecture to new dynamic cloud environments.

New Requirements and Challenges in the Cloud

In classic Three-Tier Architecture, every call from outside, usually a UI, was handled by a server that then committed the changes to a relational database in a transaction. The transaction was opened at the beginning and committed at the end.

But quality requirements and user expectations have changed. Apps need to be available at all times, globally, without any downtime. At the same time, new features need to be rolled out faster. The long maintenance windows from before are no longer tolerated. This leads to smaller teams that own small components and must be able to deploy independently and fast. Such an ecosystem needs to have a distributed state, as a single database is not sufficient: There is never time to make a database migration, and a failure of a database is not tolerated either. At the same time, a transactional database cannot handle other new requirements of parallelism, global distribution, and request speed.[1] Many old and classic Three-Tier

---

[1] As transactional databases are very common, there are quite sufficient approaches to improving their shortcomings. For example, read replicas can increase the read speed and add distribution. However, these features come at the cost of losing the transaction guarantee; most of the read replica features add a replica lag [1].

Architecture can only scale vertically (bigger instances, not multiple instances), which does not work for all workloads. Some also allow horizontal scaling; however, this often needs to be designed initially or later to be added in extensive redesigns.[2]

Microservices or serverless functions can help with these problems. Independent teams deploy smaller components faster; a failure in a component is not problematic when the service is scaled horizontally (to multiple instances).

However, such scalable microservices come with new complexity that needs to be understood and covered in the design. We will now look at some key principles that need to be considered.

The CAP Theorem and Eventual Consistency

A classic relational database (also called relational database management system (RDBMS)) has an "immediate consistency." Either a transaction is committed, and all see the data, or it is not yet committed, and nobody sees it. This is simple. However, it requires a short moment for everything to wait for that sync; multiple transactions cannot persist simultaneously, as they need an absolute order. However, this is an excellent property that comes at the cost of availability and partition tolerance, as the CAP theorem describes [2].

The CAP theorem states that any network shared-data system can have at most two of the three desired properties; see also Figure 13-1 [2–5]:

Consistency Every read operation gets the same, consistent result, in any part of the system (last committed write in a RDBMS). This can be modeled sequentially, for example, by a sequence diagram without any doubt about what will happen, so there will be no doubt about what will be read. This requires a global time in the system in which all reads can be atomic [6–8].

Availability Data are available for reads and updates without a guarantee of correctness.

Partition tolerance Data are distributed over the network. A part of the system can work without the other. This allows horizontal scaling.

---

[2] Three-Tier Architecture does not have to be bad and still serve its purposes; however, it has shown that it currently can often help in designing such systems in other ways, which now would perform better and better meet new requirements. In addition, the change to microservices away from big monoliths has a lot to do with the shift to the cloud and away from big mainframes. So, at the time, the systems were designed correctly and well suited for that usage.

New Requirements and Challenges in the Cloud

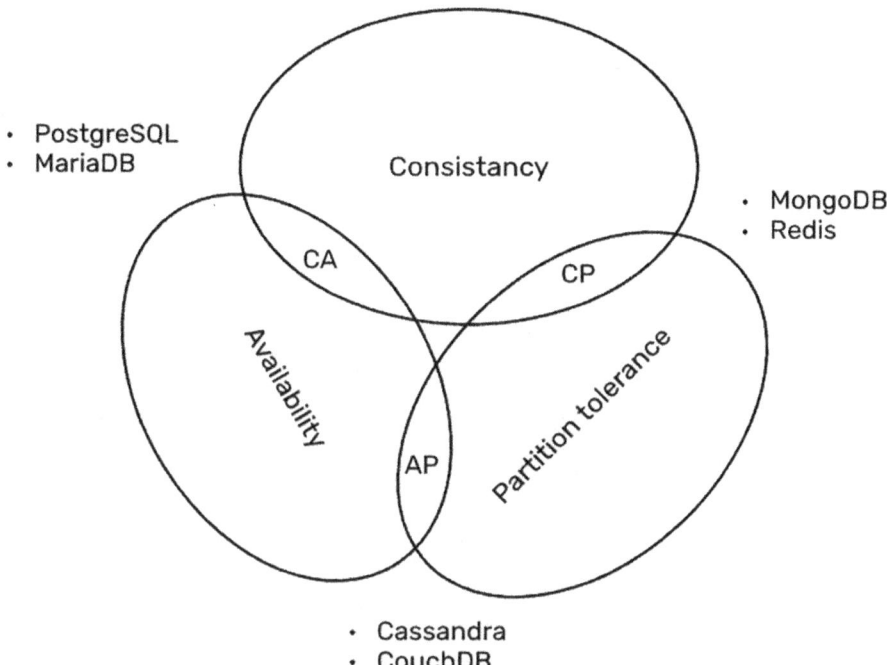

Figure 13-1 CAP theorem visualized as an Euler diagram with some classic products; however, placement depends on configuration

Figure 13-1 shows, next to the Euler diagram, some data stores in each category:

CA Consistency and availability are the default RDBMS, as described earlier. They are excellent as they are consistent and simple for businesses to understand. They implement atomicity, consistency, isolation, durability (ACID).

CP Consistency and partition tolerance are systems that can maintain consistency, even when they are separate. However, it comes at the cost of not all data being available if a node fails or is unreachable.

AP Availability and partition tolerance-supporting systems keep out the consistency of supporting writes in multiple disconnected places. This can be helpful to support offline capabilities, however, with the trade-off of diverting states.

The separation should support a mental model to see that not all data stores can help all types of use cases. In addition, it should show that a classic RDBMS does not support all cases. However, the product placement on the CAP theorem should be only exemplary. Many products have configurations that can shift the product into another corner.

> **Kafka was originally designed as a CA system**

Kafka, for example, was initially designed as a CA system, like a RDBMS. However, depending on the change in the configuration, it can also shift to another corner.[3] [9]

Databases that are not ACID but in the CP or AP corner are called BASE [10,11]:

Basically available   This means the data are available at all times to users, which results in replication of the data.

Soft state   The state of the data can change without an external trigger, as the data may be only in a transient or temporary state until the system has reconciled the final state.

Eventually consistent   The system will achieve a final consistent state after all updates are made. Then the state will be consistent. However, it is unknown when this will happen.

Eventually consistent systems are preferable to scale cheaply and deploy into multiple places, as is normal for the cloud.

It is harder to work with basically available, soft state, and eventually consistent (BASE) than with ACID, but it comes with many benefits that are needed for a microservice architecture. It is essential to understand this for API design, as asynchronous APIs are often only necessary because of the use of BASE stores and will work differently because of that.

Now we can look at where and why we need BASE systems.

Using the Cloud's Benefits

A cloud-native architecture can bring many benefits. However, building systems for the cloud is necessary, not just "lifting and shifting" them to the cloud. In Chapter 11, "Brownfield Project," we discussed how we could split a monolithic system into microservices with the strangler fig pattern [12].

Here, we want to go briefly over some issues the authors often see that are relevant to API design. At the end of the chapter, we will refer to some more complete resources for a full cloud-native architecture.

Stateless APIs

Horizontal scalable systems can be cheaper to run and recover faster from minor problems on a single machine.

It is vital to design APIs stateless as we did in the examples in Chapter 7; in this way, it is unnecessary to keep a state on the backend between calls, and a sticky

---

[3] Change Kafka configuration to use High Availability: https://kafka.apache.org/documentation/#design_ha.

New Requirements and Challenges in the Cloud

connection to the same backend between all calls is necessary. This allows for switching of backend instances without a state transfer. In addition, this allows for more straightforward tests, as each request is testable independently and repeatably. Proxies and content delivery networks (CDNs) can also be integrated to cache API calls so that the pages load faster and the user experience is enhanced.

Considerations when using a cache

Use HTTP cache headers to configure caches if needed. However, be careful before using caches for APIs.

Only opt-in for API calls that can be cached if an application has sensitive data. When configured wrong, user or session data can be leaked to other users over incorrectly configured caches.

Also be aware that these cases will bring down the 75th percentile of all API calls when configured correctly; however, it will not make the first API call faster when the cache is empty. Special setups must be developed to ensure these are delivered faster (e.g., regular calls fill the caches).

In addition to security concerns, caching is a source of outdated data. Be careful about how long a cache can be accepted or how to clear caches if data change.

Statefulness by design in PHP

Applications implemented in simple PHP often require sticky sessions, which should be avoided by the API design.

The session object ($_SESSION in PHP) is given by the programming language itself. These session data are stored in memory. They can be used for each request. This requires that sessions be handled by the same backend. To identify a session, a session identifier is then passed with every request, either by a request parameter or with an HTTP cookie.

By default, the $_SESSION is stored in a file on the server [13, 14].

Requests, however, do not need to be routed to the same backend, as the session identifier is used to keep track of the session.

When you have a stateful conversation, you need to store the whole conversation on the server side and use it on each new request from the client in that context (in PHP with the $_SESSION object).

With a stateless API, no context is given, or it needs to be loaded with each request from the database.

With a stateless server, not having local files represent a state is preferred. This makes hosting an application server simple and allows cleaner APIs that do not need to be routed always to the same backend. This mechanism is called "sticky sessions." Only when the server is stateless can each request be sent independently, allowing the same result without depending on other requests and without sticky session routing.

The session in PHP can also persist in the database, which would solve the problem; however, this always requires a request to the database and adds latency. So it is important to keep the session requirements to a minimum.

Stateless also for chatbots

Most LLM chats have a stateful-looking conversation interface. The chat gives context to AI; however, the API is mostly kept stateless (e.g., OpenAI ChatGPT[4] and Anthropic Claude[5]). You are required to resend chat messages on every request. While internal factors in LLM models support this decision, it is also necessary from a load-balancing perspective to be able to simply distribute the requests over multiple instances.

This shows that even if a context first seems like it needs a stateful API, it is often possible to change the parameters to use an elegant, stateless session. This makes testing and scaling of the deployment a lot simpler.

Eventual Consistency

Eventual consistency supports distribution in the cloud with microservices. Be careful when starting with the concept, as it is not free: Complexity will arise. Start small and increase over time.

Asynchronous Communication

Asynchronous communication is essential when applying a microservice architecture. When microservices are built only with synchronous communication in between, the blast radius will be huge. This is the radius where the systems will go down when an error happens. Decoupling systems with asynchronous APIs help to make the system fault-tolerant and resilient against these failures.

More Resources

To dive deeper into the topics, we recommend the following literature:

- The Twelve-Factor App[6]
- AWS Well-Architected Framework[7]

---

[4] https://platform.openai.com/docs/api-reference/chat/create
[5] https://docs.anthropic.com/en/api/messages
[6] https://12factor.net/
[7] https://docs.aws.amazon.com/wellarchitected/latest/framework/welcome.html

- Azure Well-Architected Framework[8]

Having covered the basics of the cloud, we will now go into the serverless world.

What Is Serverless?

By "serverless world" we mean a cloud environment that shifted from classic hardware/server provisioning to virtualization (Infrastructure as a Service (IaaS)) to containerization (Platform as a Service (PaaS)) and finally to serverless (Functions as a Service (FaaS)), with the goal of bringing changes to production faster, learning from them, and repeating [15].

Serverless can bring more significant separation, for example, if each endpoint has its independent implementation as a function. This can support designing better APIs.

Best practices like stateless and scalability are already considered and implemented by the cloud provider.

How to Implement APIs Without a Server

A common implementation on AWS is AWS Lambda. An Amazon API gateway can route REST calls to the matching AWS Lambda, which then processes an endpoint [16]. Or the lambda function is invoked directly [17].

A common implementation is shown in Figure 13-2.

The serverless approach works great not only with synchronous APIs but also with asynchronous APIs. A component calls the serverless function when an event or message is delivered to the message or event broker.

Such an implementation can be architected after the API design is done in the same way. The same principle applies: Clean API design, readable for humans, and ubiquitous language must be applied in a bounded context.

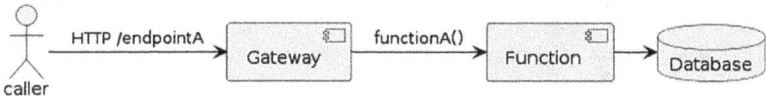

Figure 13-2 Typically synchronous API pattern with serverless functions

---

[8] https://learn.microsoft.com/en-us/azure/well-architected/

Points to Remember

- With the switch away from the large monolithic Three-Tier Architecture to smaller microservices, new requirements and options are available. These changes do not come for free.
- Consistency and availability are the default with a classic RDBMS, but they come with trade-offs that must be considered.
- Try to design APIs stateless.
- API design does not change when going serverless.

Review Questions

13.1 What is the trade-off when selecting a network storage?

(a) There is no trade-off; that is why the best database exists.
(b) A database can only support a single protocol.
(c) A database cannot have all three features: consistency, availability, and partition tolerance.

13.2 How do we design an API when implementing the API for a serverless implementation?

(a) No server means no API!
(b) We need to use Code-First.
(c) The same way: We still use API-First.

References

1. Replication with amazon aurora. Amazon Web Services. [Online] Available: https://docs.aws.amazon.com/AmazonRDS/latest/AuroraUserGuide/Aurora.Replication.html. Visited on 12 April 2025
2. Perkins L (2018) Seven Databases in Seven Weeks, a Guide to Modern Databases and the NoSQL Movement (The Pragmatic Programmers 2), Redmond E, Wilson JR (eds), 2nd edn. The Pragmatic Bookshelf, Raleigh, 1338 pp. Description based upon print version of record. ISBN: 978-16-80505-97-9
3. Brewer E (2012) Cap twelve years later: how the "rules" have changed". Computer 45(2):23–29. ISSN: 1558-0814. https://doi.org/10.1109/mc.2012.37
4. What is the cap theorem? IBM (2022). [Online] Available: https://www.ibm.com/think/topics/cap-theorem. Visited on 09 Feb 2025
5. Junker A (2023) Event-getriebene intergrationsarchitekturen. [Online] Available: https://www.informatik-aktuell.de/entwicklung/methoden/eventgetriebene-integrationsarchitekturen.html. Visited on 27 Dec 2024
6. Gharachorloo K (1995) Memory consistency models for shared-memory multiprocessors. [Online] Available: http://infolab.stanford.edu/pub/cstr/reports/csl/tr/95/685/CSL-TR-95-685.pdf. Visited on 09 Feb 2025

7. Liochon N (2015) This long run. [Online] Available: http://blog.thislongrun.com/2015/03/the-confusing-cap-and-acid-wording.html. Visited on 09 Feb 2025
8. Lamport L (1993) How to make a correct multiprocess program execute correctly on a multiprocessor. IEEE Trans Comput SRC Res Rep 96 46(7):779–782. [Online] Available: https://www.microsoft.com/en-us/research/publication/make-correct-multiprocess-program-execute-correctly-multiprocessor/. Visited on 09 Feb 2025
9. Rao J (2013) Intra-cluster replication in apache kafka. [Online] Available: https://engineering.linkedin.com/kafka/intra-cluster-replication-apache-kafka. Visited on 09 Feb 2025
10. Cox G (2024) Explanation of base terminology |baeldung on computer science. [Online] Available: https://www.baeldung.com/cs/db-base-meaning-cap. Visited on 09 Feb 2025
11. Acid vs base databases - difference between databases, AWS (2024). [Online] Available: https://aws.amazon.com/compare/the-difference-between-acid-and-base-database/. Visited on 09 Feb 2025
12. Fowler M (2024) Strangler fig. [Online] Available: https://martinfowler.com/bliki/StranglerFigApplication.html. Visited on 10 Feb 2025
13. Php: Session_start - manual (2025). [Online] Available: https://www.php.net/manual/en/function.session-start.php. Visited on 09 Feb 2025
14. Php: Runtime configuration - manual (2025). [Online] Available: https://www.php.net/manual/en/session.configuration.php. Visited on 09 Feb 2025
15. Hohpe G, Danieli M, Landreau J-F, Hashmi T (2021) Cloud Strategy: A Decision-Based Approach to Successful Cloud Migration. Architect Elevator Book Series. leanpub.com
16. Chapin J, Roberts M (2020) Programming AWS Lambda, Build and Deploy Serverless Applications with Java, 1st edn. first release. O'Reilly, Beijing, 256 pp. ISBN: 978-14-92041-00-9
17. Creating and managing lambda function urls (2025). [Online] Available: https://docs.aws.amazon.com/lambda/latest/dg/urls-configuration.html. Visited on 13 April 2025

Part IV
Summarizing

A summary of the book

Avoiding Mistakes in the Definition of Events and APIs

14

In this chapter, we will look at how we can avoid mistakes when designing APIs. First, we revisit the APIs from the first chapter. Then we point out some other problems and how they can be mitigated.

Revisiting the Examples from the Beginning

Let us start by discussing the APIs from Chapter 1 and how the content of the book can help to avoid the mistakes made there. We will reuse the APIs in the same order.

An API Directly from the Database

First we look at the API presented in section "An API Taken Directly from a Database". The API was formulated directly from the database schema. The database types correspond to well-designed API types only slightly.

This issue can be mitigated with an API-First approach, where the API is desigend without viewing the implementation. In addition, defining a ubiquitous language first will not lead to cryptic table-name-like fields, like `BPKIND`. The approach does not allow for also adding fields to an API that do not correspond to the business context. Thus, one will think not only about integrating an API when looking at the database.

An API Solely Formulated from the Backend Developer's Point of View

The second example of a bad API, in section "An API Formulated Solely from the Backend Developer's Point of View", represents an API that was built from the view of a backend developer. It used internal identifiers, like `tenant` and `partnerNumber`, to assign customers. Starting with a workshop and specifying the API before going into the implementation, having the customer of the API product in mind (AaaP) would lead to better identifiers. As the library example showed, going for a single UUID as an identifier for an object is often a good choice.

A Purely Technical API Without Business Relevance

Next, in section "A Purely Technical API Without Business Relevance", a purely technical API was presented. The API delivered a customer identifier by means of a business partner number. Such a technical translation should be part of a bounded context, not a REST API. It can be part of the implementation of an adapter, as shown in Chapter 11, "Brownfield Project."

An API from the Ivory Tower

Section "An API from the Ivory Tower" showed contrast, a nontechnical ivory tower API. Involving developers in the process of API design, as discussed in Chapter 9, "Collaborative Design and Agility," will lead to a discussion where developers will spot these problems at the design phase.

The Overloaded World-Domination API

Section "Overloaded World-Domination API" presented the authors' favorite: the world-domination API. The API tried to bring everything into a single API, leading to endpoints like `/customers/{customer-id}/account/{account-id}/transactions`, and so on.

Using DDD and designing bounded contexts before designing the APIs will eliminate the need to put everything into a single API. Like this, beautiful small APIs should be extracted from a larger business problem, which leads to the last problem, missing documentation.

APIs Without Documentation and Unexpected Behavior

The last problematic API presented in section "APIs Without Documentation and Unexpected Behavior" involved an API without documentation that demonstrates unexpected behavior.

Designing the API first (API-First) compared to Code-First will lead to more documentation in the API, as it will not be an afterthought that can be generated from the code; see the discussion in section "API-First vs. Code-First."

In combination with the understanding of standards, best practices, and conventions discussed in Chapter 6, "Interface Definitions", Chapter 8, "Developer Experience and API Implementation", and section "New Requirements and Challenges in the Cloud," this should be eliminated.

Other Problems

The examples from the beginning served as a great introduction, but they did not cover all errors that need to be avoided. Let us briefly summarize other mistakes to look for and how to avoid them.

Inconsistent naming Inconsistent naming in APIs can be a real problem for comprehension and should be avoided.
Using ubiquitous language helps avoid this.

Missing lifecycle management Not having a known lifecycle and clear versioning strategy can break production Environment of integrators. Having a clear versioning understanding and strategy in mind is crucial, as described in section "Versioning of APIs."

Wrong or unhelpful errors Wrong or unhelpful errors make the integration or debugging of an API very difficult.
Therefore, it is essential to have good error messages and formulate them as early as the design phase; see discussions in Chapter 6, "Interface Definitions."

Insecure APIs Insecure APIs can have credibility and legal issues.
It is important to secure the APIs properly, implement best practices (Chapter 6, "Interface Definitions"), and test for security vulnerabilities (section "Testing").

Performance and scalability An API can be as good as it wants to be if performance does not match the criteria of the integrator; it will probably not be used. See the discussion on caching and different types of storage in Chapter 13, "APIs and Events in a Serverless World," and the example in Chapter 11, "Brownfield Project."

Conclusion

We summarized problems associated with APIs from early in the book, as well as other ones, and showed how the book's contents helped to address these problems.

Next, we will look at some real-world beautiful APIs that did not have these problems.

Points to Remember

- Take the time to follow best practices and standards for API design described in the book to mitigate bad API design.

Review Questions

14.1 How do you mitigate the issues associated with an API that uses database fields?

(a) Reuse the schema from the database.
(b) Define a ubiquitous language in a collaborative manner.

14.2 How do you mitigate the issues associated with an API formulated from the backend developer's perspective?

(a) Use Code-First.
(b) Create a story to design an API for the backend developer.
(c) Use a collaborative workshop approach to define the API.

14.3 How do you mitigate the issues associated with an overloaded API with endpoints and methods?

(a) Use API-First.
(b) Use DDD and design bounded contexts.

A Couple of Beautiful APIs

15

In this chapter we aim to summarize what a beautiful API is and make present examples of real-world, beautiful APIs.

What Is a Beautiful API in the Eyes of the Authors?

Beautiful APIs are APIs that do not repeat the errors in section "APIs Nobody Wants to Use." They avoid those mistakes by using API-First, elegant tailored bounded contexts with a ubiquitous language. In addition, they have good documentation with examples, and perhaps also a sandbox or other tools to simplify tests and quickly integrate the API as a consumer. They are nice products.

Some Beautiful Real-World APIs

Let us next look at some really beautiful APIs and discuss why they are elegant.

Synchronous APIs

First, we look at synchronous APIs. Externally facing APIs are more often seen, so this section is longer than the one on asynchronous examples. This as API gateways and firewalls are better suited for a REST API, and they are simple to integrate

AWS S3
AWS S3 is an often used example of a an excellent REST API [1, 2]. It provides a true REST interface that manipulates objects using different HTTP methods.

The full API, however, is huge and therefore not represented as an OpenAPI,[1] but it is still well documented. SDKs also help to simplify the usage.

However, there is a greatly simplified example:[2]

- GET on API's root resource to list all Amazon S3 buckets of a caller.
- GET on a folder resource to view a list of all objects in Amazon S3 bucket.
- PUT on a folder resource to add a bucket to Amazon S3.
- DELETE on a folder resource to remove a bucket from Amazon S3.
- GET on a folder/item resource to view or download an object from an Amazon S3 bucket.
- PUT on a folder/item resource to upload an object to an Amazon S3 bucket.
- HEAD on a folder/item resource to retrieve object metadata in an Amazon S3 bucket.
- DELETE on a folder/item resource to remove an object from an Amazon S3 bucket.

Shared Mobility API

This public API, "Shared Mobility API,"[3] landed on this list because it is not only simple, comprehensible, and easy to use, but it has a great ecosystem, too. The API follows a standard that is documented and extendable and has clear governance.[4]

This standard is implemented in the API, and it is discoverable in a portal.[5] The API can be found there, as well as a demo implementation and a showcase.[6]

Next to the API, documentation with an explanation and a maturity assessment of the API[7] are also available.

Booking.com

Booking.com also has some great APIs, all divided up to serve specific functions and descriptions on how to use them. An example is the "Demand API":[8] The API has a description, including a sample integration workflow. Each endpoint is well described, and the documentation describes how to test the API in the sandbox environment.[9]

---

[1] https://docs.aws.amazon.com/pdfs/AmazonS3/latest/API/s3-api.pdf

[2] Example of exposure of resources on API Gateway https://docs.aws.amazon.com/apigateway/latest/developerguide/api-as-s3-proxy-export-swagger-with-extensions.html.

[3] https://api.sharedmobility.ch/

[4] https://github.com/MobilityData/gbfs

[5] https://opendata.swiss/de/dataset/standorte-und-verfugbarkeit-von-shared-mobility-angeboten/resource/8a6a545e-b35c-4fe0-b8ea-e609a8054822

[6] https://opendata.swiss/en/dataset/standorte-und-verfugbarkeit-von-shared-mobility-angeboten

[7] https://www.oev-info.ch/sites/default/files/2023-05/general_bikeshare_feed_specification_gbfs_eng.pdf

[8] https://developers.booking.com/demand/docs/open-api/demand-api

[9] https://developers.booking.com/demand/docs/getting-started/overview

Stripe
The API of Stripe is huge; however, it is clearly broken down into parts, and, as discussed in section "Developer Experience When Integrating an API," the teams have done a great job in providing observability and added features that allow users to jump right in and start playing around.

Asynchronous APIs

Even though most APIs that are public are synchronous, there are also some asynchronous APIs. In section "Publishing External Events", we discussed the difficulties associated with publishing events via the internet.

GitHub Webhooks
GitHub implements Webhooks to notify other applications about changes to the platform.[10] Webhooks as an implementation for asynchronous communication over the public internet with external systems is a common implementation and works well when it comes to addressing security concerns.

Slack
One public AsyncAPI known to the authors is published by Slack.[11] However, they degraded the AsyncAPI in favor of an SDKs.[12]

Conclusion

Here, we have looked at some beautiful public APIs. Not all are perfect. However, they are quite elegant and show what great APIs can look like. Some are also publicly accessible, and users can play around with them.

Points to Remember

We discussed the following points in this chapter:

- Using API-First and the process discussed in this project helps in designing beautiful APIs.
- REST APIs are used more often for public facing interfaces.

---

[10] https://docs.github.com/en/webhooks/webhook-events-and-payloads

[11] https://github.com/slackapi/slack-api-specs/blob/master/events-api/slack_events_api_async_v1.json

[12] Events on their website https://api.slack.com/events?filter=Events

Review Questions

15.1 Are all beautiful APIs public?

(a) Yes
(b) No

15.2 Does API-First with clear-cut contexts and a ubiquitous language support making a beautiful API?

(a) Yes
(b) No

References

1. Spichale K (2019) API-Design, Praxishandbuch für Java- und Webservice-Entwickler, 2. überarbeitete und erweiterte Auflage. Heidelberg: dpunkt.verlag, 1381 pp. ISBN: 978-39-60886-02-0
2. Richardson L, Ruby S (2007) Restful Web Services, 1st edn. O'Reilly, Sebastopol. ISBN: 978-05-96529-26-0

Summary   16

Having shown how mistakes in APIs can be mitigated, and having looked at some beautiful real-world APIs, we are in a position to conclude the book.

In the book we went from "what is an API?" and "what is quality?" to "how do we specify an API?" and "how can we collaboratively arrive at an elegant API" to "which processes should I use?" Here, we would like to summarize the tools and methods learned and put them in context: from tactical to strategic and from technical to business-oriented, visualized in Figure 16-1. We summarize them as ranging from business-oriented to technically oriented:

Business Model Canvas A Business Model Canvas makes it possible to model business plans collecting key users, key partners, and, especially, key business capabilities [1, 2]. We presented this topic in section "Business Model Canvas".

North Star A North Star can be used to formulate a vision statement that can be used to align all partners in product development. We discussed this in Chapter 12.

Wardley map A Wardley map is a great tool for prioritizing features and capabilities from a strong user's point of view. Users and their needs are detected, and capabilities and features are sorted from invisible to visible to the user. Based on this vertical scale, capabilities and features are sorted along a product cycle from genesis over a custom-built over product to commodity. The product or project must implement custom-built and genesis capabilities and features. Products and commodities can be purchased on the market and eventually adapted to the needs of the product. This means that capabilities with their genesis as an emerging market or as custom-built characteristics and most visible to users get the highest priority. We discussed Wardley maps in section "Qualifying Business Ideas" [3].

Capability map A capability map is used to sort the capabilities of a company into generic, supporting, and core subdomains. **Generic capabilities** mean capabilities that do not differ from company to company. **Core capabilities** are

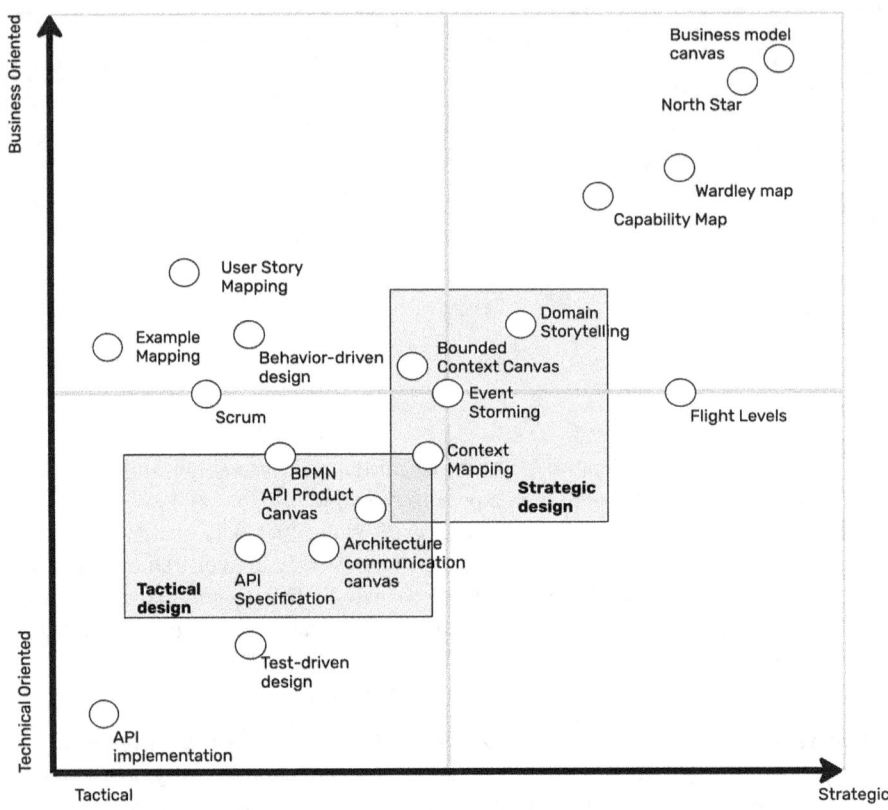

Figure 16-1 Different methods to design APIs discussed in text

capabilities that form the core of the company's business. **Supporting capabilities** bear characteristics of both. Using the capability map gives further hints on prioritizing capabilities [4]. We discussed capability maps in section "Capability Map".

User story mapping User story mapping is an agile method of sorting user stories along with user activities and tasks. They can be prioritized and releases can be defined for user stories. Such mapping can enhance the domain stories found in Domain Storytelling. We discussed User story mapping as an additional method in section "User Story Mapping".

Domain Storytelling Domain Storytelling is an interactive workshop format aimed at gathering business requirements visually [5]. Domain Storytelling was discussed in section "Domain Storytelling". We used the method throughout the book to visualize processes and business dependencies.

Behavior-Driven Design BDD is a method to overcome the gap between business and technology, especially when defining tests [6]. We discussed it directly when discussing tests (section "Testing") and in connection with agile methods along the software development cycle (section "Behavior-Driven Design").

Example Mapping Example Mapping is a tool that is used to map user stories, rules to user stories, and examples. It helps in defining business-driven tests as in BDD [7]. We discussed example mapping in section "Example Mapping".

Bounded Context Canvas A Bounded Context Canvas can be used to describe and document a bounded context [8]. We introduced Bounded Context Canvas in section "Bounded Context Canvas".

Flight Levels Flight Levels are an agile approach that is used to combine strategic and operative perspectives in project planning [9]. We discussed them as a scaled agile approach in section "Collaborative Processes in Classical Project Environments".

Event Storming Event Storming is a collaborative workshop format for defining bounded contexts. It follows the ideas of mind-storming workshops and creates overviews of business processes and insights into a bounded context with aggregates and read-only models [10]. We used Event Storming widely in this book and introduced the method in section "Event Storming".

Scrum Scrum is an agile framework for organizing a team [11]. We discussed Scrum in section "Collaborative Processes in Classical Project Environments".

Context mapping Context mapping is a method of defining context maps from Event Storming or Domain Storytelling results. It uses a context map to show the dependencies between bounded contexts [12, 13] and to show data flows [14]. We introduced context mapping in section "Context Map".

BPMN BPMN is a formal, visual standard language used to note business processes [15]. We discussed BPMN in section "Business Process Model Notation".

API Product Canvas An API Product Canvas was introduced by the authors to specify the APIs of a bounded context. The authors provided a Miro template for the canvas.[1] The API Product Canvas was introduced in section "API Product Canvas".

Architecture Communication Canvas The Architecture Communication Canvas collects the major information and decisions of an implemented bounded context [16]. It contains rough overviews of the selected architecture in a C4 model (context diagram and component diagram). We discussed the architecture communication canvas in section "Architecture Communication Canvas".

API specification An API specification is a document containing the contract formulated in an IDL. We introduced in detail OpenAPI and AsyncAPI [17, 18] (sections "Definition of Synchronous Interfaces and Definition of Asynchronous Interfaces"). We showed how OpenAPI could be used to define synchronous APIs (section "Definition of Synchronous Interfaces"). We showed additionally how AsyncAPI could be used to define asynchronous APIs (section "Definition of Asynchronous Interfaces"). We discussed GraphQL and gRPC as IDLs (sections "Definition as gRPC and Definition as GraphQL").

---

[1] https://miro.com/miroverse/api-product-canvas/

Besides the core definition of interfaces, data structures (schemas) need to be defined. We discussed how different schemas could be defined for different usages. That involved Protobuf, JSON, and Apache Avro (section "Data Formats and Their Schemas") as the most common ones.

Additionally, we presented an overview of those situations where certain protocols and schemas are most suitable (section "Advantages and Disadvantages of Different Schemas and Protocols").

Test-driven design TDD is a methodology to write tests before implementing a feature. We discussed TDD in section "Test-Driven Development".

API implementation The API implementation covers the part where the interface descriptions are implemented on the publisher and consumer sides. The implementation covers quite different aspects from

- Versioning (section "Versioning of APIs")
- Testing of APIs (section "Testing")
- Continuous integration (section "Continuous Integration")
- API Management (section "Publishing APIs")
- Developer experience (section "Developer Experience When Integrating an API")
- Data-mesh (section "Data Mesh Concept")
- Analytics and monitoring (section "Analytics and Monitoring")

In addition to those aspects, we discussed elements of APIs in a serverless world and the concept of distributed state and eventual consistency (Chapter 13).

An implementation of an API is usually not the end of development. We discussed changes in a "brownfield" example project (Chapter 11) and how extensions to software applications could be handled (Chapter 10).

Using the process and following the provided rules in API design will lead to the creation of elegant and sustainable APIs. APIs and events are the backbone of a digital, cross-linked world and must be designed collaboratively. Innovation as a whole thrives on partnership between IT and business.

References

1. Strategyzer (2024) The business model canvas. [Online] Available: https://www.strategyzer.com/library/the-business-model-canvas. Visited on 21 July 2024
2. Osterwalder A (2013) Business Model Generation: A Handbook for Visionaries, Game Changers, and Challengers, Pigneur Y (ed). Wiley & Sons, New York (2013). ISBN: 9780470876411.
3. Wardley S (2022) Wardley Maps, 2. überarbeitete und erweiterte Auflage. Simon Wardley, Heidelberg
4. Alex S, Blair M, Bryan Lail J (2022) Business capabilities, version 2, the open group. [Online] Available: https://pubs.opengroup.org/togaf-standard/business-architecture/business-capabilities.html. Visited on 27 Aug 2024
5. Hofer S, Schwentner H (2022) Domain Storytelling: A Collaborative, Visual, and Agile Way to Build Domain-Driven Software. Pearson International, London. ISBN: 978-01-37458-91-2

6. Smart J (2014) BDD in Action, Behavior-Driven Development for the Whole Software Lifecycle. Manning Publications Co. LLC, New York, 1307 pp. Description based on publisher supplied metadata and other sources.. ISBN: 978-16-38353-21-8
7. Wynne M (2015) Introducing example mapping. [Online] Available: https://cucumber.io/blog/bdd/example-mapping-introduction/. Visited on 05 Feb 2025
8. Group D (2024) The bounded context canvas. [Online] Available: https://github.com/ddd-crew/bounded-context-canvas. Visited on 28 Dec 2024
9. What is flight levels? (2025). [Online] Available: https://www.flightlevels.io/what-is-flight-levels/. Visited on 07 Feb 2025
10. Brandolini A (2024) Event storming. [Online] Available: https://www.eventstorming.com/. Visited on 30 Jun 2024
11. What is scrum? (2025). [Online] Available: https://www.scrum.org/resources/what-scrum-module. Visited on 07 Feb 2025
12. DDD Crew (2023) Context mapping. Visited on 08 Sep 2024
13. Evans E (2004) Domain-Driven Design: Tackling Complexity in the Heart of Software. Addison-Wesley, Boston
14. Junker A (2025) Mastering Domain-Driven Desgin. BPB Online. ISBN: 978-93-65892-52-9
15. Pant K, Juric MB (2008) Business Process Driven SOA using BPMN and BPEL. Packt Publishing, Birmingham. ISBN: 978-18-47191-46-5
16. Hruschka P and Starke G (2024)Arc42. [Online] Available: https://arc42.org/. Visited on 28 Dec 2024
17. OpenAPI Initiative (2021) Openapi specification v.3.1.0. [Online] Available:https://spec.openapis.org/oas/latest.html. Visited on 14 July 2024
18. Asyncapi 3.0.0 (2023). [Online] Available: https://www.asyncapi.com/docs/reference/specification/v3.0.0. Visited on 10 Nov 2024

Glossary

aggregate A referencable object which belongs to a bounded context and can be accessed via the API of a particular bounded context. 16, 210, 212

anticorruption layer A layser to secure a model from other models to leak in. 108

Apache Avro Avro is a binary-serialization format developed in 2009 as a subproject of Hadoop. 132, 143, 144, 147, 176, 177, 184, 185, 187, 190, 195, 206, 254, 255, 261, 262, 265, 370

Apache Kafka Apache Kafka is an open source event streaming platform, using a broker concept that are building up clusters. See https://kafka.apache.org/ 30, 31, 38, 141, 171, 175, 176, 181, 182, 184, 185, 189, 190, 195, 196, 211, 319, 320

API design The process of defining APIs. 22, 47, 54

API Gateway An API Gateway is a component that delivers the API to a consumer. It is a proxy that can provide security, monitoring, and routing aspects [1]. 276, 277, 286, 373

API Key Alphanumeric string to authenticate a service to an API. 166

API Management The term API Management describes the platform to provide APIs (e.g., API Marketplace, API Gateway), the associated processes, governance, and organizational mindset with APIs [1, 2]. 276, 373, *see* API Marketplace and API Gateway

API Marketplace An API Marketplace can be used to discover APIs that are published. It helps to onboard new users using self-registration, including user management; billing and rate limiting can also be integrated. It can be company internal or external. 276, 373

API Product Canvas A canvas for definition of necessary APIs of a bounded context. 216, 220, 222, 226, 237, 247, 291, 295, 306, 307, 318, 345

API specification A specification document - usually following certain standards - describing an API with access parameters and payload. 20, 22, 36, 37, 47, 49–51, 57, 114, 127, 129, 159, 170, 171, 193, 242, 253, 254, 258, 259, 263, 266, 267, 280, 369

API-First The process of design an API before writing the implementation code. 18, 20, 22, 23, 114, 125, 146, 153, 156, 247, 267, 270, 271, 275, 280, 286, 354, 359, 361–363, 365, 366, 381, 387, 388

Application Programming Interface An Application Programming Interface provides an abstraction of an application call a client can interact with in a programmable way. The API is or should be implemented by one or many solutions [3]. 121

arc42 arc42 describes an architecture documentation template used to communication and document architecture reliable https://arc42.org/overview. It used as synonym for the group defining the documentation too. 295

AsyncAPI AsyncAPI is a specification language which allows to define asynchronous messages or events independently from their implementation. 16, 126–128, 136, 171, 174–179, 181, 182, 187, 193, 195, 196, 254, 263–265, 291, 306, 307, 365, 369

at least once It means that messages are delivered one or more times. In case of a system failure, messages are never lost, but they may be delivered more than once [4]. 152, 153

at most once At most once means that the message can be delivered or not. If the message is lost, it is not redelivered [4]. 152, 153

Base64 Base 64 is a encoding of data in a subset of the American Standard Code for Information Interchange (ASCII) letters. It allows the trasfer and storage of any data in ASCII letters. The data is only using 64 different character for the transfer: `A-Z, a-z, 0-9, +, /`. The end is signalized with a = [5]. 131, 132

bounded context A bounded context is a concept of Domain-Driven Design. It allows to define the borders where a given domain model and its ubiquitous language is valid. Usually it gives the boundary of one business function. 6, 16–18, 29, 55, 82–84, 98–102, 104, 107–111, 114, 116, 205, 207–210, 212, 216, 218, 226, 227, 232, 233, 239, 240, 244, 245, 247–249, 269, 276, 291–295, 298, 301–303, 306, 307, 314, 322, 334, 336, 338, 340, 345, 360, 369, 384, 385

Bounded Context Canvas A bounded context canvas is a visual representation of a bounded context with the most important information about business decisions, ubiquitous languages, and depencies to users, external systems, and other bounded contexts. 47, 206–208, 210, 216, 233, 236, 240, 248, 249, 295, 307, 345, 369, 384

broker A broker, in integration architecture, is a compontent that provides content to other component. 35–37, *see* message broker and event broker

Brokerless In this book a communication without a broker involved. 181, *see* broker

brownfield A brownfield project is a project where earlier components are used and re-engineered. It is an analogy to a brownfield with infrastructure components that are partly abandoned. 325, *see* greenfield

business capability A capability of an organization to do its business. 16, 74

Business Model Canvas A canvas which allows to model business plans collecting key users, key partners, and especially key business capabilities [6]. 15, 18, 71–73, 77, 117, 291, 292, 297, 306, 307, 343, 344, 346, 367

Glossary

C4 Stands for context, containers, components, and code. It is model defining four hierarchical abstractions each represented with a diagram. The diagram context gives the overview and each next diagram is a zoomed in view [7]. 209, 335

camel case It is a naming convention to write words without spaces and start each word with an upper case; however, the first word needs to be lower case. Otherwise, it would be called pascal case. Example: domainDrivenApiDesign 144, 377, *see* pascal case

capability map A capability map is used to structure and detail capabilities of a business [8]. 74, 79, 88, 117, 292, 297, 306, 307, 343, 344, 346, 367

CloudEvents CloudEvents is a specification for messages proposed by the Serverless Working Group of the Cloud Native Computing Foundation (CNCF). The standard defined fields to facilitate routing, that can be applied as envelope or mapped to specific protocol fields [9, 10]. 182, 196, *see* message and event

Code-First Process of creating APIs by generate the specification out of the implementing code. 20, 22, 23, 275, 354, 361, 362

context map A context map is a tool to determine the team dependencies and the data to be exchanged between different bounded contexts. 16, 19, 71, 102, 104, 109–114, 205, 247, 248, 291, 294, 295, 297, 301, 306, 314, 334, 335, 369, 384

context mapping Context mapping is a method to determine team and service structures in a software application, using context maps. 286, 307, 345, 369

core bounded context A bounded context belonging to a core domain of an enterprise. 109

Data Product Data products are data provided by a team as a first-class citizens. It is often referred to it together with data products, where operational teams are also responsible for their data. A shift left, where data should been seen as a first-class citizens and as a product that is provided by the team as a service [11]. Sometimes also called Data as a Product (DaaP). 179, 284, 286

DevOps DevOps is the movement to bring development and operations closer together through increased collaboration, and it is often supported by tooling. It supports the agile mindset, where a team builds and runs an application, bringing the team closer to an autonomous team [14]. 35

domain event An event which is triggered by a status change of a certain business object, aggregate, or entity. The model is found by the methods of Domain-Driven Design. 16, 294, 322, 334

Domain Story A domain story is the result of a domain-storytelling-workshop. It visualizes the business requirements. 47, 82, 84, 87, 293

Domain Storytelling Domain Storytelling is a collaborative, interactive workshop format, which allows to gather business requirements in a visual way 15–19, 71, 82, 90, 93, 96, 102, 104, 114, 117, 118, 247, 286, 291–294, 297, 298, 300, 306, 307, 311, 314, 318, 337, 338, 345, 346, 368, 369

Domain-Driven Design Domain-driven design describes a methodology to define a software systems out of business perspective. 87, 374

Endpoint A defined access point of an API representing a service to read or change data. 111

entity A collection of data, which can be referenced. 210

event Events are messages with a specific semantic. They represent a state change in the past, e.g., "Book was ordered". They are not directed to a consumer and do not represent an intended action [15]. 35, 40, 207, 210, 212, 216, 318, 319, 322, 339, 345, 347, *see* eventing and message

event broker An event broker stores all messages on channels or topics. The messages, in general, only get deleted after a period of time or never at all (e.g., for Event Sourcing) [16]. 41, 42, 212, 315, 319, 336, 340, *see* broker and message broker

Event Storming Event Storming is a collaborative workshop format which allows to determine the bounded contexts of software components. Domain events are collected along the business process and with corresponding aggregates and commands, the bounded contexts can be determined. 16–19, 71, 82, 96, 99–104, 114, 118, 205, 212, 247, 248, 286, 291, 292, 294, 295, 297, 298, 300, 301, 306, 307, 314, 334, 338–340, 345, 346, 369, 378

eventing Eventing is a communicate mechanism, usually used with event-driven architecture (EDA) as an integration architecture pattern. 40, 41, *see* event and EDA

exactly once This is the behavior in that each message is delivered once and only once. Messages are never lost or read twice, even if some part of the system fails [4]. 152, 153

Example Mapping Mapping of user stories, rules, and examples so that workshop attendees understood the business by examples. 301, 369

generic bounded context A bounded context belonging to a generic domain of an enterprise. 110

Git A very popular distributed version control system. 266, 272, 280

GraphQL A data query and manipulation language that works over remote APIs. 31, 39, 41, 153, 156–158, 169, 170, 184–186, 189, 190, 195, 265, 369

greenfield A greenfield project is a project with no earlier components available. It is an analogy to a greenland without any infrastructure. 325, *see* brownfield

HTTP Authentication Authentication to an API using basic or bearer token authentication as standardized in [17]. 167

idempotency Idempotency is a mathematical term describing a function that can be applied once or multiple times with the same result: $f(x) = f(f(x))$. 147, 155, 160, *see* message

message messages get transferred from a producer to a consumer. They are transferred over a channel. The channel can be a queue, a Hypertext Transfer Protocol (HTTP) channel, or another communication channel. In messaging, messages get transfered asynchronously over a queue. 30, 40–42, 377, *see* messaging, event and eventing

message broker Message brokers sends messages to consumers that subscribe to them. The channel queues the messages for consumers; when they are read, they are deleted. Message reads must be typically acknowledged before they are deleted [16, 18]. 42, 377, *see* broker and event broker

messaging Messaging in integration architecture is a communication mechanism to transfer messages from a producer to a consumer asynchronously. The messages are transferred over a channel, here also called queue. The consumer reads them, generally by subscribing to a channel upfront, and usually deletes them by reading [18]. 40, 377, *see* queue, message, event and eventing

microservice "microservice architecture (microservices) is an architectural style that functionally decomposes an application into a set of services." [19] *see* Three-Tier Architecture

North Star Metrics to formulate a vision statement. 343, 344, 346, 367

online library The online library is an example used in this book to explain an domain-driven API design. 63, 71

open host service Service which externalized its ubiquitous language via published language and is independent from other services. 107, 109

OpenAPI OpenAPI is a specification language which allows to define REST APIs independently from their implementation. 7, 16, 20, 114, 121, 123, 125–127, 134, 136, 159, 163, 168, 170, 171, 178, 186, 187, 193, 195, 222, 254, 264, 265, 281, 291, 306, 307, 320, 364, 369, 379

pascal case It is a naming convention to write words without spaces and start each word with an upper case, including the first word; if the first word is lower case, it is called camel case. Example: DomainDrivenApiDesign. 138, 144, 375, *see* camel case

Protobuf The Protobuf is a binary serialized format developed by Google and later open sourced [11,20]. It is mainly used with gRPC Remote Procedure Calls (gRPC). 132, 141–143, 147, 154, 169, 170, 178, 187, 195, 254, 255, 261, 262, 265, 370, 377, *see* gRPC

protocol A standard way to do something in integration architecture. It describes a standard on how two or more systems interact with each other and transfer data. *see* OSI

Publish / Subscribe In Publish / Subscribe, the producer sends a message to a channel. Multiple different consumers can consume the message if they have previously subscribed to the channel [18]. 35, 41, 377, *see* message and event broker

queue A channel on a message broker. The producer sends messages onto the queue, which stores it. Consumers can than consume the queued messages [18]. A queue does not necessarily have to be on a message broker; however, this book describes only such queues. 40, *see* message broker and message

RabbitMQ RabbitMQ is an open source message broker implementing the Advanced Message Queuing Protocol (AMQP) protocol [21, 22]. 42, 184, 190, *see* message broker and AMQP

read model Read model or view is a modelling term of Event Storming. It means data which are assigned to a domain event, which serve as input data and which are not changing, when the event happens. 16

REST API An API that implements a REST architecture. 23, 216, 219, 228, 233

RESTful A RESTful-oriented service is a service that is implemented using the Representational State Transfer (REST) architecture. This resource-oriented architecture style is the most common on the web. Richardson calls them "services that look like the Web" [23]. Implementations mainly use HTTP. The style has different endpoints modeled as resources that can be accessed, changed, created, and deleted with different operations; however, to make an API really RESTful and stable, a lot of rules need to be followed [24]. 31, 32, 38, 39, 125, 126, 158, 378, *see* REST API, API, REST and HTTP

service mesh A service mesh controls all service interactions by adding a dedicated layer. It is often implemented with small, transparent service proxies between the services [26]. 276

shared kernel Two teams have parts of their model in common. 106

software development kit Is a bundle or kit of multiple tools for a developer supporting the development process. This can be something like the Java Development Kit (JDK) or a wrapper for a network-based API. 57, *see* JDK

strangler fig pattern A software pattern to gradually remove logic from a legacy system and move logic to new components around it. Which at the end make the legacy system obsolete [27, 28]. 337, 338, 350

supportive bounded context A bounded context belonging to a supportive domain of an enterprise. 109

Swagger Swagger was the predecessor of OpenAPI. Now it is a Suite for API developers from SmartBear Software. 121, 123, 125

Synergetic Blueprint Process of methodologies based on DDD approaches invented by *Junker* [29]. 4, 291, 306, 311, 314, 318, 320, 322, 325, 339, 340, 343, 345

telemetry Telemetry is the automatic measurement of data from mainly production systems. The systems emit these data, normally including logging, metrics, tracing, and security information and event management (SIEM) [30]. 264

Three-Tier Architecture A 3-tier-architecture means a layer architecure with 3 layers. Those layers are built from a persistence layer over a business layer up to a user access layer. The layer are not separated by a business perspective. They are separated by their function in a network. 347, 348, 354

ubiquitous language A language used commonly by business and IT experts. The language appears in code, APIs, and in business documents. It is valid for one bounded context. 6, 8, 17, 19, 22, 23, 50, 51, 82, 109, 115, 116, 248, 301, 384

User story mapping User story mapping is a tool to map user stories along the business process to get a deeper insight to the processes and dependencies [31]. 298, 300, 368

view Read model or view are data which are assigned to a domain event, which serve as input data and which are not changed, when the event happens. 16

vision statement Statement to align all activities to reach a formulated target. 343, 344

Visual Glossary A Visual Glossary is a visual representation of terms used in one or multiple bounded contexts using the terms and the relations in between. The terms are represented by boxes and the relations by arrows named with verbs representing the relation. [32]. 6, 16, 17, 19, 90, 92–94, 96, 102, 112, 118, 153, 170, 291, 293, 294, 297, 318, 337, 345, 346

Wardley map A Wardley map can be used to priorize the development of software components along a evolvement path of components with genesis, custom development, product, and commodity stages. 16, 74, 75, 77–79, 117, 208, 248, 291, 292, 305–307, 332, 340, 343, 344, 346, 367

Wardley mapping To create a Wardley map. 74, 75, 297

Webhook Webhooks are a not standardized way of how web servers push notifications to subscribers. They are normally implemented with an HTTP POST request [33]. Webhooks can be specified with OpenAPI 3.1 as "Callbacks"[1]. 31, 181, 315, 320, 322, *see* HTTP and OpenAPI

---

[1] https://swagger.io/docs/specification/v3_0/callbacks/

WebSockets The WebSocket protocol is a Transmission Control Protocol (TCP) based protocol (only needing HTTP for its handshake to then be upgraded by the HTTP servers). The protocol enables bi-directional communication (full-duplex) with web browsers [34]. The protocol is used by GraphQL when using Transmission Control Protocol (TCP) for subscriptions (notifications from the server to the client) [35]. 31, 171, 320, *see* protocol, and TCP

Solutions

The solutions to the review questions for each chapter are presented here.

Review Questions for Chapter 1

1.1 Ugly APIs Consider the poor and ineffective APIs you have encountered. Which of the following characteristics would you attribute to a subpar API? Multiple responses are possible.

(b) Naming a property `employee` and using it for external stakeholders.
(c) Exposing an integer database key in the API.
(d) An API that makes it possible to obtain a GUID based on a database key.
(g) One API exposing all possible resources of an enterprise

1.2 Feedback loops Reflect on a modern and collaborative development process. What are the minimum essential feedback loops required?

(c) Feedback between developers and business experts and IT specialists is necessary – but only during implementation.
(d) Customers should provide feedback to business managers when using an application.

1.3 API-First or Code-First Reflect on the development of a new web application where a production-ready version needs to be implemented. What development approach would you choose?

(a) API-First.

Review Questions for Chapter 2

2.1 What does API stand for?

(b) Application Programming Interface

2.2 What principal communication strategies do you know?

(b) Synchronous and asynchronous communication

2.3 How do you distinguish messages and events from each other?

(c) A message is deleted after reading, whereas an event can be stored forever.

Review Questions for Chapter 3

3.1 What are the quality characteristics of ISO 25010? (multiple selections possible)

- (a) Functional suitability
- (b) Performance efficiency
- (c) Maintainability
- (f) Reliability
- (g) Security
- (h) Safety
- (k) Flexibility
- (m) Interaction capability
- (n) Compatibility

3.2 Can quality requirements be contradictory?

(c) Some are contradictory and need to be balanced out.

3.3 Are quality requirements important for APIs?

(b) Yes, they are important for specification, implementation, operation, and sundown.

Review Questions for Chapter 4

4.1 How should first ideas be implemented?

(c) Sketch out with mockups.

4.2 What are different kinds of subdomains?

(a) Core, supportive, generic

Review Questions for Chapter 5

5.1 What are the main parts of a Business Model Canvas?

(a) Key partners, key activities, key resources, key propositions, customer relationships, channels, customer segments, cost structure, revenue streams

5.2 What are the evolution steps of an application in a Wardley map?

(b) Genesis, custom, product, commodity

5.3 What are the parts of a sentence in domain storytelling?

(c) Actor, action, work item, adverbial

5.4 Are quantities parts of a visual glossary?

(a) Yes

5.5 What are the steps in event storming?

(d) Event collection, event consolidation, event enhancement, definition of bounded contexts

5.6 Can you use a context map to define data exchanges?

(d) Yes

Review Questions for Chapter 6

6.1 What kind of specification language is used to define network-based interfaces?

(d) Interfaces can be defined by an Interface Description Language.

6.2 What specification language is used nowadays to define synchronous network-based interfaces?

(d) OpenAPI

6.3 What specification language is used nowadays to define asynchronous networked-based interfaces?

(b) AsyncAPI

6.4 Does AsyncAPI require a specific protocol?

(c) No

6.5 Does north–south communication typically include user interactions?

(a) Yes

6.6 What influences the decision for a data format and a protocol?

(d) All of the above

Review Questions for Chapter 7

7.1 What kind of canvas would you use to discuss a bounded context?

(a) Bounded Context Canvas

7.2 What kind of canvas would you use to discuss the architecture of a bounded context?

(b) Architecture Communication Canvas

7.3 What kind of map would you use to define a team structure?

(d) Context map

7.4 What language must be used to define an API?

(c) Ubiquitous language as published language of the bounded context

7.5 Is it helpful to change the overarching service design when designing APIs?

(a) Yes

When the service design is done from scratch, like in the example, does it make sense to change it?

7.6 What kind of canvas is helpful when designing APIs for a bounded context?

(c) API Product Canvas

7.7 Can an API be implemented after it is designed?

(a) Yes

7.8 Can an API be changed after it is designed?

(c) Yes, but good versioning needs to be applied

Versioning needs to be applied after the first version is consumed. More details can be found in section "Versioning of APIs."

Review Questions for Chapter 8

8.1 Can breaking changes be avoided in every case?

(b) No

8.2 What tests should be performed when implementing an API?

(d) Tests should be carefully considered taking into account their time intensity and the level of feedback.

8.3 What concepts can be used to specify tests?

(a) BDD and TDD

8.4 What functions can be fulfilled by an API gateway?

(c) Routing, authorization, and rate limiting, for example.

8.5 When should the use of a service mesh be considered?

(b) For internal APIs

Review Questions for Chapter 9

9.1 Do I need to shape ideas to establish a business?

(a) yes

9.2 Do I need to collect business requirements for designing APIs?

(a) Yes

9.3 Do I need context mapping to design elegant APIs?

(a) Yes, for the data flow and the team organization

9.4 Does a collaborative mindset support API design?

(a) Yes

9.5 Are collaborative and agile processes possible in large organizations?

(a) Yes

Review Questions for Chapter 10

10.1 Are enhancements of software applications supported by collaborative modeling?

(a) Yes

10.2 Can enhancements of a software application create new bounded contexts?

(a) Yes

Review Questions for Chapter 11

11.1 Can the target architecture of a modernization project be designed by the same principles as a "greenfield" project?

(a) Yes

11.2 Can certain steps of the Synergetic Blueprint be skipped?

(a) Yes, when the events and transitions are clear

11.3 How can a legacy model be decoupled from the modern model?

(c) Using an event broker, CDC, and data liberation design pattern

11.4 When modernizing a system, should all parts of the system be modernized simultaneously?

(a) No

Review Questions for Chapter 12

12.1 Does a North Star help to align partners?

(b) Yes

12.2 Can I skip domain storytelling when implementing technical enhancements?

(b) Yes

Review Questions for Chapter 13

13.1 What is the trade-off when selecting a network storage?

(c) A database cannot have all three features: consistency, availability, and partition tolerance.

13.2 How do we design an APIs when implementing the API for a serverless implementation?

(c) The same way: We still use API-First.

Review Questions for Chapter 14

14.1 How do you mitigate the issues associated with an API that uses database fields?

(b) Define a ubiquitous language in a collaborative manner.

14.2 How do you mitigate the issues associated with an API formulated from the backend developer's perspective?

(c) Use a collaborative workshop approach to define the API

14.3 How do you mitigate the issues associated with an overloaded API with endpoints and methods?

(b) Use DDD and design bounded contexts.

Review Questions for Chapter 15

15.1 Are all beautiful APIs public?

(b) No

15.2 Does API-First with clear-cut contexts and a ubiquitous language support making a beautiful API?

(a) Yes

References

1. Medjaoui M (2019) Continuous API Management, Making the Right Decisions in an Evolving Landscape, Wilde E, Mitra R, Amundsen M (eds), 1st edn. O'Reilly, Beijing, 267 pp. ISBN: 978-14-92043-55-3
2. De B (2017) API Management: An Architect's Guide to Developing and Managing APIs for Your Organization, 1st edn. Apress, New York, 11 pp. Includes bibliographical references. - Description based on print version record
3. Reddy M (2011) API Design for C++. Morgan Kaufmann, San Francisco. ISBN: 978-01-23850-04-1
4. Message delivery guarantees (2025). [Online] Available: https://docs.confluent.io/kafka/design/delivery-semantics.html. Visited on 21 Feb 2025
5. Josefsson S (2006) The Base16, Base32, and Base64 Data Encodings, RFC 4648. https://doi.org/10.17487/rfc4648. [Online] Available: https://www.rfc-editor.org/info/rfc4648. Visited on 09 Sep 2024
6. Strategyzer (2024) The business model canvas. [Online] Available: https://www.strategyzer.com/library/the-business-model-canvas. Visited on 21 July 2024
7. C4 model. [Online] Available: https://c4model.com/. Visited on 28 Dec 2024
8. Alex SM, Bryan Lail BJ (2022) Business capabilities, version 2, the open group. [Online] Available: https://pubs.opengroup.org/togaf-standard/business-architecture/business-capabilities.html. Visited on 27 Aug 2024
9. Basig L, Lazzaretti F, Aebersold R, Zimmermann O (2021) Reliable event routing in the cloud and on the edge: An internet-of-things solution in the agetech domain. In: Software Architecture. Springer International Publishing, Berlin, pp 243–259. ISBN: 978-30-30860-44-8. https://doi.org/10.1007/978-3-030-86044-8_17
10. Cloudevents - version 1.0.2 (2023). [Online] Available: https://github.com/cloudevents/spec/blob/v1.0.2/cloudevents/spec.md. Visited on 17 Dec 2024
11. Bellemare A (2023) Building an Event-Driven Data Mesh, Patterns for Designing and Building Event-Driven Architectures, 1st edn. O'Reilly, Beijing, 1 p. ISBN: 10-9812-757-9
12. Dulay H, Mooney S (2023) Streaming Data Mesh: A Model for Optimizing Real-Time Data Services, Mooney S (ed), 1st edn. O'Reilly, Beijing, 1 p. ISBN: 10-9813-069-3

13. Dehghani Z (2022) Data Mesh, Delivering Data-Driven Value at Scale, 1st edn. O'Reilly, Sebastopol, 1403 pp. ISBN: 978-14-92092-34-6
14. Wilsenach R (2015) Dev ops culture. [Online] Available: https://martinfowler.com/bliki/DevOpsCulture.html. Visited on 12 Feb 2025
15. Hohpe G (2024) Event-driven = loosely coupled? not so fast! . [Online] Available: https://www.enterpriseintegrationpatterns.com/ramblings/eventdriven_coupling.html. Visited on 12 Feb 2025
16. Bellemare A (2020) Building Event-Driven Microservices, Leveraging Organizational Data at Scale, 1st edn. O'Reilly, Beijing, 1304 pp. ISBN: 978-14-92057-86-4
17. Fielding RT, Nottingham M, Reschke J (2022) HTTP Semantics, RFC 9110. https://doi.org/10.17487/rfc9110. [Online] Available: https://www.rfc-editor.org/info/rfc9110. Visited on 29 July 2024
18. Hohpe G, Woolf B (2013) Enterprise Integration Patterns: Designing, Building, and Deploying Messaging Solutions. The Addison-Wesley Signature Series, 17. print. Addison-Wesley, Boston, 683 pp. ISBN: 032-1200-68-3
19. Richardson C (2019) Microservices Patterns: With Examples in Java. Manning Publications, Shelter Island, 11 pp. Includes bibliographical references and index. - Description based on print version record. ISBN: 978-16-17294-54-9
20. History |protocol buffers documentation. [Online] Available: https://protobuf.dev/history/. Visited on 30 Nov 2024
21. Rabbitmq/rabbitmq-server: Open source rabbitmq: Core server and tier 1 (built-in) plugins (2025). [Online]. Available: https://github.com/rabbitmq/rabbitmq-server. Visited on 12 Feb 2025
22. Rabbitmq: One broker to queue them all |rabbitmq (2025). [Online] Available: https://www.rabbitmq.com/. Visited on 12 Feb 2025
23. Richardson L (2015) RESTful Web APIs, Amundsen M (ed), 1st edn, Second Release. O'Reilly, Beijing, 373 pp. ISBN: 9781-44-9358-06-8
24. Indrasiri K (2020) gRPC: Up and Running, Building Cloud Native Applications with Go and Java for Docker and Kubernetes, Kuruppu D (Ed), 1st edn. O'Reilly, Beijing, 188 pp. Index: Seite 183–188. ISBN: 14-9205-833-5
25. Hohpe G, Danieli M, Landreau J-F, Hashmi T (2021) Cloud Strategy: A Decision-Based Approach to Successful Cloud Migration. Architect Elevator Book Series. leanpub.com
26. Sharma R (2020) Getting Started with Istio Service Mesh: Manage Microservices in Kubernetes. Springer eBook Collection, Singh A (ed). Apress, Berkeley, 1321129 pp. ISBN: 978-14-84254-58-5
27. Fowler M (2024) Strangler fig. [Online] Available: https://martinfowler.com/bliki/StranglerFigApplication.html. Visited on 10 Feb 2025
28. Tune N (2024) Architecture Modernization: Socio-Technical Alignment of Software, Strategy, and Structure, JG Perrin (ed), 1st edn. Manning Publications Co. LLC, New York, 1464 pp. Description based on publisher supplied metadata and other sources.. ISBN: 978-16-38355-84-7
29. Junker A (2025) Mastering Domain-Driven Desgin. BPB Online. ISBN: 978-93-65892-52-9
30. Riedesel J (2021) Software Telemetry: Reliable Logging and Monitoring. Manning, Shelter Island, 533 pp. Includes index. ISBN: 978-16-17298-14-1
31. Patton J (2014) User Story Mapping. O'Reilly Media, Sebastopol. ISBN: 978-14-91904-90-9
32. Zörner S (2015) Softwarearchitekturen dokumentieren und kommunizieren, Entwürfe, Entscheidungen und Lösungen nachvollziehbar und wirkungsvoll festhalten, 2., überarbeitete und erweiterte Auflage. Hanser, München, 277 pp. Literaturverz. S. [269]–272. ISBN: 978-34-46443-48-8
33. Higginbotham J (2021) Principles of Web API Design: Delivering Value with APIs and Microservices. Pearson Education, London. ISBN: 978-01-37355-63-1

34. Melnikov A, Fette I (2011) The WebSocket Protocol, RFC 6455. https://doi.org/10.17487/rfc6455. [Online] Available: https://www.rfc-editor.org/info/rfc6455. Visited on 12 Feb 2025
35. Subscriptions link - apollo graphql docs (2025), Source: https://github.com/apollographql/apollo-client/blob/main/docs/source/api/link/apollo-link-subscriptions.md. Apollo Graph. [Online] Available: https://www.apollographql.com/docs/react/api/link/apollo-link-subscriptions. Visited on 12 Feb 2025

Index

A

Aggregate, 16, 87, 93, 96, 98, 99, 112, 118, 210–212, 215, 220, 294
allOf, 214–215
AMQP, xxv, 38, 41, 171, 181, 187, 189, 195, 319, 378
Anticorruption layer (ACL), 105–108, 110, 185, 190, 232, 233, 238, 248, 295, 316, 335, 338
Anti-pattern, xxxi, 5, 50, 121, 190–192
anyOf, 230
Apache Thrift, 123, 141, 143
API, xxxi, xxxii, 6–17, 22, 25–30, 35–39, 45–58, 71–117, 125, 128, 131, 146, 160, 166–168, 170, 178, 190, 192–195, 205, 210–213, 215, 216, 220, 223, 229, 245, 249, 253–287, 297–306, 317, 320, 339, 351–353, 359–367
   composition, 219
   consumer, 35, 36, 254, 280, 281
   definition, 6, 25
   design, xxi, 18, 20, 28, 45, 47, 54, 71–118, 121, 125, 205, 212, 246–248, 264, 280, 297–305, 339, 350, 353, 360, 362
   enhancement, xxxii, 17, 215
   product canvas, 210–212, 216, 218, 220, 222, 224, 226, 227, 235, 237, 238, 247, 249, 291, 295, 306, 307, 318, 345, 369
   provider, 9, 14, 36, 266
   specification, 20, 22, 36, 37, 47, 49–51, 126, 129, 159, 170, 171, 193, 242, 253, 254, 258, 259, 263, 266, 267, 280, 301, 369

API management, xxxii, 54, 125, 275, 276, 285, 370
   platform, 267, 276, 277, 285
ApiOps, 276, 280
Architecture Communication Canvas, 206, 208–210, 216, 219, 226, 228, 234, 238, 241, 244, 248, 249, 295, 307, 345, 369
Artificial intelligence (AI), 3, 64, 114–115, 121, 147, 192–195, 331, 332, 334, 338, 352
   stateless, 350, 351
AsyncAPI, xxxi, 16, 126–128, 171, 174–182, 193, 195, 212–216, 263–265, 283, 291, 345, 365, 369
   Linter, 187, 264, 265
Asynchronous communication, xxi, xxxi, 32–35, 37, 43, 148, 170, 181, 188, 190, 191, 195, 196, 212, 216, 261, 352, 365
   distingushment to synchronous, 32
Asynchronous interface, 127–130, 170–181, 195
At least once, 152, 153
At most once, 152, 153
Avoiding mistake, xxxiii, 359–362
Avro, xxxi, 132, 143–147, 176–178, 184, 185, 187, 190, 195, 206, 254, 255, 261, 262, 265, 370
Azure Event Hubs, 38

B

Backends for frontends (BFF), 277–278, 286, 287
Backward compatibility, 17, 257, 262

Behavior driven design (BDD), 271, 300–301, 368, 369
Big Ball of Mud (BBoM), 105, 108
Bounded context, 6, 13, 16–18, 29, 55, 71, 82–84, 93, 98–102, 106–111, 181, 205, 207–210, 212, 215, 219, 227, 232, 239–241, 244, 269, 276, 291–295, 301–304, 313, 331, 336–338, 345, 353, 360, 363, 369
Bounded Context Canvas, 47, 206–208, 210, 212, 216, 220, 226, 233, 236, 240, 247–249, 295, 307, 345, 369
BPMN, 298, 299, 369
Broker, xxxi, 34–37, 41–43, 128–130, 148, 149, 152, 171, 175, 178, 181, 186, 192, 195, 205, 257, 259, 261, 319, 320, 331, 338
Brokerless, xxxi, 181, 195, 212
Brownfield project, xxxii, 55, 107, 247, 325–340, 350, 360, 361
   business analysis, 326–327
Business capability, 16, 74, 209
Business Model Canvas, 15, 18, 71–75, 77, 117, 291, 292, 297, 306, 307, 343, 344, 346, 367

C

Calendar versioning, *see* Versioning
CalVer, *see* Versioning
Capability map, 74, 75, 79, 81, 88, 117, 209, 292, 297, 306, 307, 343, 344, 346, 367, 368
CAP theorem, 152, 348–350
Cloud event, xxxii, 182–183, 196
Collaboration, xxi, xxxiii, 19, 23, 57, 108, 316, 318, 339
Collaborative development, xxx, xxxi, 14, 15, 23
COM/DCOM, 122, 123
Communication
   asynchronous, xxi, xxxi, 32–35, 37, 43, 148, 170, 181, 188, 190, 191, 195, 196, 212, 216, 261, 352, 365
   categories, xxxi, 22, 25–43, 121
   pattern, xxxii, 16
   synchronous, 33–34, 36–38, 43, 148, 149, 188, 191, 196, 210, 352
Compatibility, xxxi, 17, 48–49, 58, 142, 257, 260–263, 277
   backward, 17, 257, 262
   coexistence, 49
   forward, 257, 262

interoperability, 49
   of schemas, 257, 260
Complexity, 3–6, 35, 39, 48, 55, 175, 291, 294, 348, 352
Conformist, 105–107, 110, 248, 295
Consistancy, 153, 213, 216, 228, 283, 348–350, 352, 354, 370
Context map, 16, 19, 71, 102, 104, 109–114, 153, 205, 206, 246–248, 291, 294–295, 297, 301, 306, 314, 316, 331, 334, 335, 369
Continuous integration (CI), 52, 264–266, 272–275, 280, 286, 370
Conway's Law, 301–303, 306, 307, 314
CORBA, 123, 158
CQRS, 217–219, 221, 223, 295, 319
   considerations, 218
Cucumber, 271
Customer/supplier development teams, 105, 106

D

Day-to-day work, xxi, xxxiii, 302, 305
Delivery guarantees, 152–153, 182
Development process, xxx-xxxii, 5, 6, 8, 14–19, 23, 115, 291, 293
Distributed Computing Environment (DCE), 122
Domain, 68
   core domain, 68, 79, 81, 109
   generic domain, 68
   story, 15, 82–90, 293, 312, 313, 315, 318, 327, 328, 330
   supportive domain, 68, 79, 82
Domain-driven design (DDD), xxx, xxxi, 3, 6, 18, 20, 55, 71–118, 121, 133, 193, 205, 212, 221, 226, 232, 234, 235, 247, 271, 280, 291, 292, 303–305, 311, 322, 326, 335, 337, 339, 360
Domain storytelling, 15–19, 71, 82, 90, 93, 99, 102, 104, 114, 117, 118, 247, 291–294, 297, 298, 300, 306, 307, 311, 312, 314, 318, 337, 338, 345, 346, 368, 369
   brownfield, 327-329
Dumb pipes, 39

E

Event, xxxi, xxxiii, 27, 40–42, 96–101, 113–114, 118, 126, 132, 147, 170–172, 182, 190–192, 206, 212,

Index 393

215, 219, 221, 231, 235, 245, 259, 260, 281, 291, 318–322, 331, 336, 338, 345, 347–354, 359–362, 365, 370
domain event, 16, 96, 212, 213, 215, 217, 221, 223, 294, 322, 334, 336, 338
read model, 16, 98
sourcing, 41, 42, 147
Event broker
vs. message broker, 42
Event-driven architecture (EDA), xxi, 35, 39, 41, 126, 171, 175, 191, 212, 318, 336
Eventing, 35, 39–42, 175, 260
vs. messaging, 40
Event Storming, 16–19, 71, 82, 93–105, 109, 110, 114, 192, 205, 212, 247, 248, 291, 292, 294, 295, 297, 298, 300, 301, 306, 314, 334, 338, 339, 345, 369
Event streaming, 30, 41
Eventual consistency, 152, 153, 348–350, 352, 370
Exactly once, 152, 153
Example mapping, 301, 302, 369
Executable specification, 271

F
Feature flags, 275, 286
Feedback loops, 17, 23, 264, 268, 271, 284, 285, 306
Flexibility, 34, 56, 258, 259, 280, 339
adaptability, 56
installability, 56–57
replaceability, 57
scalability, 56
Flight levels, 304, 305, 307, 369
Forward compatibility, 257, 262
Functional suitability, 46–47
functional appropriateness, 46, 47
functional completeness, 46, 47
functional correctness, 46, 47

G
Gateway
communication, 36, 37
Git
flow, 272–273
GraphQL, xxxi, 39, 41, 153, 156–158, 169–170, 184–186, 189, 190, 195, 265, 369
Linter, 186

gRPC, xxxi, 39, 44, 141, 153–158, 169–170, 184, 185, 187, 189, 190, 195, 211, 369
linter, 187

H
HTTP
public APIs, 320

I
Inheritance, 68, 214, 215
Interaction capability, 49–53, 58, 382
appropriateness recognizability, 49, 50
inclusivity, 49, 51
learnability, 49, 50
operability, 49, 50
self-descriptiveness, 49, 51
user assistance, 49, 51
user engagement, 49–51
user error protection, 49, 50
ISO 25010, 45, 46, 58, 382

J
JBehave, 271
JSON, xxxi, 125, 130–138, 143, 146, 147, 162, 165, 166, 169, 172, 176, 178, 182–184, 187, 190, 195, 206, 254–256, 259, 261, 265, 370
convert to YAML, 166
JSON Web Token (JWT), 178, 229, 230

K
Kafka, xxxi, 30, 38, 141, 171, 172, 175, 176, 181, 182, 184–187, 189, 190, 195, 196, 205, 211, 216, 319, 320, 350
Protocol, 38
Keystone interface, 275, 286
API-First, 275

L
Legacy system, 12, 55, 107, 325–329, 335–337, 340
Linter, 186, 187, 264, 265, 280
Logging, 284–286

M
Maintainability, xxxi, 54–56, 58, 382
analyzability, 54–56

modifiability, 54, 55
modularity, 54–55
reusability, 54, 55
testability, 54–56
Message broker
  vs. event broker, 42
Message queuing telemetry transport (MQTT), xxxi, 41, 171, 182
Messaging, 35, 39–43, 175
  vs. eventing, 35, 40
Metrics, 48, 53, 208, 277, 282, 284–286, 343
Modernization
  architectural transition, 335–339
  change data capture, 336
  data liberation, 336
  decoupling layer, 335–338
  modernization layer, 335–339
  strangler fig, 337, 338

N
North Star, 343, 344, 346, 367

O
oneOf, 230
Online library, xxxi, xxxii, 14, 55, 63–75, 77–81, 88, 90, 99–104, 108–114, 132, 170, 185, 195, 205–249, 253, 258, 291, 292, 294, 295, 301–303, 307, 311–315
  implementation ideas, 65–67
  prioritization, 68–69
  requirements, 63–65
OpenAPI, xxxi, 7, 16, 20, 38, 50, 114–116, 121, 123, 125–127, 134, 136, 158–163, 167, 168, 170, 171, 178, 186, 187, 193–196, 216, 220, 222, 254, 264, 265, 281, 291, 306, 307, 320, 345, 364, 369
  linter, 186, 187, 265
  parameters, 163
  request bodies, 162, 163
  responses, 50, 163–165
  security schemes, 166
    API Keys, 166, 167, 178
    HTTP Authentication, 166, 167
    Mutual TLS, 166, 167
    OAuth 2.0, 166–168
    OpenID Connect, 166, 168
Open host service, 105, 107–109, 212, 295, 335

OpenID Connect, 49, 57, 166, 168, 240, 243
Open Network Computing Remote Procedure Call (ONC RPC), 122

P
Partnership, 105, 108, 311, 370
Performance efficiency, 47–48, 58, 382
  capacity, 47, 48
  resource utilization, 47, 48
  time behavior, 47, 48
Polymorphism, 55, 230, 231
Protobuf, xxxi, 132, 141–143, 146, 147, 154, 155, 169, 170, 178, 179, 184, 187, 195, 254–256, 261, 262, 265, 370
Protocol, xxxi, xxxii, 30–32, 34, 35, 37–40, 42, 107, 108, 116, 121, 124, 126, 127, 149, 153, 158, 171, 172, 180, 182–190, 195–197, 210, 263, 317, 319, 320, 354, 370

Q
Quality requirements, xxxi, 45-58, 182, 208, 209, 347, 382
Query-based API styles, 39

R
RAML, 121, 123–125
Redpanda, 38
Reliability, 52, 58, 269, 272, 382
  availability, 52–53
  fault tolerance, 52, 53
  faultlessness, 52
  recoverability, 52, 53
Remote APIs, xxxi, 25, 30, 53, 266
Request-and response-based, 38–40
Request-Reply, 191
Resilience, xxxi, 53, 272, 278
REST, xxxi, xxxii, 8, 38, 153, 156, 158–169, 184–186, 189, 190, 195, 196, 205, 219, 221, 258, 317, 336, 353, 363
RESTful-oriented, 38
Review questions, xxix, 23, 43, 58, 70, 117–118, 196–197, 248–249, 286–287, 307–308, 322, 340, 346, 354, 362, 366, 381–388
RPC-style, 38, 39

S

Safety, 57
    fail safe, 57
    hazard warning, 57
    operational constraint, 57
    risk identification, 7
    safe integration, 57
Scaled agile, 307
Security, xxxi, 53
    accountability, 53
    authenticity, 54
    confidentiality, 53
    integrity, 53
    nonrepudiation, 53
    resistance, 54
Semantic versioning, *see* Versioning
SemVer, *see* Versioning
Separate ways, 105, 107
Serverless, xxxii, 182, 347–354, 370, 387
Service mesh, xxxii, 54, 179, 276, 278–279, 286, 287, 385
Service-oriented architecture, 39, 335
Shared kernel, 105, 106
Sidecar pattern, 278
Small project, xxxiii, 264
Smart pipes, 39
Solutions, xxix, 3, 11, 19, 153, 191, 192, 213, 218, 219, 231, 232, 234, 236, 238, 240, 242, 254, 257, 260, 266, 277, 278, 281, 285, 295, 312, 318, 320, 321
Spotify Model, 304, 305, 307
Strangler fig, 337, 338, 350
Strategic design, 292
Swagger, 121, 123, 125
Synchronous communication, 33, 36–38, 43, 148, 149, 170, 188, 196, 210, 352
    distinguishment to asynchronous, 32
Synchronous interface, xxxi
    specification, 156
Synergetic Blueprint, 4, 291, 295–297, 306, 311, 314, 318, 320, 322, 325, 339, 340, 343–345, 386
    shortcuts, 343
System Object Model, 122

T

Tactical design, 205, 206, 295, 345, 346
TDD, 266, 270, 271, 275, 286, 370
Team topologies, 295, 303, 335

Test
    acceptance, 269, 271, 301
    chaos, 269
    compiler, 265–266
    contract, 266–268, 280, 286
    e2e, 269–270, 272
    integration, 266, 268–269, 272
    linter, 186, 264, 265, 280
    load, 269
    penetration, 269
    performance, 269
    reliability, 269
    security, 269
    static code, 187, 264–265, 280
    unit, 266–268, 270, 286
Thrift, 123, 141, 143
Tracing, 29, 154, 164, 284–286

U

Ubiquitous language, 6, 8, 9, 17, 19, 22, 23, 50, 51, 82, 105, 109, 110, 115, 116, 208, 232, 247, 248, 271, 301, 353, 359, 361–363, 366, 384, 387, 388
User error protection, 49, 50, 131, 146, 280
User story mapping, 298–300, 368

V

Versioning, 28, 49, 139, 142, 147, 159, 170, 171, 182, 194, 249, 253–263, 283, 306, 361, 370, 385
    calendar versioning, 254
    endpoint, 258
    header, 258
    media-type, 258–259
    semantic versioning, 159, 253, 254, 283
    synchronous APIs, 257–259
Vision statement, 343, 344, 367
Visual Glossary, 6, 16, 17, 19, 88, 90–96, 101, 102, 112–114, 118, 153, 154, 170, 194, 208, 291, 293, 294, 297, 318, 337, 345, 346, 383
VUCA, 3–6, 22, 291

W

Wardley map, 15, 16, 74–80, 110, 117, 208, 248, 291, 292, 297, 305–307, 332–335, 340, 343, 344, 346, 367, 383
Web APIs, 25
WSDL, 121, 123–125, 265

X

XML, xxxi, 123, 124, 130–132, 137–141, 146, 169, 182, 184, 195, 196, 265
XPCOM, 123

Y

YAML, 125, 131, 132, 136–137, 146, 165, 166, 178, 195, 264
    convert to JSON, 166, 178
yq, 166

Z

Zero trust, 278

GPSR Compliance

The European Union's (EU) General Product Safety Regulation (GPSR) is a set of rules that requires consumer products to be safe and our obligations to ensure this.

If you have any concerns about our products, you can contact us on

ProductSafety@springernature.com

In case Publisher is established outside the EU, the EU authorized representative is:

Springer Nature Customer Service Center GmbH
Europaplatz 3
69115 Heidelberg, Germany

www.ingramcontent.com/pod-product-compliance
Lightning Source LLC
LaVergne TN
LVHW010333260326
834688LV00036B/693